By the same author

Alternating Currents: An Analytical and Graphical Treatment for Students and Engineers

By FREDERICK BEDELL and ALBERT C. CREHORE

Translated into German by Alfred H. Bucherer

Theorie der Wechselströme: in analytischer und graphisher Darstellung

Translated into French by J. Berthon

Étude Analytique & Graphique des Courants Alternatifs

The Principles of the Transformer

By FREDERICK BEDELL

A Laboratory Manual of Physics and Applied Electricity

Arranged and Edited by EDWARD L. NICHOLS

VOL. II—PART II

Experiments with Alternating Currents

By FREDERICK BEDELL

DIRECT AND ALTERNATING CURRENT MANUAL

WITH DIRECTIONS FOR TESTING AND A DISCUSSION OF THE THEORY OF ELECTRICAL APPARATUS

BY

FREDERICK BEDELL, Ph.D.

PROFESSOR OF APPLIED ELECTRICITY IN CORNELL UNIVERSITY

ASSISTED BY

CLARENCE A. PIERCE, Ph.D.

SECOND EDITION, ENLARGED AND REVISED

NEW YORK

D. VAN NOSTRAND COMPANY

25 Park Place

LONDON: CONSTABLE & CO., Ltd.

1915

First Edition, September, 1909; November, 1909

Second Edition (enlarged), July, 1911; (revised), December, 1911
August, 1912; November, 1912; March, 1914, August, 1915

Note.—The first edition was issued under the title
"Direct and Alternating Current Testing."

Press of
The New Era Printing Company
Lancaster, Pa.

DIRECT AND ALTERNATING CURRENT MANUAL

WITH DIRECTIONS FOR TESTING AND A DISCUSSION OF THE THEORY OF ELECTRICAL APPARATUS

BY

FREDERICK BEDELL, Ph.D.

PROFESSOR OF APPLIED ELECTRICITY IN CORNELL UNIVERSITY

ASSISTED BY

CLARENCE A. PIERCE, Ph.D.

SECOND EDITION, ENLARGED AND REVISED

NEW YORK

D. VAN NOSTRAND COMPANY

25 Park Place

LONDON: CONSTABLE & CO., Ltd.

1915

First Edition, September, 1909; November, 1909

Second Edition (enlarged), July, 1911; (revised), December, 1911
August, 1912; November, 1912; March, 1914, August, 1915

Note.—The first edition was issued under the title
"Direct and Alternating Current Testing."

Press of
The New Era Printing Company
Lancaster, Pa.

PREFACE.

This manual consists of a series of tests on direct and alternating current apparatus, selected with reference to their practical usefulness and instructive value. While the book has been prepared primarily for students, it is hoped that it may prove helpful to others. The presentation is in the form of a laboratory manual; the author, however, has not restricted himself to a mere statement of instructions for conducting tests but has directed the reader's attention to the principles that underlie the various experiments and to the significance of the results. Experience has shown that theory is more readily grasped when it is thus combined with its application and that the application is more intelligently made when its broader bearings are understood. The material has been systematically arranged and it is believed that the book may be found useful for reference or as a text, aside from its use in testing.

From the text proper are excluded specialized tests and those that are of limited application or require unusual testing facilities, such tests being described in the appendices to the several experiments. These appendices thus permit a fuller discussion of some of the details of the tests and various modifications than could be included to advantage in the text proper. The tests in general are those that can be performed in any college laboratory.

No attempt has been made to make the work exhaustive or complete; on the contrary every effort has been made to eliminate matter of secondary importance and that which is of questionable technical or pedagogical value.

The aim has been to arrange an introductory series of experiments of a comprehensive nature, so that in a reasonable time and with a reasonable amount of effort the student may acquire the power to proceed to problems requiring a continually increasing initiative and originality. Although standardized tests afford the quickest way for obtaining certain desired results and, in the case of a student, for obtaining a knowledge of testing methods, the ability to conduct such tests with full instructions given is soon acquired. Beyond this point the exclusive use of standardized tests should be avoided. Standards in electricity serve best as new points of departure. The student who is to become more than the "ordinary slide-rule engineer" or "mental mechanic" will have sufficient intellectual curiosity to desire more than any standardized tests can give him and should be encouraged in every way to seek new results and to devise ways and means for obtaining them with the facilities at hand. To attempt to formulate such work would at once deprive it of its freshness. The student may well be referred to the current technical press and to the transactions of the engineering societies for suggestions as to subject matter for further study and also as to methods to be adopted.

With reference to prepared blanks and forms, the writer believes that their use can be, and often is, carried too far, leading perhaps to good technical but not to good pedagogical results. In a certain sense the one who prepares the forms and lays out the work is the one who really performs the experiment, the tabulators of data being assistants who, for commercial work, require only a common school education.

Progress undoubtedly results from the development of individualism and if room for such development is to be given in a college course—specifically in a college laboratory course—the more or less standardized instruction must needs be reduced so as not to fill the entire available time. The natural tendency has been quite the reverse. Two decades ago, the study of electrical engineering meant, practically, the study of direct currents, there being little else. Laboratory courses were developed in which the whole available time was well filled with test after test upon direct current generators and motors. The transformer and alternator were then added, with extensive time-consuming tests, with the apparent assumption that the full development of alternating currents was reached. In succeeding years came the general development of polyphase currents, the rotary converter, induction motor, etc., these subjects being added to a crowded course by a process of compression rather than judicious elimination. The student was given more than he could possibly assimilate. As types of machines have multiplied, it would take years to perform all the permutations and combinations of tests on all the different types. But is this necessary for a student? Why not develop a student's powers by a few typical experiments on a few typical kinds of apparatus?

With this end in view the writer has made selection from material which has long been collecting in the form of typewritten outlines. These have been in a process of continual evolution, frequently rewritten and used by many classes. By a process of elimination and survival, experiments consisting of a large amount of mechanical data-taking and tabulation and a relatively small amount of technical content have been dropped in favor of those experiments which have proved most effective in student development. Various demands upon the writer's time prevented his preparing for the press a book on testing a number of years ago and the present appearance of the book is due in no small way to the valuable assistance of Dr. Pierce. Meanwhile several admirable manuals have appeared, which differ, however, in aims or scope from the present work. The author hopes to find leisure, in the near future, to make good some of the omissions of the present volume and to include in a later edition additional chapters on alternating current motors and converters.

The present work is self-contained and requires only such preliminary courses in physical and electrical measurements as are usually given in colleges. The book may be used to advantage in conjunction with standard texts on electrical engineering, as those by Franklin and Esty, S. P. Thompson, and Samuel Sheldon, or with an introductory text such as that by H. H. Norris. The experiments given in the book may be supplemented by others of an elementary, intermediate or advanced nature, as circumstances may require. The division of experiments into parts and sections will be found to add materially to the flexibility of the book.

The author desires to express his appreciation of the initial instruction and inspiration of Professor H. J. Ryan and of the continuous coöperation for many years of Professor G. S. Moler. He wishes also to express his indebtedness to many who have been associated with him in laboratory instruction, in particular to Dr. A. S. McAllister, as many references in the present text bear evidence. He likewise desires to express his appreciation of the spirit of coöperation shown by Professors H. H. Norris and V. Karapetoff and other engineering colleagues. The author is indebted, furthermore, to various professors and students, who have used and corrected this book in proof during the last year and to a number of engineers who have looked over the proof sheets and have made valuable suggestions. For all shortcomings the author alone is responsible.

ITHACA, N. Y., July 1, 1909.

PREFACE TO THE SECOND EDITION.

The preparation of this edition for the press has given the author an opportunity to make good certain omissions in the first edition and to include discussions of the induction motor, the induction generator, frequency changers, the synchronous motor, the synchronous converter, wave analysis and a selection of special problems. In a restricted sense the book is now complete. It is not exhaustive and has many short-comings, but it is believed that the reader who has become familiar with the principles of testing herein contained can proceed to conduct such further tests as he desires and to manipulate and use new types of apparatus without special instruction.

In order to present the discussion of the circle diagram for an induction motor with the utmost conciseness and clearness, the graphical constructions for determining slip, efficiency and power factor—which are not essential to the understanding of the circle diagram proper—have been omitted from the main discussion and placed in an appendix. For the same purpose, in the discussion of the theory of the synchronous motor (Experiment 10–B), particular emphasis has been laid upon the essential principles which govern the operation of the motor and matters of secondary importance have been sub-ordinated.

Although much has been written on the subject of wave analysis, common experience has been that it is indeed a laborious task to familiarize one's self with the usual methods and to use them for accurately analyzing a wave. In an endeavor to eliminate all unnecessary labor, the author has given explicit instructions for wave analysis that occupy only two pages (pp. 335–6), a numerical example for determining the old harmonics up to the seventeenth occupying two pages more. It is believed that this method, which is based upon the work of Runge, will be found generally useful.

While the problems in Chapter XII. may prove useful, they should be looked upon as chiefly suggestive; if they inspire the reader to undertake other than standardized tests, they will have served their purpose.

ITHACA, N. Y.,
May 1, 1911.

CONTENTS.

CHAPTER I.

Direct Current Generators.

CHAPTER II.

Direct Current Motors.

CHAPTER III.

Synchronous Alternators.

CHAPTER IV.

Single-Phase Currents.

PAGE.

CHAPTER V.

TRANSFORMERS.

CHAPTER VI.

POLYPHASE CURRENTS.

CHAPTER VII.

PHASE CHANGERS, POTENTIAL REGULATORS, ETC.

CHAPTER VIII.

INDUCTION MOTORS.

CHAPTER IX.

INDUCTION MACHINES: FREQUENCY CHANGERS AND INDUCTION GENERATORS.

CHAPTER X.

SYNCHRONOUS MACHINES.

CHAPTER XI.

WAVE ANALYSIS.

CHAPTER XII.

CHAPTER I.

DIRECT CURRENT GENERATORS.

Experiment 1–A. Generator Study and Characteristics of a Series* Generator.

PART I. GENERATOR STUDY.

§ 1. Faraday discovered (1831) that when a conductor cuts lines of force an electromotive force is generated in the conductor proportional to the rate at which lines are cut, and all dynamos (or generators as they are now commonly called) operate on this principle. To generate an electromotive force, it is essential therefore to have a conductor (or several conductors combined by various winding schemes) forming the *armature* as one member; and to have lines of force or magnetic flux set up by *field magnets* which form the second member. For operation it is necessary also to have a source of mechanical power† by which either one of these members can be given a motion with respect to the other. A generator may have‡ a stationary field and revolving armature; or, a revolving field and stationary armature, designated as the *revolving field type.* Although the latter is useful for large alternators, serious objections to it have been found in direct current machines, for with the commutator stationary the brushes must revolve, which leads to difficulties in construction and operation. It is the custom, therefore, to build all

* Where a series generator is not available, this study may be taken without experimental work or in connection with Part I. of Exp. 1–B.

† Power is required to overcome friction and other losses, and to overcome a counter torque (§§ 1–3, Exp. 2–A) which varies with the load.

‡ Outside of this classification is the *inductor alternator* which has a stationary armature, a stationary field winding and a revolving inductor of iron; its study should be taken up later under alternators.

direct current generators and motors with a stationary field and revolving armature.

§ 2. The student should consult any of many excellent treatises for a detailed discussion of different types of generators and if possible should note one or two machines in the laboratory or elsewhere which are examples of each important type. Machines noted should illustrate the following terms, some of which are briefly explained in later paragraphs: *Stationary field, revolving field, bipolar, multipolar, separately-excited, self-excited, series wound, shunt wound, compound wound, magneto-generator, open and closed coil armature, drum armature, Gramme (or ring) armature.*

Only the general structure of the various machines need be noted. Observe particularly the magnetic circuit of each machine and the disposition of the field winding. Keep in mind that magnetic flux is proportional to magnetomotive force (field ampere-turns) divided by the reluctance of the complete magnetic circuit, *i. e.,* the sum of the reluctance of each part (air gap, core. etc.). As the reluctance of any part of a magnetic circuit is equal to the length divided by the product of the cross-section and permeability, it is obvious that an unnecessarily long magnetic circuit should be avoided, a fact neglected in some early machines.

§ 3. In a *bipolar* generator, one pole is north and the other south; in a *multipolar* generator (with 4, 6, 8, etc., poles) the poles are alternately north and south.

Each armature conductor accordingly passes first underneath a north and then underneath a south pole and has induced in it an electromotive force first in one direction* and then in the

* (§ 3a). An exception is the so-called unipolar, homopolar or acyclic dynamo, which has a unidirectional electromotive force generated in the armature conductor; it accordingly delivers direct current to the line without commutation. Faraday's disk dynamo (one of the earliest dynamos) was of this type. For years it was the dream of zealous electricians to make this type of machine practicable, but it was considered only as an interesting freak, for at ordinary speeds the voltage generated is too low

other, *i. e.*, an alternating electromotive force. The simplest form of generator is therefore the alternator, the current being taken from the armature to the line without any commutation. If the armature is stationary, the alternating current from the armature is taken directly to the line; if the armature is revolving, the armature windings are connected to collector rings (or slip rings) from which the current is taken to the line by means of brushes.

In a direct-current generator the armature windings are connected* to the several segments or bars of a *commutator*, from which the current is taken by brushes to the line. The alternating electromotive force generated in each coil is thus commutated, or reversed in its connection to the line, at or near the time of zero value of the electromotive force of the coil.

The electromotive force in each coil increases from zero to a maximum and back to zero, and at any instant the electromotive forces in the various individual coils have different values ranging from zero to a maximum, according to the positions of the coils. The sum of these coil-voltages, as impressed upon the line as *terminal voltage*, is however practically constant.

for most purposes. But changed conditions have made it a practical and important machine (1) driven at high speed by the steam turbine, or (2) driven at moderate speed to generate large currents at low voltage for electrochemical work. Dynamos of this class are not included in this study. For further information, see "Acyclic Homopolar Dynamos," by Noeggerath, A. I. E. E., Jan., 1905; also, *Standard Handbook*, or Franklin and Esty's *Electrical Engineering*. For description of some structural improvements, see pp. 560 and 574, *Electrical World*, Sept. 12, 1908.

* (§ 3b). The details of armature windings will not be here discussed; they are amply treated in many text and handbooks. In almost all machines a *closed coil* winding is used. (The Brush and *T–H* arc dynamos and a few special machines use *open coil* winding.) In a closed winding, the armature coils are connected in series and the ends closed. There are two ways of connecting the coils in series: *wave winding* and *lap winding*. In the wave or series winding there are always two brushes and two paths for the current from brush to brush, irrespective of the number of poles. In the lap or parallel winding, generally used in large generators, there are as many paths (and brushes) as poles. The two schemes are essentially the same in a bipolar machine.

§ 4. Field magnets are usually* energized by direct current passed through the field windings; permanent magnets being used only in small machines, called *magneto-generators,* used for bell-ringers, etc. A generator is *separately-excited* or *self-excited* according to whether the current for the field is supplied by an outside source or by the machine itself. Alternators are separately excited; direct current generators are usually self-excited.

§ 5. A direct current machine (either generator or motor) may be: (1) *Series wound,* with the field winding of coarse wire in series with the armature and carrying the whole armature current; (2) *Shunt wound,* with a field winding of fine wire in shunt with the armature and carrying only a small part of the whole current; (3) *Compound wound,* with two field windings, the principal one in shunt and an auxiliary one in series with the armature.

The compound generator is in most general use, being best suited for all kinds of constant potential service, both power and lighting; the shunt generator performs similar service but not so well. The characteristics of these machines will be studied fully in Exp. 1–B.

The series generator is of interest because: (1) It is one of the earliest types and of historical importance; (2) It is the simplest type and illustrates in a simple manner the principles which underlie all dynamo-electric machinery, both generators and motors; (3) In a compound wound generator or motor, the series winding is an important factor in the regulation of potential or speed. In itself the series generator is of relatively small importance, because neither current or voltage stay constant; it is used only in some forms of arc light machines with regulating devices for constant current.

In direct current motors, all three types of winding are employed: series wound motors for variable speed service in traction, crane work, etc.; shunt and compound wound (including differ-

* The *induction generator,* to be studied in a later experiment, does not come under this classification.

ential wound) motors for more or less constant speed service (Exp. 2-A).

PART II. CHARACTERISTICS OF A SERIES GENERATOR.

The characteristic curves to be obtained are: the *magnetization curve,* with the machine separately excited; the *external series characteristic,* with the machine self excited; and the *total characteristic,* which is computed.

§ 6. **Magnetization Curve.**—This curve shows the armature voltage (on open circuit) corresponding to different field currents when the generator is separately excited from an outside source, as in Fig. 1. No load is put upon the machine. Means for varying the field current must be provided; see Appendix, § 14.

§ 7. *Data.*—Readings are taken of field current, armature voltage and speed; the first reading is taken with field current zero, showing the voltage due to residual magnetism. The field current is then increased by steps from zero to the maximum* rating of the machine, the readings taken at each step giving the "ascending" curve. The descending curve is then obtained by decreasing the field current by steps again to zero. In Fig. 3, only the ascending curve is shown; see also Fig. 2, of Exp. 1-B.

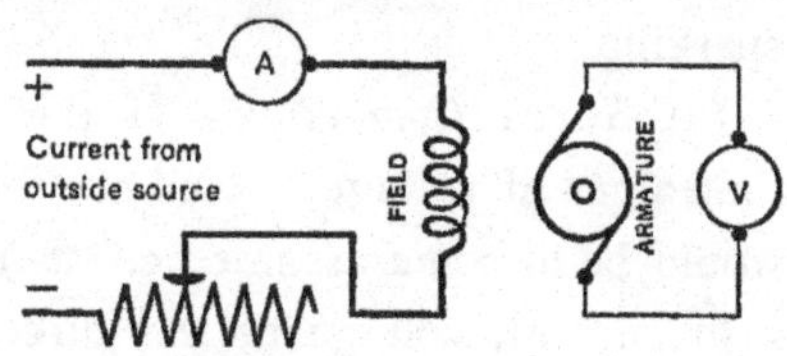

FIG. 1. Connections for magnetization curve,—separately excited.

To obtain a smooth curve, the field current must be increased or decreased continuously; there will be a break in the curve if a step is taken backwards or if the field circuit is broken during

* (§ 7a). *Current Density.*—For field windings an allowable current density is 800-1,000 amp. per sq. in. (1,600-1,275 circ. mils per amp.); for armatures, 2,000-3,000 amp. per sq. in. (640-425 c. mils per amp.). For a short time these limits can be much exceeded. The sectional area of a wire in circular mils is the square of its diameter in thousandths of an inch.

a run. This is true of all characteristics or other curves involving the saturation of iron.

§ 8. *Brush Position.*—During the run the brushes are kept in one position; if for any reason they are changed, the amount should be noted. For a generator the best position is the position of least sparking and of maximum voltage, which locates the brushes on the "diameter" or "line" of commutation. Under load this line is shifted forward from its position at no load, on account of field distortion caused by armature reaction,* and the brushes are accordingly advanced a little to avoid sparking. As it is desirable to keep the brushes in the same position in taking all the curves, with load or without load, it is well to give the brushes at no load a little lead, but not enough to cause much sparking.

§ 9. *Speed Correction.*—If the speed varied during the run, the values of voltage as read are to be corrected to the values they would be at some assumed constant speed. Since, for any given field current, voltage varies† directly with speed, this correction is simply made by direct proportion; each voltmeter reading is

* (§ 8a). *Armature Reactions.*—Armature current has a *demagnetizing* effect and a *cross-magnetizing* effect, the two effects together being called *armature reaction,* as discussed in various text books. The demagnetizing effect due to back ampere-turns weakens the field; the cross-magnetizing effect due to cross ampere-turns distorts the field (weakening it on one side and strengthening it on the other) and shifts forward the line of commutation. In many early machines this made it necessary to shift the brushes forward or back with change of load to avoid sparking; in modern machines the armature reactions are not sufficient to make this necessary and the brushes are kept in one position at all loads.

If a very accurate determination of the neutral position of the brushes is desired, it can be found by a voltmeter connected to two sliding points which are the exact width of a commutator bar apart. The neutral position is the position of zero voltage between adjacent commutator bars, and this is shown by the voltmeter.

† (§ 9a). If the speed can be varied at will, this can be verified for one field excitation. A peripheral speed of 3,000 feet per minute is permissible with the ordinary drum or ring armature.

multiplied by the assumed constant speed and divided by the observed speed.

§ 10. *Curve.*—After the speed correction is applied, the magnetization curve is plotted as in Fig. 3. The abscissæ of this curve, field amperes, are proportional to field ampere-turns or magnetomotive force; the ordinates, volts generated at constant speed, are (by Faraday's principle, § 1) proportional to magnetic flux. The curve, therefore, is a magnetization curve (showing the relation between magnetic flux and magnetomotive force) for the magnetic circuit of the generator, which is an iron circuit with an air gap. The bend in the curve indicates the saturation of the iron.

§ 11. **External* Series Characteristic.**—This characteristic, which is the operating or load characteristic of the machine, shows the variation in terminal voltage for different currents, when the machine is self excited and the external resistance is varied. The armature, field and external circuit are in series, as in Fig. 2; readings are taken of current, voltage and speed, for an ascending curve as in Fig. 3. The descending curve may be taken if desired.

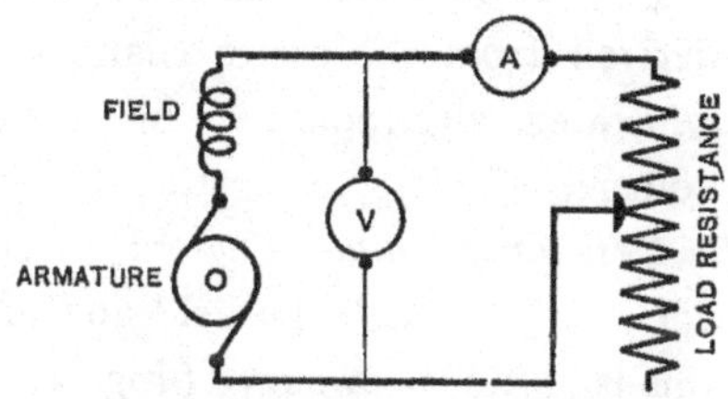

FIG. 2. Connections for series characteristic,—self excited.

For any point on the curve, the resistance of the external circuit is $R = E \div I$, or the tangent of the angle between the I-axis and a line drawn from the point to the origin. Below the knee of the curve, it will be seen that a small change in the external resistance will make a large change in current and voltage.

* (§ 11a.) In any characteristic the term "external" indicates that the values of current and voltage external to the machine are plotted; the term "total" indicates that the total generated armature current and voltage are the quantities used.

If the speed varied during the run, the external characteristic should be corrected* for speed as before (§ 9).

The watts output, for any point on the external characteristic is given by the product of current and voltage, and may be plotted as a curve.

§ 12. If the field coil is connected so that the current from the armature flows through it in the wrong direction, so as to demagnetize instead of building up the residual magnetism, the machine will not "pick up." For each direction of rotation, the proper connection of the field will be found to be independent of the direction of the residual magnetism. Note the effect of previous magnetization (from an outside source) first in one and then in the other direction, and the effect in each case of reversing the field connections.

§ 13. **Total Series Characteristic.**—The total characteristic is derived from the series characteristic, so as to show the total generated electromotive force instead of the terminal brush voltage.

Resistance Data.—The only additional data needed are the voltage drops through the field and armature for different currents; this is plotted as a curve (Fig. 3) which is practically a straight† line. With the armature stationary, current from an outside source is passed through the field or armature (separately); the current is measured and the difference of potential at the terminals. The ratio $E \div I$ gives the resistance. This is called meas-

* (§ 11b). This correction is applied to the external characteristic and not to the total characteristic for convenience. Inasmuch as it is the *generated* electromotive force which is proportional to speed, to be accurate the correction should be applied to the total and not to the external characteristic.

† (§ 13a). This would be a straight line if the resistance were constant. The resistance varies with temperature; see Appendix, § 15. The armature resistance also varies with current since it includes the resistance of brushes and brush contact, which depends upon current density. The hot resistance is to be measured after the machine has run awhile, and is to be considered constant.

uring resistance by "fall of potential" method; see Appendix, § 17.

Curve.—By adding to the external characteristic the RI drop for field and armature, we have the generated voltage or total characteristic Fig. 3.

Interpretation.—The total characteristic falls below the magnetization curve on account of armature reaction, that is, the demagnetizing effect of the armature current which weakens the field and hence reduces the generated voltage; for, in taking the magnetization curve, there was no armature current and hence no armature reaction. The external characteristic falls below the total series characteristic, on account of resistance drop.

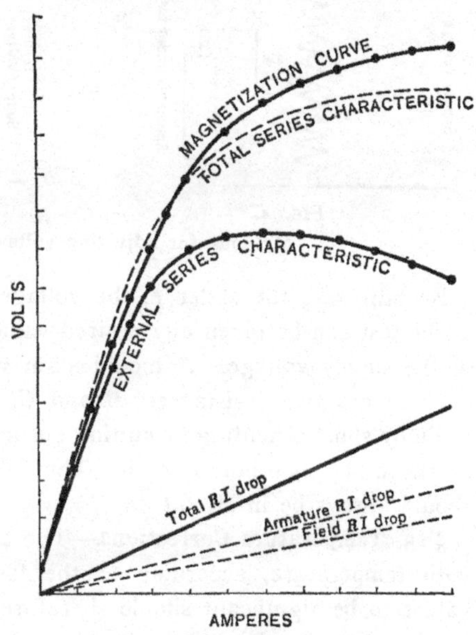

FIG. 3. Characteristics of a series generator.

The magnetization curve would be higher than the total characteristic for all currents, if in taking it the brushes were given no lead, that is were in the position of maximum voltage. Giving the brushes a lead lowers the magnetization curve so that for small values of the current it may fall below the total characteristic.

APPENDIX I.

MISCELLANEOUS NOTES.

§ 14. **Current and Voltage Adjustment.**—For currents of small values, when a wide range of adjustment is desired, a series resistance (Fig. 1) is frequently inadequate and it is better to shunt off current from a resistance R, as in Fig. 4.

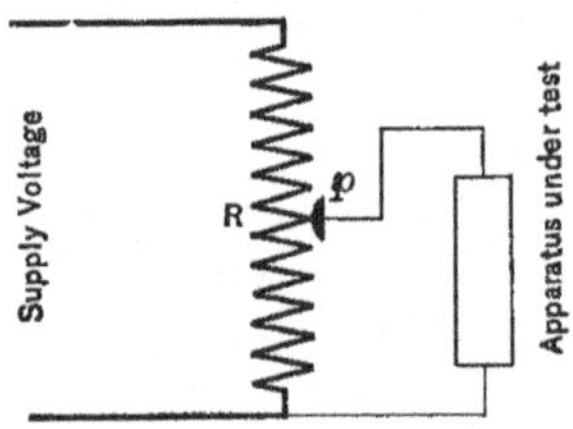

FIG. 4.

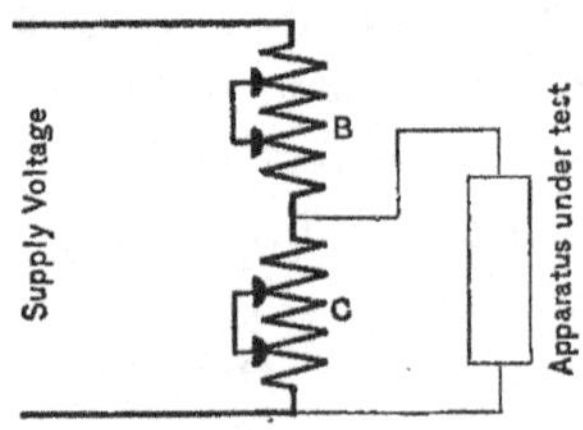

FIG. 5.

Methods for adjusting voltage or current.

By adjusting the slider p, the voltage delivered to the apparatus under test can be given any desired value from zero up to the value of the supply voltage. A modification which is sometimes convenient employs two resistances, B and C, Fig. 5. The adjustment is made by short circuiting or cutting out more or less of one resistance or the other, but not of both. The full amount of one resistance should always be in circuit.

§ 15. **Temperature Corrections.**—The conductivity of copper varies with temperature, according to the law given below. Resistance values to be significant should therefore be for some specified temperature; known for one temperature they can be computed for any other. Temperature rise can be computed from increase in resistance. In all cases where accuracy of numerical results is important, as in commercial tests for efficiency, regulation, etc., definite temperature conditions should be obtained; for this the detailed recommendations of the A. I. E. E. Standardization Rules should be consulted. To meet standard requirements, a run of several hours is commonly required. In practice work this is not necessary, it being usually sufficient to specify resistances as *cold* when taken at the beginning and *hot* when taken at the close of the test.

Let R_t be the resistance of a copper conductor at a temperature

$t°$ C. At a higher temperature the resistance will be greater and experiment shows that the increase in resistance will be in direct proportion to the temperature rise.

At a temperature $(t+\theta)$ °C. the resistance is accordingly

$$R_{t+\theta} = R_t(1 + a\theta).$$

The temperature coefficient a (per degree C.) depends upon the initial temperature t (degrees C.), or the temperature for which the resistance is taken as 100 per cent., and has for copper the following values:*—

t	0°	6°	12°	18°	25°	32°	40°	48°
a	.0042	.0041	.0040	.0039	.0038	.0037	.0036	.0035

From the formula given above, if the resistance is known for one temperature, the resistance can be computed for any other temperature or for any temperature rise.

§ 16. From this formula we can also compute the temperature rise θ, above the initial temperature t, corresponding to a known increase in resistance. By transformation the formula becomes

$$\theta = [(R_{t+\theta}/R_t) - 1] \div a.$$

The temperature rise above an initial temperature t is accordingly equal to the *per cent. increase in resistance divided by* a.

§ 17. **Fall of Potential Method for Measuring Resistance.**—This method is based upon the fact that the fall of potential through a resistance R carrying a current I is $E = RI$ (Ohm's Law). The resistance R which is to be determined may be the resistance of any conductor whatever (transformer coil, field winding, armature, etc.) which will carry a measurable current without undue heating and is not itself a source of electromotive force. An armature, therefore, must be stationary while its resistance is being measured by this method.

Connect the unknown resistance to a source of direct current through a regulating resistance, Fig. 6 (see also § 14), so that the current will not unduly heat the resistance or exceed the range of instruments. Take readings of the two instruments simultaneously,

* A. I. E. E. Standardization Rules; also, A. E. Kennelly, *Electrical World*, June 30, 1906.

and without delay so as to minimize the effect of heating. The resistance R is equal to $E \div I$.

Fig. 6 shows the usual arrangement of apparatus, in case the voltmeter current is but a small part of the total current. The voltmeter leads should be connected directly to the resistance to be measured (not including unnecessary connectors, etc.) or should be pressed firmly against its terminals. The resistance of an armature winding is taken by pressing the voltmeter leads against the proper bars 180° or 90° apart; if resistance of brushes and connections is to be included, the voltmeter is connected outside of these connections.

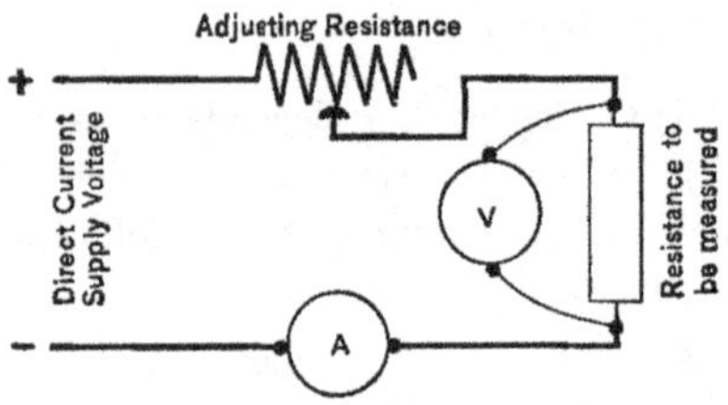

Fig. 6. Measurement of resistance by fall-of-potential method.

In case the ammeter current is very small, so that the voltmeter current is a considerable part of the total current, the voltmeter should be connected outside the ammeter so as to measure the combined drop of potential through the ammeter and unknown resistance. With the voltmeter connected either way, an error is introduced which may often be neglected but can be corrected for when particular accuracy is desired.

§ 18. The voltmeter should always be disconnected before the circuit is made or broken, or any sudden change is made in the current, to avoid damage to the instrument.

If the resistance being measured is highly inductive, not only the instrument but also the insulation of the apparatus under test may be damaged by suddenly breaking the current through it on account of the high electromotive force induced by the sudden collapse of the magnetic field. This may be avoided by gradually reducing the current before breaking the circuit.

§ 19. The value of an unknown resistance can be found in terms of a known resistance placed in series with it by comparing the drops in potential around the two resistances, the current in each having the same value.

EXPERIMENT 1–B. Characteristics of a Compound* Generator.

§ 1. **Introductory.**—A compound generator is made for the purpose of delivering current at constant potential either at the terminals of the machine or at some distant receiving point on the line. In the former case the machine is *flat compounded*, the ideal being the same terminal voltage at full load as at no load, giving a practically horizontal voltage characteristic. In the latter case the machine is *over compounded,* giving a terminal voltage which rises from no load to full load to compensate for line drop, so that at the receiving end of the line the voltage will be constant at all loads. Constant potential service is used both for power and for lighting. Constant delivered voltage is essential in lighting for steadiness of illumination and in power for constant speed.

§ 2. For such service, the series generator is not at all adapted, its voltage being exceedingly low at no load and, for a certain range, increasing greatly with load.

§ 3. A shunt generator almost meets the conditions, generating a voltage which is nearly constant but decreasing slightly with load (Figs. 4 and 6). Obviously by increasing the field excitation (field ampere-turns) when the machine is loaded, the voltage can be increased to the desired value; this is true, however, only in case the iron is not saturated and it is accordingly possible for the increase in field ampere-turns to produce a corresponding increase in the magnetic flux. (Compare Fig. 2.) In a shunt machine this increase in field excitation can be obtained by an increase in field current produced either by an attendant who adjusts the field rheostat or by an automatic† regulator.

* (§ 1a). This experiment can be applied to a *Shunt* generator by omitting §§ 20–25.

† (§ 3a). *Tirrell Regulator.*—Many older forms of regulators, which operated by varying field resistance, are superseded by the Tirrell Regulator. This regulator operates through a relay as follows: (1) When the voltage is too low, it momentarily short circuits the field rheostat, causing the

§ 4. In a compound generator, the necessary increase of field excitation with load is simply and effectively obtained by means of an auxillary series winding. Since the current in the series winding is the load current, the magnetizing action of the series winding (that is its ampere-turns or magnetomotive force) increases in direct proportion to the load. This increases the magnetic flux and hence the generated voltage by an amount dependent upon the degree of saturation of the iron.

Looked at in another way, a shunt winding (which alone gives a falling characteristic) and a series winding (which alone gives a rising characteristic) are combined so as to give the desired flat compounding or a certain degree of over-compounding. As the shunt winding alone gives very nearly the desired characteristic, the shunt is the principal winding, the series winding being supplementary and of relatively few ampere-turns.

The characteristic curves for a shunt or compound generator may be classed as *no-load characteristics*, and *load characteristics*.

PART I. NO-LOAD CHARACTERISTIC.

§ 5. There is one no-load characteristic, the *saturation curve*, which shows the saturation of the iron for different field excitations; for this the generator is usually self-excited but may be separately excited when so desired.

§ 6. (*a*) **No-load Saturation Curve.***—This curve shows the terminal voltage for different values of field current.

§ 7. *Data.*—The machine is connected as a self-excited shunt

voltage to rise; (2) when the voltage is too high, it momentarily removes the short circuit, causing the voltage to fall. The voltage would be much too high or too low, if the short circuit were permanently made or broken. The short circuit is, however, rapidly made and broken. and of a varying duration, a nearly constant voltage being thus secured. It may be applied directly to a generator (D.C. or A.C.) or to its exciter. It may be used advantageously in connection with a compound winding, and may be arranged so as to cause the voltage to rise with load in the same manner as in an over compounded generator.

* Also called *excitation characteristic*, or *internal shunt characteristic*.

generator, Fig. 1, and is driven without load at constant speed. Readings are taken of field current, terminal voltage and speed. The field current is varied by adjusting the field rheostat by steps from its position of maximum to minimum resistance. This gives the ascending curve; the resistance is then increased again to its maximum for the descending curve. If the rheostat, with resistance all in, does not sufficiently reduce the field current, a second rheostat may be placed in series with it. The machine "builds up" from its residual magnetism as does the series generator; if the field winding is connected to the armature in the wrong direction, the machine will not pick up but will tend to become demagnetized. Should the direction of rotation be reversed, the field connection should be reversed.

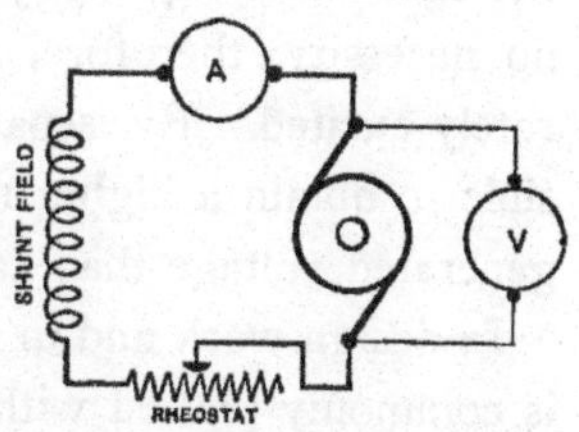

FIG. 1. Connections for no-load saturation curve.

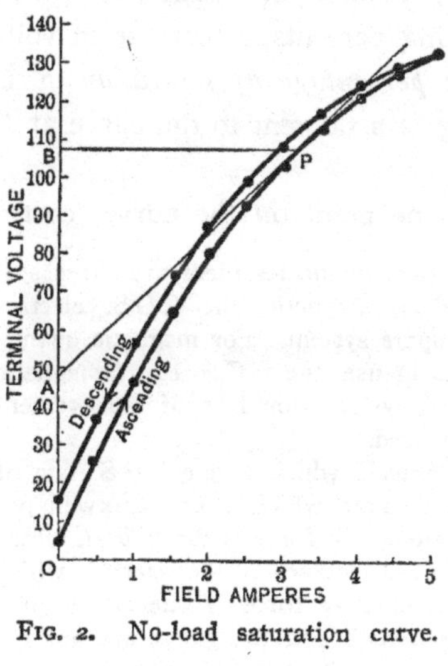

FIG. 2. No-load saturation curve.

§ 8. *Curves.*—Voltage readings are corrected by proportion for any variation in speed (§ 9, Exp. I–A), and the curves plotted as in Fig. 2.

§ 9. *Interpretation of Curves.*—The curves in Fig. 2 show the saturation of the iron and are much the same as the characteristic of a series dynamo. The current through the armature is small, being only a few per cent. of full-load current; the resistance drop through the armature may accordingly be neglected and the measured terminal voltage be taken as (practically) equal to the total generated voltage. Likewise, the armature current is so small that armature reactions are negligible, and the curve is practically the same as a separately-excited magnetization curve. There is no necessity, therefore, for taking curves both self- and separately-excited. By separately exciting a generator, it is possible to obtain a higher magnetization and consequently a higher generated voltage than can be obtained by self-excitation.

In design work and in manufacturing tests, the saturation curve is commonly plotted with field ampere-turns, instead of amperes, as abscissæ. However plotted, the abscissæ are proportional to magneto-motive force and the ordinates to magnetic flux.*

§ 10. *Saturation Factor and Percentage of Saturation.*—There are two ways for expressing† numerically the amount of saturation for any point P on the working part of the curve. (1) The *saturation factor, f,* is the ratio of a small percentage increase in field excitation to the corresponding percentage increase in voltage thereby produced. (2) The *percentage of saturation, p,* is the ratio $OA \div OB$, when in Fig. 2 a tangent to the curve at P is extended to A.

Compute these two for some one point on the curve, corre-

* (§ 9a). *Magnetic Units.*—For electrical quantities there are three systems of units in use—the C.G.S. electromagnetic, the C.G.S. electrostatic and the practical or volt-ohm-ampere system. For magnetic quantities there is only one system of units in use, the C.G.S. electromagnetic system; magnetic units of the practical system would be of inconvenient size, they have no names and are never used.

The unit of magnetic flux is the *maxwell,* which is one C.G.S. line of force. The unit of flux density is the *gauss,* which is one maxwell per square centimeter. The unit of magnetomotive force is the *gilbert,* which is $(10 \div 4\pi)$ ampere-turn. The unit of reluctance is the *oersted,* which is a reluctance through which a magnetomotive force of one *gilbert* produces a flux of one *maxwell.* The *maxwell* and the *gauss* are authorized by International Electrical Congress, but not the *gilbert* and the *oersted.*

Analogous to Ohm's Law (current = electromotive force ÷ resistance), we have the corresponding law for the magnetic circuit: flux (maxwells) = magnetomotive force (gilberts) ÷ reluctance (oersteds).

† A. I. E. E. Standardization Rules, 57, 58.

sponding say to normal voltage, and check by the relation $p = 1 - 1/f$.

These terms are useful because they make possible an exact numerical statement of the degree of saturation of a machine, under working conditions, without the reproduction of a saturation curve. For a more complete study, compute p and f for different points and plot.

PART II. LOAD CHARACTERISTICS.

§ 11. The usual load characteristics are the *shunt, compound* and *armature* characteristics.

In taking the *shunt* and *compound* characteristics, the machine is left to itself with the field rheostat in one position during the run, the curve showing the variation in terminal voltage with load.

In taking the *armature* characteristic the field rheostat is constantly adjusted; the curve shows the variation in excitation necessary to maintain a constant terminal voltage at different loads.

The *differential* and *series* characteristics are not commercial characteristics but are included to show more fully the operation of the series winding. (For *full-load saturation curve,* see § 33.)

§ 12. (*b*) **Shunt Characteristic.**—This is the working characteristic of the machine when operated at normal speed as a shunt-wound generator and shows the variation in terminal voltage with load (Curve *A*, Fig. 4).

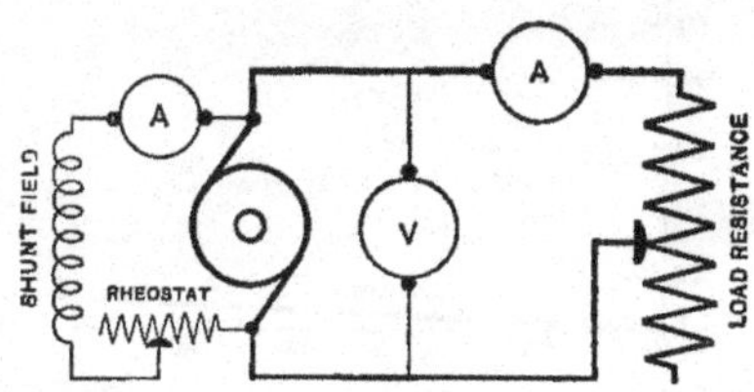

FIG. 3. Connections for shunt characteristic.

§ 13. *Data.*—The connections are shown in Fig. 3. Readings are taken of terminal voltage, field current, line current and speed. No speed correction is made, there being none which is simple and accurate. The field rheostat is set in one position and no change is made in it during the run.

§ 14. The setting of the rheostat for commercial testing (§ 21a) is made for normal voltage at full load. For the purposes of this experiment, it is usually preferable to set the rheostat for normal voltage (or for any selected value of voltage) at no load; in this case the shunt, compound and differential curves, Fig. 6, all start from the same no load voltage. The load current is then increased from no load up to about 25 per cent. overload and then decreased, if so desired, back to no load. The return curve will fall a little below, on account of hysteresis.

Data are also to be taken for a characteristic starting at no load with a voltage below normal (§ 18).

Armature resistance is measured by the fall-of-potential method, (§ 17, Exp. 1–A).

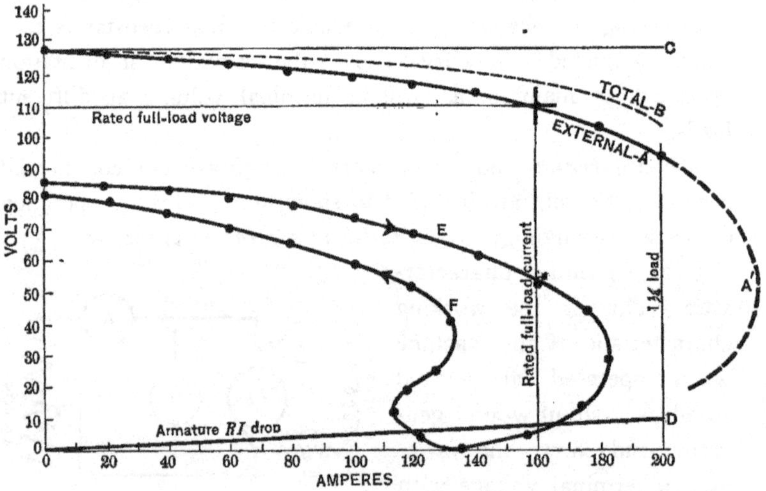

FIG. 4. Shunt characteristics.

§ 15. *Curves.*—The armature *RI* drop is plotted as Curve *D* in Fig. 4.

For the *external* shunt characteristic (Curve *A*, Fig. 4), plot observed line current as abscissæ and observed terminal voltage as ordinates.

For the *total* shunt characteristic (Curve *B*), plot total armature current* (line current plus field current) as abscissæ, and total generated voltage (terminal voltage plus armature *RI* drop) as ordinates.

§ 16. *Interpretation* (*Armature Reactions and Regulation*).— An ideal characteristic would be the straight horizontal line, Curve *C*, indicating a constant voltage at all loads. As a matter of fact the terminal voltage (Curve *A*) decreases with load. There are, at constant speed,† three causes for this: (1) *armature resistance drop*, (2) *armature reactions which reduce the magnetic flux* and (3) *decreased field excitation as the voltage decreases.*

The difference between Curves *A* and *B* shows the effect of (1) resistance drop; the difference between *B* and *C* shows the effect of (2) armature reaction and (3) decreased excitation, and of (4) if speed varies.

The difference between *B* and *C* will show the effect of armature reactions (2) alone‡ if a run is made at constant excitation and constant speed, thus eliminating (3) and (4). This is the practical *method for determining armature reactions.* The machine may be self excited or (preferably) separately excited.

§ 17. The *regulation* of the generator is shown by the drop in Curve *A*. To express regulation numerically as a per cent., the rated voltage at full load is taken as 100 per cent. In a commercial test, therefore, the curve is taken by beginning at full load at rated voltage (100 per cent.) and proceeding to open circuit. The regulation|| is the per cent. variation from normal

* The difference between line and armature currents is so small that for many practical purposes the distinction between them can be neglected.

† (§ 16a). Should the generator slow down under load, as when driven by an induction motor, this would constitute a fourth cause (4).

‡ (§ 16b). Included, as a part of armature reaction, is the effect of local self-induction of the armature conductors, when traversed by the armature current which (in any one conductor) is rapidly reversing in direction.

|| A. I. E. E. Standardization Rules 187, *et seq.*

full-load voltage (in this case the per cent. increase) in going from full load to no load.

§ 18. *Characteristics taken with Low Field Excitation.*—On short circuit a shunt generator has no field excitation and the short-circuit current (depending on residual magnetism) is commonly less than normal full-load current. The current, however, is much greater before short circuit is reached. On account of this excessive current, the complete characteristic curve cannot be obtained with the field rheostat in its normal setting. To show the form of the complete shunt characteristic, set the field rheostat for a no-load voltage much below normal, and take Curve *E* (Fig. 4) from open circuit to short circuit, and Curve *F* returning from short circuit to open circuit. The form of these curves should be interpreted.

§ 19. With a weak field, armature reactions cause the terminal voltage to fall off with load more rapidly than with a strong field. This is seen by comparing Curves *E* and *F* with Curve *A*. The effect of armature reactions is least when the iron is highly saturated, for then any decrease in magnetomotive force (due to armature ampere-turns) does not cause a corresponding decrease in magnetic flux.

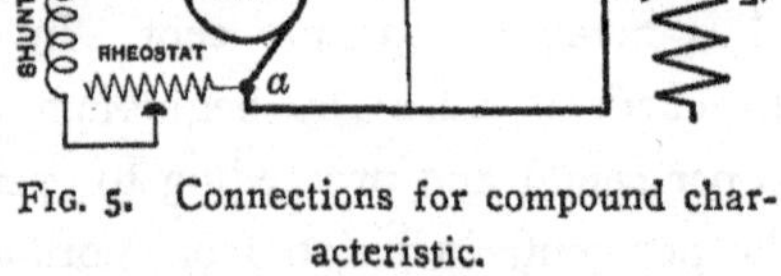

FIG. 5. Connections for compound characteristic.

(Compare Fig. 2.) It follows, therefore, that a ***shunt generator*** gives the best regulation when worked ***above the knee*** of the saturation curve. It will be found (§ 22) that this is not so for a compound generator.

§ 20. (*c*) **Compound Characteristic.**—The connections for taking the compound characteristic, Fig. 5, are the same as for the shunt characteristic, Fig. 3, with the addition of the series field winding which is in series* with the armature. The same readings of terminal voltage, field current, line current and speed are taken as for the shunt characteristic and no speed† corrections are made.

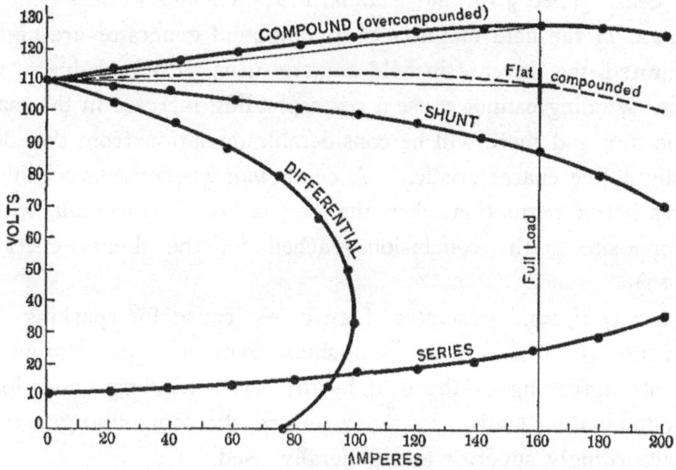

FIG. 6. Series, shunt, compound and differential characteristics.

§ 21. In Fig. 6 are plotted shunt, compound and differential characteristics, beginning with the same no-load voltage.‡

The compound characteristic cannot be made a perfectly straight line from no load to full load. What can be done is to have the terminal voltage at full load the same as the no-load

* (§ 20a). *Short Shunt and Long Shunt.*—The connection shown in Fig. 5 is *short shunt;* it would be *long shunt* if the shunt field were connected to the line terminals *ac,* instead of to the armature terminals *ab.* Both methods of connection are used commercially, the difference between them being slight.

† (§ 20b). A generator is compounded for the particular speed at which it is to operate. When it is to be driven by an induction motor, it may be compounded so as to take into account the slip of the motor, *i. e.,* its slowing down under load.

‡ (§ 21a). In commercial testing, the compound and shunt characteristics would be taken with the same normal voltage at full load (§ 14). The differential characteristic would not be taken.

voltage (flat compounding) or a definite percentage higher (over compounding). In either case the regulation is the maximum percentage deviation from the ideal straight line at any part of the characteristic, rated full-load voltage being taken as 100 per cent. (See § 17, and Standardization Rules.)

§ 22. If the field magnets of a compound generator are highly saturated, the increase in field ampere-turns with load due to the series winding cannot cause a corresponding increase in the magnetic flux and there will be considerable deviation from the ideal straight line characteristic. A *compound generator* accordingly gives better regulation when the iron is *below saturation,* which is opposite to the conclusion reached for the shunt generator (§ 19).

In a compound generator there is less cause for sparking and shifting of brushes than in a shunt generator, on account of the strengthening of the field by the series winding under load. For fluctuating loads, as railway service, the compound generator is accordingly superior and generally used.

Obviously, on account of the series winding, it is much worse to overload or short circuit a compound than a shunt generator.

§ 23. *Shunt for Adjusting Compounding.*—If the characteristic of a compound generator rises more than is desired, there are too many series ampere-turns. These can be reduced without changing the number of turns by reducing the current which flows through them. This is done by a shunt resistance in parallel with the series winding. A generator is usually given more series turns than are necessary, the desired amount of compounding being obtained by adjusting the shunt resistance. This is much easier than changing the number of series turns and makes it possible to change the amount of compounding at any time, even after the machine is in use.

§ 24. (*d*) **Differential Characteristic.**—The connections for this are the same as for the compound characteristic (Fig. 5) except that the series field winding is reversed so as to be in opposition to the shunt winding. The effect of the series winding is now to decrease (instead of increase) the magnetization of the iron, as the armature current increases, causing the voltage to fall off with load more rapidly than with the shunt field alone. As there is no demand for this, generators with differential winding are not used. (In a motor, Exp. 2–A, a differential winding is useful in giving constant speed).

§ 25. (*e*) **Series Characteristic.**—This characteristic shows the effect of the series winding alone, with the shunt winding not connected. The procedure is the same as in testing a series generator, the connections being as in Fig. 2, Exp. 1–A.

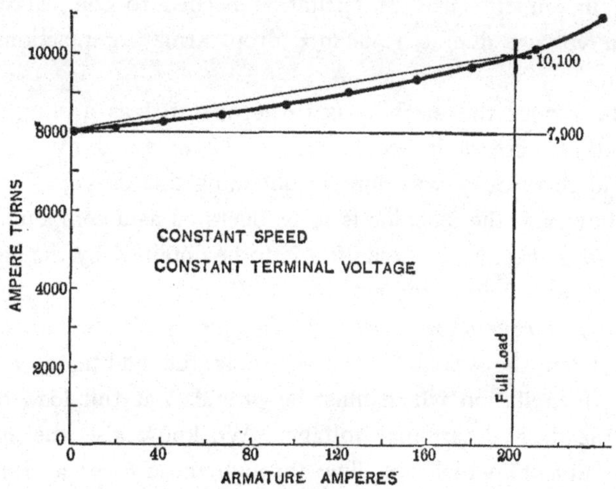

FIG. 7. Armature characteristic or field compounding curve, showing that at full load 2,200 more ampere-turns are needed than at no load for constant terminal voltage.

§ 26. (*f*) **Armature Characteristic.**—This curve is used in determining the proper number of series turns for compounding a generator; and is therefore frequently called a *field compounding curve.** It shows, Fig. 7, the variation in field excitation

* This has also been called an "excitation characteristic," a name which is ambiguous since it may be taken to mean the saturation curve, §6.

(amperes or ampere-turns*) necessary at different loads to maintain a constant voltage at the terminals of a shunt generator driven at constant speed.† The connections are shown in Fig. 3, readings being taken of field current, line current, terminal voltage and speed. Separate excitation may be used when a higher excitation is wanted than can be obtained by self-excitation. The load current is increased from no load to about 25 per cent. overload. At each load, before readings are taken, the voltage is brought to the desired constant‡ value by adjusting the field rheostat.

§ 27. The rise in the armature characteristic shows the increase in ampere-turns of excitation needed to compensate for loss in voltage due to resistance drop, armature reactions, etc. (§ 16).

If in service the machine is to be operated as a shunt generator, this increase in excitation can be obtained by adjusting the field rheostat as was done in obtaining this curve.

If, however, the machine is to be operated as a compound generator, this increase in excitation is to be obtained by the ampere-turns of the series winding.

§ 28. *Determination of Proper Number of Series Turns.*—We know from the armature characteristic the additional ampere-turns of excitation which must be provided at full load to produce the desired terminal voltage. We know also the amperes (load current) which will flow through these turns at full load. The necessary number of turns is accordingly readily found by dividing ampere-turns by amperes. Thus in Fig. 7, we note that

* To plot in ampere-turns, the number of turns in the shunt field must be known; see Appendix I. The number of turns multiplied by field current gives the number of field ampere-turns.

† (§ 26a). In case the generator is to be normally driven by an induction motor, with speed decreasing with load, it should be so operated in taking the armature characteristic. (See §§ 16a, 20b.)

‡ (§ 26b). The curve may be taken for a voltage which increases with load; such a curve would show the series ampere-turns to be added for over-compounding.

for full load (200 amperes) there are needed 2200 more ampere-turns excitation than at no load. The series winding will be traversed by the current of 200 amperes, and must accordingly have 11 turns in order to make the required 2,200 ampere-turns.

If the armature characteristic were a straight line, the series turns calculated as above would be the same for all loads and the generator could be compounded so as to have perfect regulation and give an exactly constant voltage at all loads. But the armature characteristic always curves, bending more after saturation is reached. The series turns are, therefore, calculated for one definite load (full load); for other loads the compounding will be only approximately correct (§ 21).

The armature characteristic and hence the proper number of series turns for correct compounding, will differ for different speeds and terminal voltage,—an interesting subject for further investigation.

APPENDIX I.

MISCELLANEOUS NOTES.

§ 29. **Determining the Number of Shunt Turns.**—The number of shunt turns on a generator can be more or less accurately determined, if the machine has a series winding or a temporary auxiliary winding with a known number of turns.

With the machine separately excited, take an ascending no-load saturation curve, using the shunt field winding of unknown turns; take a second similar curve, using the series or auxiliary field winding of known turns. A comparison of the two curves shows that the shunt winding requires a much smaller current than does the auxiliary winding to give the same generated armature voltage. Determine this ratio of currents for equal terminal voltage (found for several voltages and averaged) and suppose it to be 1 : 40. The shunt turns are then 40 times as many as the auxiliary turns, the ampere-turns for equal terminal voltage being the same. If for example the auxiliary turns are 10, the shunt turns are accordingly 400.

§ 30. The number of turns in two field windings can be compared by the use of a ballistic galvanometer (or voltmeter or ammeter used as a ballistic galvanometer); the chief advantage of the method is that it does not require facilities for running the machine. With the armature stationary and the galvanometer connected to the terminals of one field, break a certain armature current and note the throw of the galvanometer. Repeat, breaking the same armature current with the galvanometer connected to the other field. The ratio of galvanometer throws gives the desired ratio of field turns. It is best to take a series of readings and average the results.

§ 31. An estimate of the number of turns in a coil can be made from its measured resistance, size of wire and mean length of turn. This can be used as a check, but the method is commonly only approximate on account of the uncertainty of the data.

§ 32. **To Compound a Generator by Testing with Added Turns.**—The proper number of series turns required to compound a generator can be ascertained by trial by means of temporary auxiliary turns. With the generator running at full load, pass current from an independent source through these auxiliary turns and adjust this current until the terminal voltage of the generator has the desired full-load voltage. This current (say 220 amperes), multiplied by the number of auxiliary turns (say 10) through which it flows, shows that 2,200 extra ampere-turns are needed at full load. If the full-load current is 200 amperes, the generator would accordingly require 11 series turns.

§ 33. **Full-load Saturation Curve.**—For obtaining this curve, the field excitation is varied and the load adjusted at each reading, so that the external current remains constant at its full-load value. Field currents are plotted as abscissae and terminal voltages as ordinates. Such a curve is to be taken later (Exp. 3–A) on an alternator; it may accordingly be omitted, in the present experiment, if time is limited.

CHAPTER II.

DIRECT CURRENT MOTORS.

EXPERIMENT 2–A. Operation and Speed Characteristics of a Direct Current Motor, (Shunt, Compound and Differential).

PART I. INTRODUCTORY.

§ 1. Generators and Motors Compared.—Structurally a direct current generator and a direct current motor are alike,* the essential elements being the field and the armature. The same machine may accordingly be operated either as a generator or as a motor.

Operating as a generator, the machine is supplied with mechanical power which causes the armature to rotate against a counter† or opposing torque; this rotation of the armature generates an electromotive force which causes current to flow and electrical power to be delivered to the receiving circuit.

Operating as a motor, the machine is supplied with electrical power which causes current to flow in the armature against a counter† or opposing electromotive force; this current creates a torque which causes the armature to rotate and mechanical power to be delivered to the shaft or pulley.

* (§ 1a). Since generators are built in much larger sizes than motors, one generator being capable of supplying power for many motors, there may be a difference in design due to size. Moderate size machines, generators or motors, are built with few poles,—four being common in small motors. On the other hand, very large machines—that is generators—are built with many poles.

In all direct current machines,—generators or motors—it is common practice to use a stationary field and a revolving armature (§ 1, Exp. 1–A).

† (§ 1b). There is no counter torque in a generator until current flows in the armature; there is no counter electromotive force in a motor until there is rotation of the armature.

It is seen that the operation, either as a generator or as a motor, involves (1) the generation of an electromotive force and (2) the creation of a torque, both of which depend upon fundamental laws of electromagnetism.

§ 2. **Generation of Electromotive Force.**—An electromotive force is generated in a generator or in a motor due to the cutting of lines of force, this electromotive force being proportional to the rate at which the lines of force or flux are cut, as already discussed in § 1, Exp. 1–A.

In a generator this electromotive force causes (or tends to cause) a current to flow; in a motor, it is a counter electromotive force and opposes the flow of current.

§ 3. **Creation of Torque.**—A torque is created in a generator or motor due to the forces acting upon a conductor carrying current in a magnetic field. In a motor this torque causes (or tends to cause) a rotation of the armature with respect to the field; in a generator, it is a counter torque and opposes the rotation of the armature.

The creation of torque depends upon the following fundamental principle:—When a conductor carrying current is located in a magnetic field, it is acted upon by a force that tends to move the conductor in a direction at right angles to itself and to the magnetic flux, the force being proportional* to the current and to the flux density.

This force creates a torque,—that is a turning moment or couple—equal† to the product of the force and the length of the

* (§ 3a). In C.G.S. units this force is equal to the product of the current, flux density, length of conductor and sine of the angle between the conductor and direction of flux. This sine is unity when the conductor and flux are at right angles, as in most electrical machinery. When there are a number of conductors, each conductor is subject to this force; the total torque of a motor is therefore proportional to the total number of armature conductors.

† (§ 3b). Torque may be expressed as *pounds at one foot radius, pound-feet, kilogram-meters,* etc. Power is proportional to the product of torque

radius or lever arm to which the force is applied. It accordingly follows that: *torque is proportional to armature current and to the flux density of the field;* this is irrespective of whether the armature is rotating* or not. A reversal of either the current or the flux alone reverses the direction of the torque.

Of the *total* torque, part is used in overcoming friction, windage and core loss; the remainder is *useful* torque and is available at the pulley.

§ 4. **Automatic Increase of Current with Load.**—The counter-electromotive force E' is always a few per cent. less than the supply voltage E. The difference is due to the resistance drop in the motor armature,—including brushes, brush contact and connections, and the series field (if any); that is

$$E' = E - RI. \tag{1}$$

---

and speed; thus, if R.P.M. is revolutions per minute and T is torque in pound-feet

$$\text{H.P.} = \frac{2\pi \times \text{R.P.M.}}{33{,}000} T.$$

If power is known, torque may be found by dividing power by speed. In pound-feet, torque is

$$T = \frac{33{,}000}{2\pi} \times \frac{\text{H.P.}}{\text{R.P.M.}}.$$

When power is in watts, it is frequently convenient to express torque in "*synchronous watts*"; thus,

$$T = \frac{\text{Watts}}{\text{R.P.M.}}.$$

(One synchronous watt = 7.04 pound-feet.)
(One pound-foot = 0.142 synchronous watt.)
Torque is also expressed in "watts at 1,000 R.P.M."; thus,

$$T = 1{,}000 \times \frac{\text{Watts}}{\text{R.P.M.}}.$$

* (§ 3c). Torque with the armature at rest (*static* torque) can be determined for various field currents and for various armature currents by means of a lever arm attached to the armature and a spring balance or platform scales.

Within the usual range of operation, this RI drop for a commercial motor is only a few per cent. of the total line voltage. Good design does not permit more, inasmuch as the output and efficiency are decreased by the same percentage.

The current which flows in the armature is seen to be

$$I = \frac{E - E'}{R}. \tag{2}$$

If under running conditions the current I is not sufficient to give the motor enough torque (which is proportional to current and flux) to do its work at the speed at which it is running, the motor will begin to slow down, thus decreasing the counter-electromotive force E' (which is proportional to speed and flux). As E' decreases I increases, until the torque is sufficient to meet the demands upon the motor. The current accordingly increases automatically with the load, and this increase can be continued until the safe* limit, determined by heating, is reached.

On the other hand, if the current I is more than is needed to give the torque required for the load at a certain running speed, the surplus torque will cause the armature to accelerate, thus increasing E' and decreasing I to a value which gives the proper torque for the load and speed.

It will be seen that a small change in E' is sufficient to cause a large change in I and therefore in the torque. As an example, suppose $E' = 100$, $E = 104$; if an increase in speed causes E' to increase 2 per cent., that is to 102, the current I will be reduced 50 per cent.

§ 5. **Relations between Speed, Flux and Counter-electromotive Force.**—Counter-electromotive force is proportional to speed (S) and flux (ϕ); that is

$$E' \propto \phi S. \tag{3}$$

Hence, speed varies directly as the counter-electromotive force and inversely with the flux; that is

* A motor is usually rated so that it can be run for several hours at 25 per cent. over its rated load.

$$S \propto \frac{E'}{\phi}; \tag{4}$$

or,

$$S \propto \frac{E - RI}{\phi}. \tag{5}$$

This is the speed equation for a motor. It is seen that if ϕ is reduced the speed will increase. The equation shows the definite numerical relations of the quantities involved. *How* an increase in speed is brought about by a decrease in flux is made more clear in § 7.

§ 6. **Speed of a Shunt Motor.**—A shunt motor with constant supply voltage has a constant field current and therefore a constant flux. It accordingly follows that the speed is nearly constant. The *RI* drop causes it to decrease with load (compare equation 5); this is partially offset, however, by the effect of armature reactions, as seen later (§ 8).

§ 7. It is seen from equation (5) that the speed may be increased or decreased by weakening or strengthening the field. The process is explained as follows:—

When the field is weakened the counter-electromotive force is reduced; this permits more current to flow in the armature, thus giving greater torque* and speed. The speed accordingly increases until E' has increased so as to limit the current (and hence the torque) to a value which will give no further acceleration.

The *cause* for increase of speed is *surplus torque*.

§ 8. **Armature Reactions and Brush Position.**—If the brushes are given a backward lead (which is usual in motors running in one direction, in order to obtain better commutation) the field is

* (§ 7a). As an example, suppose the field is weakened so that the flux is reduced 2 per cent. and E' the same amount; and suppose the armature current increases 50 per cent. Torque is proportional to flux and armature current and in this assumed case is increased 47 per cent.; for .98 x 1.50 = 1.47. This increase is only temporary, for the armature current and the torque decrease as higher speeds are reached.

weakened by the demagnetizing effect of armature reactions (§ 8a, Exp. 1–A). This causes the flux to decrease with load, so that the speed does not decrease as much as it would with the brushes in the neutral position. On account of armature reactions, therefore, the speed regulation of a motor is better; the voltage regulation of a generator is worse (§ 16, Exp. 1–B).

The proper brush position for best commutation is the position which gives minimum speed.

§ 9. If the backward lead of the brushes is increased, the speed of the motor under load can be increased until it equals or exceeds the speed at no load. Such a control of speed by brush adjustment is not practicable, however, on account of bad commutation and destructive sparking; the brushes should be given the position of best commutation. A *small* variation of speed can be made, if desired, by shifting the brushes, provided it is not enough to cause much sparking.

§ 10. **Speed Control.**—From equation (5) it is seen that the speed of a motor can be varied: by changing the impressed voltage, E; by varying resistance, R (series controller); or, by varying flux, ϕ. Each of these methods is in use for operating variable speed motors.

(*a*) Varying line voltage. Several line voltages can be obtained by using a number of line wires. Such a system is called a *multiple-voltage* system.

(*b*) Varying resistance. The series controller is in common use with series motors; it is used occasionally with shunt motors of small size.

(*c*) Varying flux. This can be accomplished either by a change in *excitation* (magnetomotive force) or a change in *reluctance;* for flux = magnetomotive force ÷ reluctance.

(1) Speed control by varying excitation is obtained in a shunt motor by a rheostat in series with the field (§ 7); in a series motor, by an adjustable resistance in parallel with the field.

(The possible method of control by brush-shifting, § 8, is not used.)

(2) Speed control by varying reluctance is obtained in certain shunt motors by varying the air-gap.

A limit to speed control by a variation in flux (by varying either excitation or reluctance) is reached on account of armature reactions; a considerable reduction in flux causes bad commutation. For varying the speed through a wide range, therefore, these methods can only be used if the effects of armature reactions are overcome. This was first satisfactorily accomplished by the *compensated winding* of Prof. H. J. Ryan, which was placed in slots in the pole faces. This compensation is now generally accomplished by the more easily constructed interpoles or *commutating poles* of the *interpole motor.*

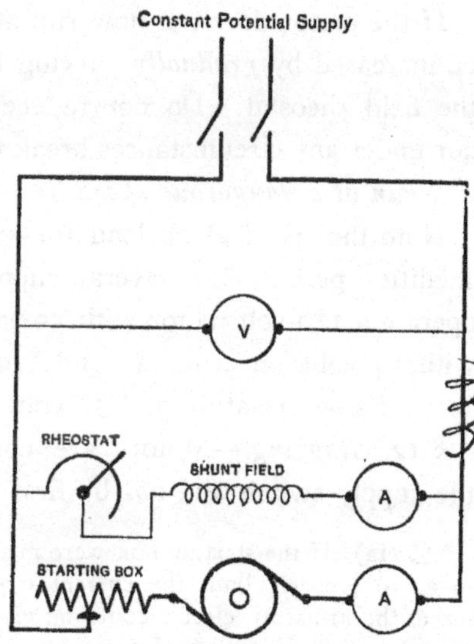

FIG. 1. Connections for operating a shunt motor.

PART II. OPERATION.

§ 11. **Shunt Motor.**—If the motor is compound, cut the series coil out of the circuit. Connect the supply lines to the main terminals of the motor and complete the connections, as in Fig. 1. Note the queries, § 15.

To start the motor, have all the starting box resistance in circuit and all the field rheostat out of circuit; make sure that the field circuit is complete. The circuits should be so arranged that closing the supply circuit will excite the field (which takes an appreciable time) *before** the armature circuit is closed. The armature circuit is then closed and this is commonly done by the starting box lever.

Bring the motor, unloaded, up to speed by cutting the starting box resistance *slowly*† out of circuit until the whole resistance is cut out. Note the ammeter during the process and the increase of speed as indicated by the hum of the motor. The starting box should be kept in circuit only during starting, for (except in special cases) it is not designed for continuous operation.

If the motor does not now run at normal speed, the speed can be increased by *gradually* varying the field current by means of the field rheostat. Do not reduce the field current too much, nor under any circumstances break the field circuit,‡ or *the motor will run at a dangerous speed.*

Note the speed at no load for several excitations; also, when facilities permit, for several supply voltages. (For example, operate a 110-volt motor with 55 volts on armature and on field; with 55 volts on armature and 110 volts on field; but *not* with 110 volts on armature and 55 volts on field.)

§ 12. *Stopping.*—Motors are commonly stopped by opening the supply switch and *not* by first opening the armature circuit.

* (§ 11a). If the starting box were made with sufficiently high resistance, so as to properly limit the current irrespective of counter-electromotive force, the armature circuit could be closed simultaneously with the field. This, however, is not usual practice.

† (§ 11b). Starting boxes are sometimes made so that it is impossible to manipulate them too rapidly. The "multiple-switch" motor starter, used particularly in starting large motors, has a number of switches, thrown successively by hand; these give good contact for large currents and require time for cutting out the successive sections of the resistance.

‡ (§ 11c). *Automatic Release.*—This danger is commonly guarded against by a solenoid on the starting box which releases the lever and allows it to spring back to the starting position when there is no current in the field circuit. This also acts as a "*no-voltage*" release, giving protection against damage which might occur were the current supply cut off and put on again with the starting box resistance all out.

There is then no sudden discharge of field magnetism and consequent liability to damage; for, as the armature slows down it generates a gradually decreasing electromotive force which maintains the field excitation so that it too decreases gradually. (If there is an automatic release on the starting box, it opens the armature and field circuits after the field excitation has decreased to a low value.)

The effects of induced electromotive force caused by sudden field discharge can be reduced by absorbing its energy in a high resistance shunt in parallel with the field circuit, or in a short-circuited secondary circuit around the field core. A brass field-spool will act in this way.

Throwing power suddenly off the line, by opening the supply switch, may cause fluctuations in line voltage,—particularly in case of large motors under load. To avoid this, before the supply switch is opened, the starting resistance may *first* be gradually introduced into the armature circuit, which, however, is not to be opened; *then* the supply switch is opened.

§ 13. **Compound Motor.**—In a compound* motor, the series winding strengthens the field as the armature current increases. On starting or under heavy load (*i. e.*, at times when the armature current is large) the motor is accordingly given a very strong field and therefore has—for a given armature current—a greater torque than it would have with the shunt† winding only.

* (§ 13a). To tell whether a series winding is connected "compound" or "differentially," throw off the belt and start the motor (for a moment) with the series coil only. If the motor tends to start in the same direction as it does with the shunt coil, the winding is "compound" or "cumulative;" if in the reverse direction, the winding is "differential."

† (§ 13b). This means a greater torque than it would have with the *same* shunt winding only. The motor could be given a different shunt winding which would give as strong a field and as great a torque as is obtained by means of the compound winding. Such a shunt winding, however, would give the strong field at *all* times; whereas the compound winding gives the strong field only at *particular* times,—*i. e.*, at starting and under load.

Under load the compound winding, by strengthening the field, causes the motor to slow down. For certain kinds of service—as in operating rolling mills, cranes, elevators, etc.—this is desirable in that the motor can work at great overload without the excessive demand for power which would be made by a constant speed motor. As compared with a shunt motor, it works under load at greater torque and less speed, and can stand a greater overload. In this respect it is similar to the series motor (see § 18). It differs from the series motor in that at light load there is still a certain strength of field due to the shunt winding, and the speed, therefore, cannot exceed a certain value, whereas a series motor will attain a dangerous speed if the load is thrown off. Under some operating conditions the compound motor can accordingly be used where neither the shunt nor the series motor would be suitable.

If slowing down with load is not wanted and a constant speed is desired at all loads, together with a large torque at starting, the series winding is used during starting only and is then cut out or short-circuited.

§ 14. **Differential Motor.**—Since a differential winding weakens the field as the load increases, such a winding makes possible a speed which *increases* with load. This is practically not desirable. In some cases, however, it is desirable to have the *same* speed at full load as at no load and to use a series winding just sufficient to overcome the tendency which a shunt motor has to slow down with load. If the series turns are too many for this, their effect can be cut down by a *shunt* of proper resistance connected in parallel with the series winding.

The starting torque of a differential motor is poor, particularly under load, inasmuch as the large starting current in the differential winding greatly weakens the field. For this reason, when a differential winding is used, it is usually cut out of circuit or short-circuited during starting.

If there are many series turns and no shunt is used, the current taken by a differential motor may become excessive as the load increases, thus weakening the field so that the motor races, or even reversing the field so that the motor suddenly reverses.

§ 15. **Queries.**—For increasing the speed, is the field current increased or decreased? Why? What is the use of the starting box? In starting, why do you not close the field and armature circuits simultaneously? Why is the starting box connected in series with the armature and not in series with the line? Why is a strong field needed for starting? Does this become of more or of less importance when starting under load? Would an added series winding be an advantage or a disadvantage in starting? Why would it be dangerous to break the field circuit? What is the effect of shifting the brushes? What is the proper position for the brushes? What is effect of interchanging positive and negative supply lines? What changes in connections are necessary to reverse the direction of rotation of the armature? (Be careful not to run more than a moment in the reverse direction, if the brushes would thus be damaged.)

PART III. SPEED CHARACTERISTICS.

§ 16. **Shunt, Compound and Differential Motor.**—It is the purpose of the experiment to determine the variation of speed with load for the same motor connected in three ways,—shunt, compound and differential; the line voltage is constant throughout the three runs. The brushes should be in one position during all the runs (§ 8), or the amount of any change noted.

With the motor connected as a shunt machine, Fig. 1, adjust the field current by means of the field rheostat so that the motor runs, on no load, at the speed for which it is designed, and keep the field current constant at this value during the run. For the other two runs, compound and differential, adjust the field current for this same no-load speed* and keep the field current constant during each run.

* (§ 16a). Starting with the same no-load speed, and making runs from no load to full load, gives the three speed characteristics of Fig. 2 coinciding at no load; this is the best procedure for instruction purposes.

In commercial testing, the field should be adjusted so that the motor runs at rated speed at full load. The curve is then taken from full load to no load; the maximum per cent. variation in speed from its full load value is the per cent. speed regulation. (Standardization Rules, 195.)

Vary* the load on the motor by steps between no load and 25 per cent. overload, reading line voltage, field current, armature† current (or else line current) and speed, for each step. Make runs with the motor connected shunt, compound and differential.

With current as abscissae (either line current or armature current) and speed as ordinates, plot speed characteristics for the three runs as in Fig. 2.

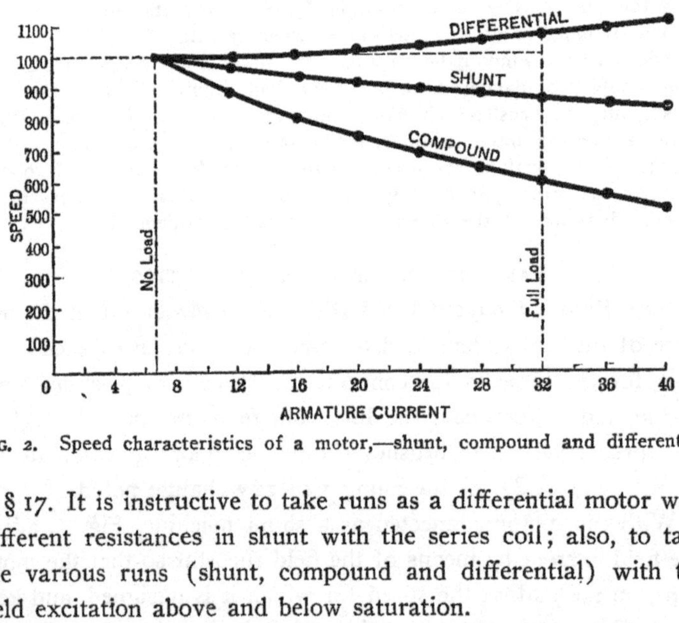

Fig. 2. Speed characteristics of a motor,—shunt, compound and differential.

§ 17. It is instructive to take runs as a differential motor with different resistances in shunt with the series coil; also, to take the various runs (shunt, compound and differential) with the field excitation above and below saturation.

* (§ 16b). This may be done by means of a brake, a blower, a belted generator or other convenient load; if a generator is used, its output may be absorbed in resistance or pumped back into the line (§ 26, Exp. 2-B).

† (§ 16c). If the armature current is measured, the field current is added to give the line current; if the line current is measured, the field current is subtracted to give the armature current.

APPENDIX I.

SERIES MOTOR.

§ 18. *Operation.*—A series* motor is distinctly a variable speed motor. Its characteristics are shown in Fig. 3. The speed increases rapidly as the load is decreased, becoming dangerously† great if the load is removed or reduced too much. The series motor, therefore, cannot be run at no load and normal voltage; it can be run at no load with a series resistance in circuit.

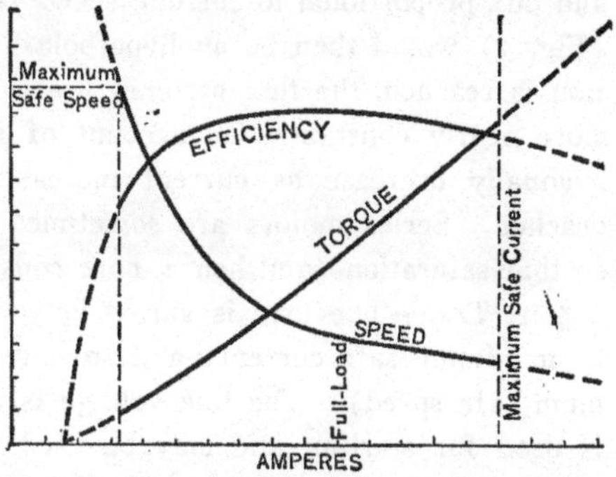

FIG. 3. Characteristics of a series motor, operated at constant voltage.

The series motor, besides being used for traction,‡ is used for hoists, etc. For such service it is well adapted. The important characteristic is that by slowing down under heavy load, it can increase its torque without requiring a corresponding increase in power; for *torque* = *power* ÷ *speed* (§ 3b). If the speed did not decrease with load, it is seen that the power would have to be greatly increased to give the same torque. This would require a much larger motor.

* (§ 18a). For the purpose of comparison with the shunt, compound and differential motor, the characteristics of the series motor are here described, although its test is not usually to be included as a part of the present experiment. When the test is made, it is well to combine it with efficiency measurements, § 33, Exp. 2-B.

† (§ 18b). In the laboratory, be prepared to shut down quickly if excessive speed is reached. With a belted load, there is danger of the belt flying off; with a brake, there is danger of an unintentional sudden decrease in load.

‡ (§ 18c). In traction, the controller is usually so arranged that two motors can be connected in series or in parallel with each other for speed control, thus giving each motor *half* or *full* voltage. The series resistance is likewise used for control and for starting. In starting, the resistance and both motors are all in series.

§ 19. *Torque.*—Since torque varies as *flux* $\times$ *current*, the torque would vary as I^2, if flux were proportional to current. For small currents—below saturation—this is more or less true. For large currents—after saturation—the flux is practically constant and the torque increases directly as I. The torque curve, Fig. 3, is therefore at first more or less parabolic and then becomes a straight line.

§ 20. *Speed.*—From equation (5) it is seen that speed varies inversely with flux. For small currents, if we consider RI negligible and flux proportional to current, speed varies as $1/I$; the speed curve (Fig. 3) would then be an hyperbola. For larger currents saturation is reached, the flux becomes practically constant and the speed more nearly constant. On account of RI drop, speed continues to gradually decrease as current increases, even after saturation is reached. Series motors are sometimes *overwound*, that is, wound so that saturation (and hence more constant speed) is soon reached.

§ 21. *Test.*—The load is varied between an overload (determined by maximum safe current) and an underload (determined by maximum safe speed). The line voltage is constant; a series resistance is used for starting and may be used for adjusting voltage. Any method for loading can be used. If a shunt generator is used as a load, its output may be absorbed in resistance or pumped back into the line. (See § 26, Exp. 2–B.) The pumping back method has been modified by A. S. McAllister, so as to form a convenient method for determining the torque of *any* kind of motor, direct or alternating (*Standard Handbook*, 3–239 and 8–151; McAllister's *Alternating Current Motors*, p. 185).

§ 22. *Power.*—Power is equal to EI and, when E is constant, power is directly proportional to I. In Fig. 3, power would be represented by a straight line passing through the origin. It will be seen, therefore, that the power required does not increase as rapidly as does the torque.

EXPERIMENT 2–B. Efficiency of a Direct Current Motor* (or Generator) by the Measurement of Losses.

§ 1. **Introductory.**—Efficiency is the ratio of output to input. The obvious and direct method for determining the efficiency of a motor is, therefore, to measure the output† and the input and take their ratio. An indirect method, known as the *method of losses* or *stray power* method, avoids the measurement of output. In this method the losses are measured and the output obtained by subtracting the losses from the input; the efficiency is then determined.

This method of losses possesses several advantages over methods that involve the measurement of output. The motor output is in some cases a troublesome quantity to measure, especially if accuracy is essential; but, even with the same degree of accuracy in the measurement of output and of losses, the efficiency cannot be as accurately determined‡ from the former as from the latter.

* With the appendices, this experiment covers the main features of the usual methods for determining the efficiency of any machine, direct or alternating. The main experiment is explicit for determining the efficiency of a shunt motor, and it is suggested that the student, without reference to the Appendices, first performs this main experiment. The Appendices should then be read and, if desired, a second experiment made (either now or later) under some of the special conditions which are there treated.

† (§ 1a). *Direct Measurement of Output.*—The output of a motor can be determined directly by *electrical* measurement (using for a load a calibrated generator, § 24), or by *mechanical* measurement (measuring torque by means of a Prony brake, Brackett cradle dynamometer, etc.). Power can be readily computed when torque and speed are known (§ 3b, Exp. 2–A). There are various forms of absorption and transmission dynamometers conveniently arranged for the direct measurement of power. For description of Prony brake, see Flather's *Dynamometers and the Measurement of Power* and the usual hand and text books; also *Electric Journal,* I., 419. For the cradle dynamometer, see Nichols' *Laboratory Manual,* Vol. II., and elsewhere.

‡ (§ 1b). Let us suppose that the error in measuring the input, output or losses is one per cent., due to inaccuracies in the instruments or in

A further advantage of this method is that a load run is not an essential, as will be seen later, and hence may be omitted. Conditions often arise, as in testing large machines, when a load test is impossible and this advantage then becomes important. It is always best, however, to make the load run when this can be done.

The method of losses is general and can be applied for determining the losses, and hence the efficiency, of a shunt, compound, differential or series wound motor or generator. In the following paragraphs the directions are full and explicit for testing a shunt-wound motor. Modifications are outlined in the Appendices for applying the method to other types of motors and generators.

§ 2. For testing any machine two runs are made: a *load run* to ascertain working conditions, and a *no-load run* (or runs) to determine losses under these same conditions.

In making the no-load run for losses the machine can be driven *electrically* as a motor or mechanically as a generator. The former method is used in this experiment (§ 7); the latter method is described in § 21 of Appendix I.

The resistance* of the armature is to be found by the fall of potential method both before and after the load run, in order that it may be determined both cold and hot (see § 17, Exp. 1–A). Since this includes the resistance of the brushes and of brush contact, which varies with current, to be exact it would be necessary to measure the armature resistance for each load

their reading. Assume the true output to be 95 when the true input is 100. The output, as measured, might vary from 94.05 to 95.95 and the input, from 99 to 101; hence the efficiency, determined from output, might vary from 93.1 to 96.9 per cent. On the other hand with the same percentage error in their determination, the measured losses might vary from 4.95 to 5.05 and the measured input from 99 to 101; hence the efficiency, determined from losses, could only vary from 94.9 per cent. to 95.1 per cent.

* In measuring armature resistance the voltmeter is to be connected to the same points as in the load run.

current. No account will be taken of a possible difference between the contact resistance with machine running and that measured with armature stationary.

§ 3. **Load Run (Shunt Motor).**—This run is made* to ascertain the working conditions for which the losses are to be determined, that is, to ascertain the load current and hot resistances for calculating copper losses and to ascertain the normal speed and excitation for which the iron and friction losses are to be determined in the no-load run. (The load run is a repetition of the run made in Exp. 2–A for obtaining speed characteristics.)

§ 4. Connect the motor to the supply lines, the voltage of which should remain practically constant during the run. (See Fig. 1 of Exp. 2–A.) Adjust the field current by means of the field rheostat so that the motor runs at its rated full-load† speed (or the speed for which its efficiency is desired) and *keep the field current constant* at this value during the run. Care in keeping the field current constant will increase the accuracy of the results; it is not sufficient to leave the rheostat in one position and assume the field current constant because it is very nearly so.

* (§ 3a). *Omission of Load Run.*—It will be seen that the load run is not essential and that the method may be employed even when the load run is impossible. Whenever it is possible, however, the load run should be taken, since it serves to get the machine "down to its bearings," that is, down to its working condition of friction as well as of temperature.

When the load run is omitted, cold resistances are measured and hot resistances determined by suitable temperature corrections or assumptions. Values of field current and speed are determined for no load; values are assumed for full load which it is believed will most nearly represent the operating conditions for which the efficiency is to be obtained. In a motor, for example, we may assume a constant excitation and a constant speed, or a speed which is say 5 per cent. lower at full load, etc. In a generator we may assume a constant speed and a constant excitation, or an excitation which is a certain amount lower (shunt generator) or higher (compound generator) at full load.

† (§ 4a). For commercial testing the speed should be adjusted to its rated value at full load; in laboratory practice the adjustment, when desired, may be made at no load.

§ 5. Beginning at about 25 per cent. overload, as estimated from the input, vary the load by steps from overload to no load or *vice versa;* at each step measure the line voltage, armature current,* field current and speed.

§ 6. The motor may be loaded in any manner that is convenient. A brake may be used for this, but it is frequently more convenient to load with a generator and to absorb† the output of the generator by suitable resistances.

§ 7. **No-load Run (Shunt Motor); Machine Driven Electrically.**‡—For a shunt machine one‖ no-load run is made; the machine is operated as a motor at the same constant excitation as in the load run. The object is to determine the losses for different speeds at this constant excitation. Before taking readings the motor should be run awhile so as to attain its normal working condition of lubrication, temperature, etc.

With the motor running unloaded, adjust the field current to the same value as during the load run and *hold constant* at this value during the no-load run. By varying the electromotive force impressed on the armature terminals, vary the speed of the motor by steps so as to cover as wide a range of speed as possible; this will give more accurate results than if only the speed range of the load run is covered. At each step measure the

* See § 16c, Exp. 2-A.

† If a direct current generator of suitable voltage is used, the current from the generator may be "pumped back" into the motor supply line (§ 26).

‡ (§ 7a). This run can be made with the machine driven mechanically (§ 21) instead of electrically.

‖ (§ 7b). Although a run at only one excitation is necessary for determining the efficiency of a shunt motor, runs at other excitations are recommended. These additional runs may be taken by the two voltage method (§ 7d). They are necessary if hysteresis loss is to be separated (Appendix I.) or if flux density is variable (Appendix III.). If a run is wanted at a very high saturation, a higher voltage may be supplied to the field than the rated voltage supplied to the armature.

electromotive force impressed on the *armature* terminals, armature current,* field current and speed.

By using two resistances, *B* and *C*, arranged as in Fig. 1, the electromotive force impressed on the armature may be varied by short circuiting more or less of *B* or of *C*. A single series resistance *B* may suffice, but the adjustment in many cases can be better made with two. An independent generator can be used as a supply to obtain variable voltages for the armature circuit, or the two voltages† of a three-wire system.

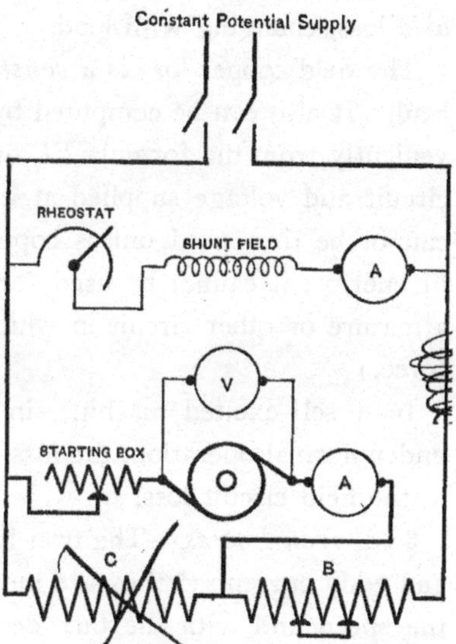

FIG. 1. Connection for no-load run as a shunt motor for determining losses.

§ 8. **Results.**—The losses of the motor include:

(1) Copper losses of field and armature;

(2) Iron losses of armature;

(3) Friction and Windage, or air resistance.

Losses (2) and (3) are *rotation losses* and are independent of load.

* (§ 7c). For the no load run the armature current is small; if a low reading ammeter is used, it should be short-circuited at starting to avoid damage by the initial rush of starting current.

† (§ 7d). *Two-voltage Method.*—For instruction purposes a complete series of armature voltages and corresponding speeds is desirable. Where two supply voltages (as 110 and 220 volts on a 3-wire system) are available, accurate results may be obtained by a two-voltage method, by taking 8 or 10 readings and averaging first with say 220 and then with 110 volts impressed on the armature of a 220 volt motor. These points, accurately determined, are sufficient for working up results by the straight line method of Fig. 2, in which they are represented by black dots *p* and *q*. By this method the trouble of adjusting armature voltage is avoided.

§ 9. *Copper Losses.*—The copper losses for any circuit can be computed, if the current and resistance through which it flows are known, being equal to RI^2 where R is resistance and I is current. The armature copper loss is thus computed; it is a *variable* loss, changing with load.

The field copper loss is a *constant* loss and does not vary with load. It also can be computed by the formula RI^2, or more conveniently from the formula EI, the product of current in the field circuit and voltage supplied at its terminals. (The formula EI cannot be thus used unless copper loss is the only expenditure of energy; it cannot be used for determining copper loss of an armature or other circuit in which there is a back electromotive force.)

In a self-excited machine, in which a field rheostat is used under normal operation, the loss in the rheostat is to be included in the field circuit loss.

§ 10. *Iron Losses.*—The iron losses are losses due to hysteresis and eddy currents;* they are independent of load, but vary with the speed and with the flux density in the armature. At constant speed, hysteresis loss (within the usual working range) varies approximately as the 1.6 power of the flux density; eddy currents as the square of the flux density. At constant flux density, hysteresis loss varies directly with the speed and eddy currents with the square of the speed. If the field current of the motor is held constant, the flux density in the armature will be practically constant for all loads. It will be modified under load† to a small extent by armature reaction, the effect of which will be neglected. Hence in a shunt motor run with constant

* This includes eddy currents in the pole pieces and in armature copper as well as in armature iron.

† (§ 10a). *Load Losses.*—Losses which occur under load in addition to copper losses and to the no-load iron, friction and windage losses are termed load losses. Any loss due to field distortion constitutes such a loss. Load loses are usually neglected as small or are estimated. See Standardization Rules 114–7.

field current, the iron losses are independent of load and depend upon speed alone.

§ 11. *Friction and Windage.*—The friction and windage losses are also independent of load and depend alone upon speed, being (for all practical purposes) directly proportional* to speed. Friction includes frictions of brushes as well as of bearings.

§ 12. *Rotation Losses W_0 (Combined Iron Losses, Friction and Windage).*—In the no-load run the power supplied to the armature (product of armature voltage and current) gives the rotation losses plus a small armature copper loss. This copper loss is subtracted (or neglected as small) to get the rotation losses. These losses are sometimes termed *stray power.*†

The combined rotation losses W_0, thus determined at no load, will be present at all loads and will have the same value for the same speed and excitation. If the speed of the motor is very nearly constant, the W_0 losses will be correspondingly constant. Rotation losses are commonly classed among the *constant* losses,‡ inasmuch as they are independent of load and the variation due to any small change of speed is small.

For determining efficiency there is no necessity for ascertaining the separate losses due to hysteresis, eddy currents, friction and windage, their combined value W_0 being sufficient.

§ 13. A curve should be plotted showing the rotation losses W_0 for constant field current at different speeds. To plot this curve accurately, it is best|| to first plot for various speeds the

* (§ 11a). Windage increases more rapidly than the first power of the speed; but windage loss is comparatively small and does not, at usual speeds, materially affect the law of variation of the *combined* friction and windage losses.

† The term stray power applies to any loss except copper loss.

‡ (§ 12a). The *no-load losses* are the rotation losses *plus* the copper loss of the field circuit (and the practically negligible copper loss of the armature); the no-load losses are therefore termed "constant."

|| (§ 13a). This is advantageous because a straight line can be drawn more accurately than a curved one, when the observed data are few or irregular; two accurate points are sufficient, but three are better as a

values of $W_o \div$ speed, which will give the straight line *ac* in Fig. 2. At very low speeds, there may be a deviation from a straight line, due possibly to errors in assumptions as to friction, etc., at these speeds. This, however, does not affect the accuracy of the construction; the straight part of the curve is to be extended back to *a*.

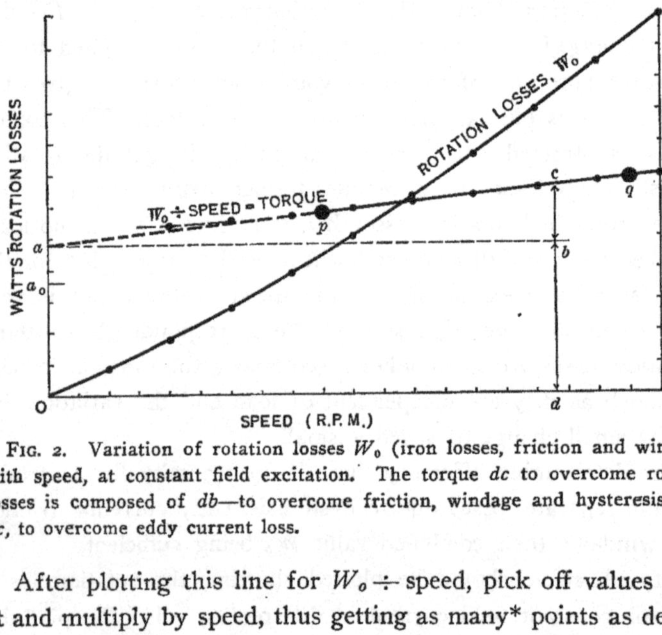

FIG. 2. Variation of rotation losses W_0 (iron losses, friction and windage) with speed, at constant field excitation. The torque *dc* to overcome rotation losses is composed of *db*—to overcome friction, windage and hysteresis—and *bc*, to overcome eddy current loss.

After plotting this line for $W_o \div$ speed, pick off values from it and multiply by speed, thus getting as many* points as desired for plotting the W_o curve. For fuller treatment, see Appendix I.

§ 14. *Efficiency.*—For any load (corresponding to readings in the load run, or assumed), the *input* is equal to the product of line current and voltage.

The *losses* are: the (variable) RI^2 loss in the armature for the particular armature current I; the (constant) copper loss of

check. It is always desirable to plot the results of any experiment, if possible, as a straight line, arc of circle, or as some curve whose law is known. The arc of a circle is much used in alternating current testing.

* Obtained in this way, more points may be used in plotting W_0 than the number of observations.

the field; and the (almost constant) rotation loss W_0, obtained from the curve in Fig. 2 for the particular speed and excitation.

The *output* is found by subtracting these losses from the input; the *efficiency* is output divided by input.

§ 15. Curves should be plotted with power output (or more simply with armature current) as abscissae, showing separate and total losses, input, output, efficiency, total current and speed; also useful torque (watts output ÷ speed); see Fig. 3. Compare the curves of Fig. 3 with the curves for a transformer, Fig. 4, Exp. 5–A, and Fig. 8, Exp. 5–B.

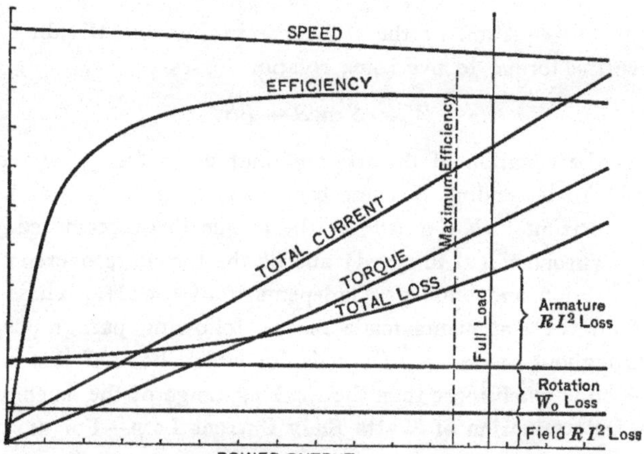

FIG. 3. Losses and efficiency of a shunt motor.

Maximum efficiency occurs when the variable loss (armature RI^2) equals the constant losses; see § 28.

It is seen that efficiency at light loads is low; this is true of both generators and motors. For this reason several generators are commonly run in parallel in a central station; as the load on the station decreases, the generators are cut out one at a time, so that the remaining generators will be more or less fully loaded and will run nearer the point of maximum efficiency.

APPENDIX I.

INTERPRETATION OF METHOD; AND SEPARATION OF LOSSES.

§ 16. **Interpretation* of Figure 2.**—For constant flux density (constant field excitation in a shunt machine), the losses due to hysteresis, friction and windage are proportional to speed (§§ 10, 11) and may be expressed as AS, where A is some constant and S is speed. Eddy current loss being proportional to the square of the speed may be expressed as BS^2, in which B is some constant. The total rotation loss is accordingly the sum

$$W_0 = AS + BS^2,$$

which is the equation of the W_0 curve in Fig. 2. Dividing by S, we have the torque to overcome rotation losses

$$W_0 \div S = A + BS,$$

which is the equation of the straight line† *ac* in Fig. 2. (See § 3b, Exp. 2–A.) Extending this line back to zero speed at *a* and drawing the horizontal *ab*, we have *bc* the torque to overcome eddy current loss (proportional to speed) and *db* the torque to overcome hysteresis, friction and windage (independent of speed). These statements and the statements made in the following paragraphs, hold true throughout the range of speeds for which $W_0 \div S$ is a straight line, which is much more than the working range of the machine.

§ 17. **Determination of Watts Eddy Current Loss.**—For any speed,

* (§ 16a). The *principle* of the graphical method which is here used was brought out by R. H. Housman and by G. Kapp, independently, in 1891 (London Electrician, Vol. XXVI., pp. 699 and 700); each made use of a straight line relation for plotting data obtained by running a motor at constant excitation and varying armature voltage. The *details*, as here given, have been modified by the writer with a view to making the method simpler and more useful. The original papers are excellent, but their method has been made unnecessarily cumbersome by writers who have followed them. Earlier, Mordey had used equations similar to those of § 16 for analytical separation of losses.

† (§ 16b). Since, at constant excitation, armature voltage (or more strictly counter-electromotive force) is proportional to speed, the W_0 curve can be drawn with E' as abscissae instead of speed. We then divide by E' (instead of S) and get the straight line *ac*, the ordinates of which ($W_0 \div E'$) are amperes.

multiplying *bc* by S, gives watts eddy current loss; multiplying *db* by S gives watts loss in hysteresis, friction and windage.

If the eddy current loss were zero, *ac* would coincide with the horizontal line *ab;* the first equation in § 16 would become $W_0 = AS$, showing that W_0 would be proportional to speed and the W_0 curve in Fig. 2 would become a straight line.

§ 18. *A Convenient Approximation.*—Since the eddy current loss is commonly only a small part of the total rotation losses, for *small* changes in speed it is nearly correct and often very convenient to say the rotation losses W_0 are *directly* proportional to speed.

§ 19. **Further Separation of Losses.**—Hysteresis loss can be approximately separated from friction and windage by additional runs at other field excitations. Friction and windage can not be separated from each other by any simple means and hence are considered together. There are various graphical and analytical methods for separating losses, all based on the following facts: friction and windage losses vary as first power of speed and are independent of flux density; eddy current loss varies as square of speed and square of flux density; hysteresis loss varies as first power of speed and 1.6 power of flux density. At any one speed, armature voltage is taken as a measure of flux density. In any of these methods it is necessary to make some assumption or approximation; for this reason the graphical methods are superior. (In the graphical method given below the approximation consists in obtaining Oa_0 by extrapolation to zero excitation.)

The analytical methods will not be taken up here; they consist in obtaining several equations (based upon the above relations) and eliminating between them after substituting numerical values obtained from observation of W_0 at various speeds and flux densities.

§ 20. *Graphical Method.*—Various graphical methods for separating losses differ chiefly in detail; the following procedure (either *a* or *b*) is suggested:

(*a*) Make a series of no-load runs, as already described, at various field excitations, extending these to as low a field excitation as possible. Plot results as in Fig. 2, obtaining a series of curves (straight lines) *ac* with intercepts Oa_1, Oa_2, Oa_3, etc., corresponding to various field currents. It is desired to find a value for an intercept Oa_0 for the supposed case of zero field current, for which of course no run can be made. To obtain this, plot a curve showing Oa_1, Oa_2, Oa_3, etc., for various field currents and continue the curve back to zero field current so as to get a value for Oa_0 by extrapolation.

(*b*) It will be found by experience that the value of Oa found by a run at a very low field excitation will differ but little from the desired value Oa_0 for zero excitation; that is, the iron losses at very low excitation are negligible. Instead of a series of no-load runs and extrapolation, *one* no-load run is taken at as low an excitation as possible; the value Oa obtained from this run is taken as the value of Oa_0 which would be obtained at zero excitation.

Referring to Fig. 2, Oa_0 obtained by either procedure just described is the torque to overcome friction and windage, for at zero excitation there is no hysteresis loss. To obtain watts loss in friction and windage at any speed, multiply Oa_0 by S; this is independent of excitation. To obtain watts loss in hysteresis at any speed for some particular excitation, multiply a_0a (for that excitation) by S.

§ 21. **Determination and Separation of Losses; Machine Driven Mechanically by an Auxiliary Driving Motor.**—This method, with the machine driven mechanically, is not limited to testing direct current machines; it can be used in testing alternators, synchronous motors, etc. By this method separate values are found for the *iron losses* and for the *mechanical losses;* that is, for hysteresis and eddy currents combined and for friction and windage combined.

The preceding method, with the machine driven electrically (§ 7), gave directly the eddy current loss and the combined hysteresis, friction and windage (§ 17). Each method has its advantages; in the one hysteresis is combined with eddy current, in the other with friction and windage.

The procedure is as follows: (1) The machine to be tested is separately excited and is driven as a generator* on no load at normal speed and excitation by means of a shunt motor; compare § 25. The motor input is measured. (2) The generator field circuit is broken and motor input again measured; the diminution† in motor input

* The armature winding is idle; this test therefore can be made for finding iron loss and friction of a machine with armature unwound.

† (§ 21a). This assumes that the motor losses remain constant. The small change in armature RI^2 loss will usually be negligible; if not

gives the iron losses (hysteresis and eddy current) of the generator. (3) The brushes of the generator are lifted, the diminution in motor input giving brush friction. (4) The belt is next thrown off, the diminution in motor input now giving the generator journal friction, windage and the belt loss.

The iron losses may be found for various excitations at normal speed. These losses should be determined for an *increasing* excitation; the losses with a decreasing excitation would be more.

For obtaining iron losses alone, this method with the machine driven mechanically is better than the method (§ 7) with the machine driven electrically; for it gives iron losses directly, separate from friction, and it is not necessary to go through any separation of losses as in § 20. This avoids error due to extrapolation and makes no assumption that friction and windage are directly proportional to speed.

On account of belt tension, journal friction will be more than in the no-load test with the belt off (§ 7). Belt losses are also included with friction and windage. This may sometimes be desirable, since it is the usual condition of operation. In a test of the motor *per se*, these losses ought not to be included, but they cannot be simply separated (§ 24a).

If the loss found by lifting the brushes is more when the machine is excited than when not excited, the brushes are not in the neutral position, thus causing additional loss by current circulating through an armature coil and brush.

If it is desired to separate the iron losses into components, hysteresis loss and eddy current loss, runs are made with varying speed and a constant excitation for each run. For each run plot *iron-loss* $\div S$ as a straight line, similar to *ac* in Fig. 2. For any speed, the product $bc \times S$ gives watts eddy current loss for the particular excitation; $ab \times S$ gives watts hysteresis loss.

negligible, it should be taken into account. Belt loss cancels out and does not enter into the determination of iron losses or brush loss.

APPENDIX II.

MISCELLANEOUS NOTES.

§ 22. **Efficiency of a Generator.**—To find the efficiency of a machine as a generator, a load run is made as a generator to ascertain the working conditions of speed, excitation and voltage. A no-load run as a motor is then made under these same conditions. The load run should be made whenever possible, but it can be omitted (§ 3a). In the load run, the field rheostat may be kept in one position (§ 12, Exp. 1–B) or changed so as to maintain the desired terminal voltage (§ 26, Exp. 1–B), according to what may be taken as the working conditions of the machine. Commercially the latter is more usual. For a compound generator, see § 31.

§ 23. **Efficiency of a Motor Generator.***—A load run is to be made when possible and measurements made of the various currents and voltages for both motor and generator. (See § 3a.) A no-load run is to be made if possible with the generator uncoupled; this determines the motor losses. Next make a run with the generator coupled but not excited, the increase in losses over the no-load run showing the friction and windage of the generator. Follow this with a run in which the generator has its proper excitation, the increase in losses over the preceding run showing the iron losses of the generator after copper losses have been taken into account. This last run gives the combined rotation losses for both machines. The copper losses are computed and added to these to get the total losses; knowing these, the efficiencies are readily computed for the two machines, combined and separately. As in the case of a generator or motor, due care is to be taken in all the no-load runs to have the proper speed and flux density in both machines. If the flux density in either machine was not constant in all the runs (as would be the case in a compound or differential machine), take note of Appendix III. The test may be made by reversing the set, that is, running the generator as a motor; this makes it possible to determine the friction and windage of the motor separate from iron losses.

§ 24. **Calibrated Generator for Measuring Motor Output.**—The out-

*The details of this test can be modified according to circumstance; see § 21.

put of a motor can be determined if for a load it drives a shunt generator whose losses are known; it is best to have the generator separately excited.

The motor output is equal to the power taken to drive the generator, that is, to the measured generator output (EI) *plus* generator losses. The losses are the copper losses and the rotation losses picked from curves (as in Fig. 2) for the particular speed and excitation; to this should be added the belt losses,*—a small but uncertain quantity. If the generator is separately excited, no account need be taken of field copper loss.

§ 25. **Calibrated Motor for Measuring Power to Drive a Generator.**—The power used in driving a generator can be determined if it be driven by a shunt motor whose losses are known. The power taken to drive the generator is equal to the motor input (EI for the armature) less armature RI^2, less W_0 for the particular speed and excitation, less belt loss (§24a).

§ 26. **Return of Power to Line by "Loading Back."**—If a direct current generator of suitable voltage is used as a load for a direct current motor, the current from the generator may be "pumped back" into the motor supply line (or into any other supply line). Used as a method of loading, it saves power, avoids the necessity of providing load resistances for the generator and introduces little complication.

The variation in load put upon the motor in driving the generator is obtained by varying the generator field current. First let us suppose that this is adjusted until the generator generates a voltage equal to the line voltage. When connected to the line (the positive terminal to the positive line), the generator will now neither give nor receive current, that is, will neither give power to nor receive power from the line. (At a lower excitation, it will receive power as a motor.) If the field current of the generator is now increased, it will generate a voltage higher than that of the line and will supply power to the line. This power can be increased by a further increase

* (§ 24a). *Belt Losses.*—Cotterill (*Applied Mechanics,* p. 265) says: "In ordinary belting this loss is small, not exceeding 2 per cent." The belt, on account of its tension, also increases the journal friction of both motor and generator.

in excitation, thus increasing the load on the driving motor as desired.

When the loading back method is thus used simply as a *loading* method and not as a testing method (§ 27), no measurements are made on the generator; measurements are made on the motor the same as though the generator were loaded with resistances.

Since one machine takes power as a motor and the other returns it as a generator, the net power taken from the supply line is only that which is required to supply the losses in the two machines.

§ 27. *Opposition Method for Testing Two Similar Machines.*—If two similar machines are operated as in the preceding paragraph and measurements are taken on both, they can be tested by Kapp's* opposition method and their combined losses determined.

There are various other opposition methods for accomplishing the same object; in each of these two similar machines are run, one as motor and the other as generator under load conditions. The two machines are connected both electrically and mechanically, so that power circulates between them and the only outside power taken is that necessary to supply the combined losses. These losses may be all supplied by the line (Kapp's method) or either partly or wholly by an auxiliary motor or by an auxiliary booster, giving rise to the various methods† of Hopkinson, Potier, Hutchinson and Blondel.

Although opposition methods are economical of power, they are not economical of time or apparatus; they are accordingly limited to testing pairs of large machines which could not be tested under load conditions in any other way. Temperature runs, regulation and efficiency tests are made in this way. Kapp's method is the simplest, but (on account of the different field excitation of the two machines) theoretically is not so accurate as some of the other methods.

§ 28. **Point of Maximum Efficiency.**—Consider that a machine has a certain constant loss (W_0 + field copper loss) and a variable loss (armature RI^2) which varies as the square of the load current I and

* This method and a modification by Prof. W. L. Puffer is fully described in Foster's *Electrical Engineering Pocketbook;* see also § 27a.

† (§ 27a). For full description and complete references, see Swenson and Frankenfield's *Testing of Electromagnetic Machinery;* see also R. E. Workman, *Electric Journal,* Vol. I., 1904, pp. 244, 289, 363; Karapetoff's *Exp. Elect. Eng.;* and various text and handbooks.

hence as the square of EI (the line voltage E being constant). These are shown in Fig. 4, in which the curve for total losses is a parabola.

At any point P on the total loss curve, the loss PA, expressed as a percentage of EI, is $PA \div OA$, which is the tangent of the angle POA. It is clear that this percentage loss is a minimum (and the efficiency a maximum) for the point P' where the line OP' is tangent to the total loss curve. But at this point P', we have $A'B' = B'P'$. (From the properties of a parabola, Od is bisected at c.) Hence:—*For any apparatus having a constant loss and a variable loss proportional to load current, maximum efficiency occurs at such a load that the constant loss and variable loss are equal.* The same result can be shown analytically by obtaining an expression for efficiency, differentiating and equating to zero (See Franklin and Esty's *Electrical Engineering*, I., 137).

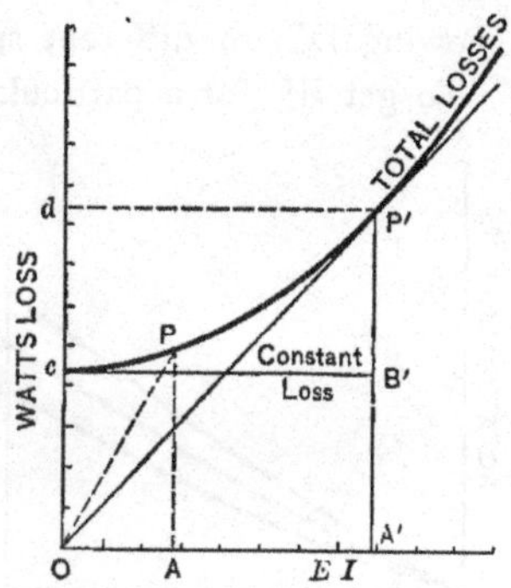

FIG. 4. Total losses represented by a parabola; P' is point of maximum efficiency.

This is true for any apparatus; thus, in a transformer, the efficiency is a maximum when the copper loss and constant core loss are equal. Within limits the designer may make the efficiency a maximum at the particular load he desires, giving due consideration to expense and to the uses to which the apparatus is to be put.

APPENDIX III.

MODIFICATION FOR VARYING FLUX DENSITIES.

§ 29. In the foregoing tests, the load run was made with constant field excitation, and hence at constant flux density; the no-load run was made at this same constant flux density. In cases where the flux density varies during the load run (due to a variation in the shunt field current or due to the action of the series field coil in a compound, differential or series wound machine), three (or more) no-load runs should be made at three different flux densities.

The following is suggested as a method for conducting the test.

§ 30. **Varying Excitation, Shunt Machine.**—First let us consider the case of a shunt machine, in which the excitation varied during the load run. Make three no-load runs at three excitations covering the range of excitations used in the load run. From these no-load runs, after plotting $W_0 \div S$, plot three curves (A, B, C in Fig. 5) showing W_0 for different speeds as before.

To get W_0 for a particular speed, erect a perpendicular in Fig. 5, corresponding to that speed. This perpendicular intersects the three curves A, B, C, giving (for a particular speed) the values of W_0 for different field currents. For each speed a derived curve may now be plotted giving W_0 for different field currents.

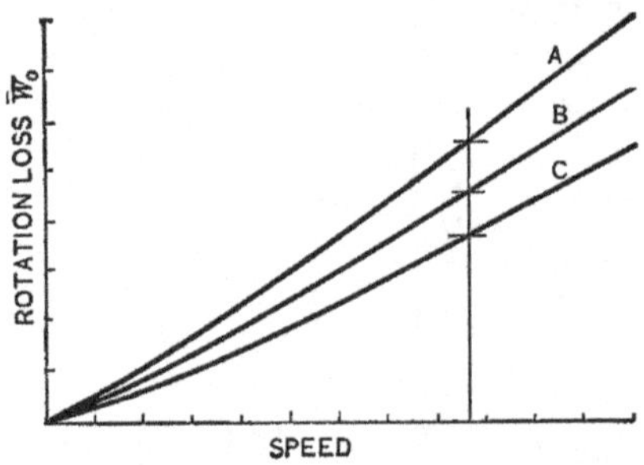

Fig. 5. Rotation losses W_0, for different excitations.

§ 31. **Compound Generator.**—In testing a compound generator, first make a load run to ascertain the *equivalent shunt excitation* and then make no-load runs as a shunt motor.

Load Run.—Make a load run as a compound generator, and note the values of terminal voltage and speed at three (or more) different loads; in each case ascertain the equivalent shunt excitation, *i. e.*, the field current which would give the same terminal voltage (and hence the same flux density) with the machine run as a shunt* generator at the same speed.

No-load Runs.—Knowing this equivalent shunt excitation, make the three corresponding no-load runs as a shunt motor at constant excitation, in each run using one of the three equivalent shunt field currents just determined.

* (§ 31a). This equivalent shunt excitation may be determined after each reading: without stopping the machine, the series winding should be first short circuited and then opened; or, the machine may be stopped and started again. Instead of this the equivalent excitation can be found from a separate shunt run (like an armature characteristic § 26, Exp. 1–B) in which is determined the field current which will give for each load the same terminal voltage as in the compound run.

Results.—Results are worked up as in the preceding paragraph. Curves are plotted as in Fig. 5 and derived curves found showing the variation of W_0 with field current for any speed. Such a derived curve is plotted for each speed observed in the load run.

§ 32. **Compound or Differential Motor.**—A load run is first made to find the equivalent shunt excitation; no-load runs are then made as a shunt motor.

Load Run.—Make a load run as a compound or differential motor, and note the speed at three (or more) different loads so chosen as to cover the speed variation of the run. In each case ascertain the equivalent shunt excitation, *i. e.,* the field current which would give the same speed* (and hence the same flux density) with the machine run as a shunt motor,—the load and the line voltage being the same as before.

No-load Runs.—Knowing this equivalent shunt excitation, make the three corresponding no-load runs as a shunt motor at constant excitation, each run using one of the three equivalent shunt field currents just determined.

Results.—The results are worked up as in the preceding paragraphs. From the three no-load runs three curves are plotted, as in Fig. 5, showing W_0 for varying speed at different excitations. From these curves a derived curve may be plotted showing the variation of W_0 with field excitation for any speed. Such a derived curve is plotted for each speed observed in the load run, and from it the value of W_0 obtained for the corresponding excitation.

* (§ 32a). The equivalent shunt excitation may be determined after each reading by cutting out the series coil as in § 31a.

The adjustment to a definite speed is, however, difficult without some particularly sensitive tachometer. To avoid this adjustment, proceed as follows:

Determine say five shunt speed characteristics, that is make five runs at different constant shunt excitations, determining speed for different loads. For each excitation plot speed as ordinates and armature current as abscissæ. By interpolating between these curves, we can find the shunt excitation that gives a particular speed for a particular armature current. This will give the equivalent shunt excitation corresponding to any speed and armature current found in the load run as a differential or compound motor. Knowing the equivalent shunt excitation, the corresponding no-load runs are made.

§ 33. **Series Motor.**—A series motor may be tested for losses in substantially the same manner as a shunt motor. So far as losses are concerned, a series motor is like a shunt motor in that the losses are the copper losses, which can be computed, and the rotation losses W_0 which depend only upon speed and excitation. In a series motor, however, speed and excitation vary greatly with load.

Load Run.—A load run as a series motor is taken to obtain speed and current for different loads; see Appendix I., Exp. 2-A. (If the no-load run is to be taken as in § 36, the load run is not a necessity.)

§ 34. *No-load Runs for Obtaining Rotation Losses; General Procedure.*—No-load runs may then be made at different constant excitations and W_0 found for different speeds by varying the armature voltage and measuring armature input in the usual manner. Readings are taken of field current, armature current and armature voltage. (The procedure is sometimes to take runs with constant armature voltage and varying excitation.) Any convenient means may be employed for obtaining the proper constant excitation and the desired armature voltage; the armature and field can best be supplied separately and not in series (see also § 37). Curves are plotted for each excitation, as in Fig. 5, showing W_0 for different speeds. Instead of speeds, armature voltage is commonly plotted as abscissae.

§ 35. *No-load Run for Obtaining Rotation Losses; Special Procedure.*—No-load runs, taken as in § 34, gives curves (Fig. 5) which tell the *complete* story, giving rotation losses for different speeds and field currents. As a matter of fact such complete information is often unnecessary; for, with constant potential supply, a series motor has a definite counter-electromotive force and a definite speed for any particular current (see Fig. 3 of Exp. 2-A). It is necessary, therefore, to get the rotation losses with each field current for the one corresponding speed only, this speed being obtained by supplying the armature with the proper voltage.

§ 36. This proper voltage to supply the armature could be found by trial (being adjusted until the speed in the no-load run for a particular field current is the same as in the load run for the same current). It is easier, however, to compute this voltage *without making a load run.*

We know that in any run (load or no load) speed is proportional to counter-electromotive force for the same excitation. For a particular field current I we will, therefore, have the same speed in the no-load run as in an assumed load run with current I, if in the no-load run the counter-electromotive force (which in this case is the impressed* armature voltage) is equal to the counter-electromotive force of the assumed load run. But for the load run we can compute the counter-electromotive force, $E' = E - RI$, for any assumed load current I. (Here E is the rated or assumed constant line voltage for which the losses are desired; R is the *hot* resistance of the armature and field, including brushes, etc.) Hence this is the proper voltage to supply the armature in the no-load run when the field current is I.

In testing a series motor by this method the field is excited with current I, which is given successive values, and the armature is supplied with the corresponding proper voltage, $E - RI$. (Or the armature can be given successive voltages and I adjusted to correspond.)

§ 37. A convenient method sometimes used for adjusting field current and armature voltage to their proper corresponding values is to connect the field and armature in series as a series motor with one regulating resistance in series with the line and one in shunt with the armature. For the first reading the series resistance is adjusted; after that, adjusting the shunt resistance alone will tend to cause the field current and armature voltage to assume automatically their correct relative values. (For this condition the series resistance is made equal to the armature resistance.) For modified ways of conducting the test, see R. E. Workman, *Electric Journal*, I., 169.

§ 38. *No-load Run for Friction.*—When the field current is very small, hysteresis and eddy current losses are so small that W_0 gives practically the friction and windage loss; compare paragraph (*b*), § 20. A run at low field excitation can be made as in § 34. This run, however, can most conveniently be made with the field and armature in series, the motor being run as a series motor on no load at a low voltage. The voltage and the speed are controlled by a series resistance; no shunt resistance is used. At no load the current through the field is so small that iron losses in the armature are negligible.

* The copper drop due to armature resistance at no load can be neglected or a small correction made.

CHAPTER III.

SYNCHRONOUS ALTERNATORS.

EXPERIMENT 3–A. Alternator Characteristics.*

§ 1 **Introductory.**—Alternating current generators are usually synchronous. Any machine—generator, motor or converter—is said to be synchronous when the current which it delivers or receives has a frequency proportional to the speed of the machine; otherwise it is asynchronous† or non-synchronous.

In a synchronous machine, the current or electromotive force has one half-wave or *alternation*—first positive and then negative—for each pole passed by a given armature conductor. A *cycle* is a complete wave of two alternations. In a synchronous machine, there is, therefore, one cycle for each pair of poles passed; the *frequency* (cycles per second) is, accordingly, equal to the speed (in revolutions per second) multiplied by the number of pairs of poles.

To deliver current with a frequency of 60 cycles per second (7,200 alternations per minute), a bipolar alternator would have to be driven at 60 revolutions per second, or at 3,600 revolutions per minute; a 4-pole machine, at 1,800 revolutions per minute, etc. Alternators are commonly made multipolar, and usually with many‡ poles, so as to avoid excessive speed.

* The curves used to illustrate this experiment and Exp. 3–B all relate to the same machine.

† (§ 1a). The induction motor and the induction generator are asynchronous. An induction motor must run *below* synchronous speed, *i. e.*, there must be a certain slip, in order to produce power. An induction generator, on the other hand, must be driven *above* synchronous speed in order to generate an electromotive force.

‡ (§ 1b). The high speed of the steam turbine has made possible, in fact has made necessary, large alternators with only few poles; for example, a *bipolar* 10,000 K. W. turbo-alternator, 1,500 revolutions per minute, is men-

§ 2. **Types of Alternators.**—Synchronous alternators are of the following three types (compare Part I., Exp. 1–A);

1. Alternators having a *revolving armature* and stationary field, used only for small machines.

2. Alternators having a *revolving field* and stationary armature, the most common type.

3. *Inductor alternators,* having a stationary armature and stationary field, the revolving part or inductor consisting only of iron.

The first type corresponds to the nearly universal type of direct current generator; there is, however, no commutator and alternating current is delivered from the armature winding to the line by means of collector (or slip) rings and brushes. In the second and third types, the armature is stationary and current is delivered directly to the line without collector rings. The continuity in insulation, thus made possible, is an important advantage in high potential machines. In revolving field alternators, the field current is introduced through slip rings.

Each type is made in several forms which may be studied by reference to standard works,* or better by examination of actual machines. The form most desirable depends upon conditions of operation, character and speed of prime-mover, etc. In some cases, it is desirable to make the moving mass as small as possible; in other cases—as in direct-connected engine-driven generators—a certain fly-wheel effect is advantageous. Alternators of the second type usually have an *internal* revolving field, a conspicuous exception being the umbrella form of *external* revolving field in the vertical-shaft alternators at Niagara. In the old Mordey and Brush form of machine, the stationary armature coils were in a vertical plane between the two parts of the revolv-

tioned in *Electric Journal,* p. 550, October, 1908. The steam turbine has thus modified both alternating and direct current generators (§ 3a, Exp. 1–A).

* See also "The Mechanical Construction of Revolving-field Alternators," by D. B. Rushmore, *Transactions A. I. E. E.,* Vol. XXIII., p. 253.

ing field. The inductor alternator, although possessing obvious mechanical advantages, is handicapped by large magnetic leakage and consequent poor regulation—unless built in an expensive manner with much material.

§ 3. **Choice of Frequency.**—In the early applications of alternating current, when power transmission was not developed and current was used for lighting only, the common frequencies in America were 125 and 133⅓ cycles per second, and these frequencies were satisfactory for the service. The efficiency of a transformer increases with the frequency (Exp. 5–B), and from this consideration even a higher frequency would be desirable; but as frequency is increased, we have greater inductive drop and poorer regulation in generator, line and transformer.

The rotary converter, introduced in the early nineties, required a lower frequency. The highest frequency at which it can operate is practically 60 cycles, and 25 cycles is better. With its advent, the higher frequencies were abandoned; 25 and 60 cycles became standard, the former for power alone, and the latter for lighting and (usually) for combined power and lighting. The induction motor has its best* operation within this same range.

Below 25 cycles, or thereabouts, the flicker of incandescent lamps of the usual types becomes prohibitive. On account of the high speed of the steam turbine, it is not adapted for driving generators below 25 cycles. The series alternating current motor, which is more economical the lower the frequency, is practically the only apparatus for which a frequency lower than 25 cycles is desirable. As the art progresses, it is possible that some new application may be developed which will demand a frequency much higher or lower than the frequencies now recognized as standard.

* (§ 3a). In a discussion on the choice of frequency, A. I. E. E., Vol. XXVI., p. 1400, June, 1907, Dr. Steinmetz stated that the most efficient frequency for the induction motor is 40 cycles, the best frequency for small motors being higher and for large motors lower. He also states that, for converters, 25 cycles is better than either a higher or lower frequency.

§ 4. **Characteristics.**—Four characteristics are to be taken:

One no-load characteristic—the *no-load saturation curve.*

Three load characteristics—the *external characteristic,* the *full-load saturation curve,* and the *armature characteristic.*

These characteristics are similar to the corresponding characteristics of a direct current generator, Exp. 1–B.

§ 5. **No-load Saturation Curve.***—This curve shows the terminal voltage for different values of field current, the machine being driven at constant speed† without load. The connections are shown in Fig. 1. Data are taken in the same way as for the no-load saturation curve of a direct current generator (§§ 5–10, Exp. 1–B); the alternator, however, is necessarily separately excited.

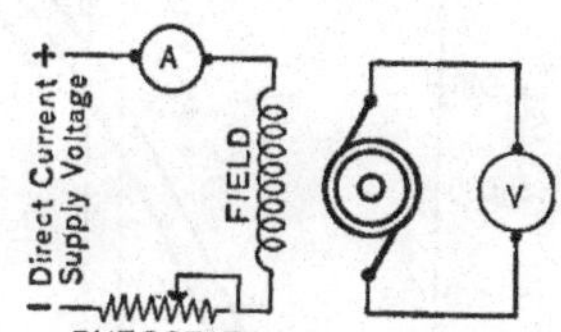

FIG. 1. Connections for no-load saturation curve.

The field current can be varied‡ by a field rheostat in series, as in Fig. 1 (with a second rheostat in series with it to give greater range, if necessary); or by an arrangement of resistances as in Figs. 4 and 5, of Exp. 1–A.

Voltage readings are corrected by proportion for any variation of speed, and plotted as in Fig. 2. Descending, as well as ascending, values may be plotted when desired. The saturation factor and per cent. saturation are determined as in Fig. 2, of Exp. 1–B.

§ 6. **External Characteristic.**—This curve shows the variation in terminal voltage with load. The alternator is driven at nor-

* (§ 5a). If the alternator is motor-driven, it is commercial practice to determine its core loss and friction at the same time that the no-load saturation curve is taken. See § 15.

† (§ 5b). Speed and frequency are proportional; with a good frequency meter at hand, it may be more convenient to observe frequency than speed. If the speed can be varied, note that voltage is proportional to speed.

‡ (§ 5c). As in all such curves, the variation should be made continuously and no back steps should be taken (§ 7, Exp. 1–A).

mal speed and excitation, readings being taken of speed and field current to see that they are constant. The connections are shown in Fig. 3.

The characteristic is to be obtained for unity* power factor, a non-inductive variable resistance being used for a load. Readings are taken of terminal voltage and external current from 0 to 25 per cent. overload.

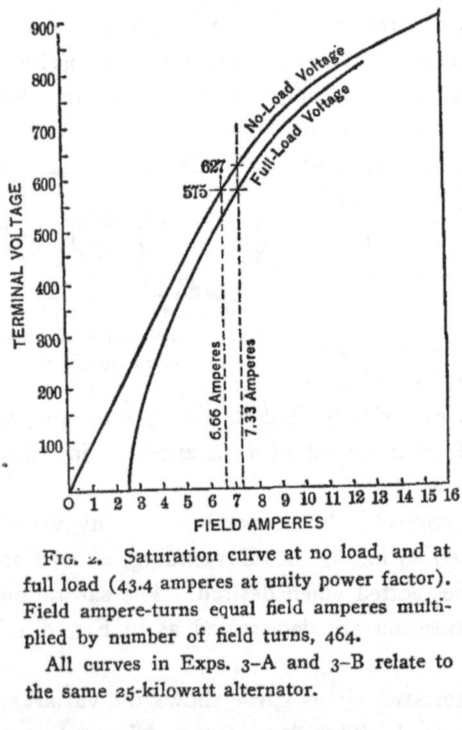

Fig. 4. Saturation curve at no load, and at full load (43.4 amperes at unity power factor). Field ampere-turns equal field amperes multiplied by number of field turns, 464.

All curves in Exps. 3-A and 3-B relate to the same 25-kilowatt alternator.

In commercial testing, the excitation is adjusted for normal voltage at full load. Fig. 4 shows the characteristic of a 25 K.W. alternator in which the voltage increases from 575 at full load to 627 at no load—a regulation of 9 per cent. (See §§ 14, 17, Exp. 1-B.) It is desirable to have the regulation as "close" as possible, *i. e.*, with the smallest possible variation in the voltage from no load to full load. Since with non-inductive load the power factor is unity, the power output is found by taking the product of terminal voltage and external current.

§ 7. The causes for the decrease in terminal voltage with load are impedance drop in the armature (due to its resistance and inductance) and armature reactions, discussed more fully in the

* For other power factors, see § 13; take data as in § 14. Also see Fig. 7, Exp. 3-B.

next experiment. Compare also § 16, Exp. 1–B. As in the case of a shunt generator (§19, Exp. 1–B), when the iron is highly saturated, the demagnetizing effect of armature reaction is the least and the regulation the best.

§ 8. The external characteristic, Fig. 4, is practically an ellipse.* At one end of the characteristic (near open circuit), an alternator tends to regulate for constant voltage; at the present day, this is the usual working part of the characteristic. At the other end (near short circuit), an alternator tends to regulate for constant current. The earliest alternators were constructed for such operation. Constant current alternators are used (less now than formerly) for series arc lighting. For this service an alternator should have high armature reaction so as to limit the current on short circuit to the desired value; a reactance external to the armature will serve equally well.

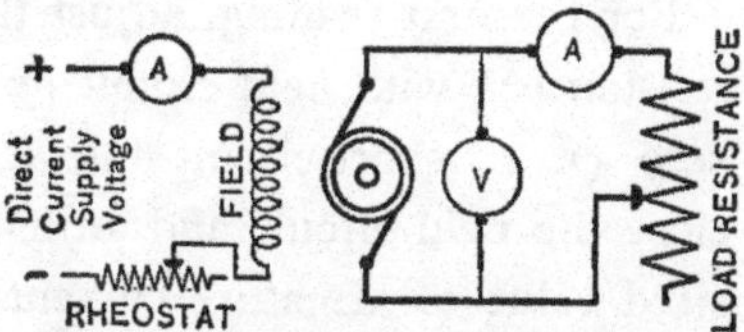

FIG. 3. Connections for loading an alternator.

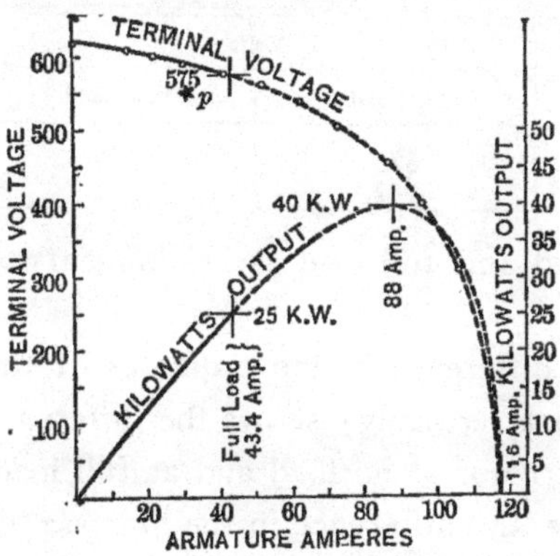

FIG. 4. External characteristic of an alternator at unity power factor. (The dotted parts of these curves were calculated according to Exp. 3–B.)

§ 9. **Full-load Saturation Curve.**—The machine is run at constant speed so as to give its normal full-load† current at different field excitations. The connections are as in Fig. 3. To obtain the curve for unity power factor, a non-inductive resistance

* See discussion of Fig. 7, Exp. 3–B.

† (§ 9a). Curves taken at intermediate loads (one fourth, one half and three fourths full load) would lie between the no-load and full-load

is used as load; with constant armature current, readings are taken of terminal voltage for different field currents, and plotted as in Fig. 2.

For the first reading, adjust the field rheostat to its maximum resistance;* with field circuit open, reduce the load resistance to zero (*i. e.*, short-circuit the armature through the ammeter); close the field circuit and adjust the field rheostat until the desired value of armature current is obtained. For each succeeding reading, increase the load resistance by a small step and readjust the field rheostat until the desired value of armature current is again obtained, taking care that the increase or decrease in excitation is continuous.

§ 10. In Fig. 2, the excitation data are as follows:

Excitation.		Volts.	
Amperes.	Ampere Turns.	No Load.	Full Load.
6.66	3090	575	525
7.33	3401	627	575

A comparison of the no-load and full-load saturation curves, Fig. 2, shows the following:

At constant excitation, the difference in the ordinates of the two curves (their distance apart vertically) shows the difference in terminal voltage of the alternator at no load and at full load.

At constant terminal voltage, the difference in the abscissæ of the two curves (their distance apart horizontally) shows the difference in excitation (magnetomotive force) required at no load and full load in order to maintain the voltage constant.

At constant excitation, a voltage of 575 at full load increases to 627 when the load is thrown off, giving a regulation of 9 per

curves of Fig. 2. To take these is unnecessary, unless some special object is in view. For inductive load, the full-load saturation curve will be lower than with non-inductive load (as shown in Fig. 1, Exp. 3–B, for zero power factor). For different power factors, see § 13, and take data as in § 14.

* This resistance should be sufficient to reduce the field current to but a small fraction of its normal value.

cent., the same as already obtained from the external characteristic, Fig. 4.

For constant terminal voltage of 575, the excitation must be increased from 6.66 amperes at no load to 7.33 amperes, at full load. This will be found to check approximately with the armature characteristic, Fig. 5; an exact check can not be expected.

Fig. 2 shows that, as we go above saturation, there is less difference between the no-load and full-load voltages, *i. e.*, the regulation is better (§ 7).

§ 11. **Armature Characteristic or Field Compounding Curve.**—This curve is taken for an alternator* in the same way as for a direct current generator (§ 26, Exp. 1–B). The curve in Fig. 5, taken for a constant terminal voltage of 575 at unity power factor, shows that in going from no load to full load (43.4 amperes) the excitation is increased from 6.6 to 7.25 amperes. This checks with the increase 6.66 to 7.33 amperes in Fig. 2.

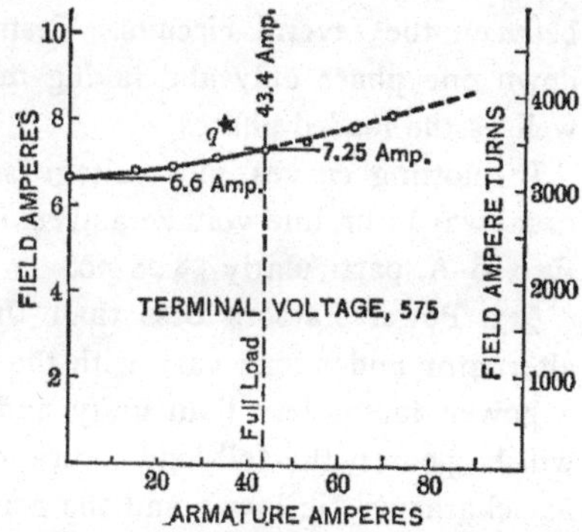

Fig. 5. Armature characteristic, or field compounding curve; unity power factor; speed constant.

Armature characteristics for lower power factors than unity will rise more rapidly (§ 13).

* (§ 11a). *Composite Winding.*—Although an alternator can not be compounded by a series winding carrying the line or armature current, as in the case of a direct current generator (since the field winding requires a direct current and the line or armature current is alternating), the result can be accomplished by rectifying part of the alternating current and passing it through what is called an *auxiliary* field winding. Such an alternator is said to be *composite* wound. The alternating current to be rectified is commonly derived from the secondary of a transformer, through the primary of which flows the line or armature current; for the core of this transformer the armature frame or spider is used. The

APPENDIX I.

MISCELLANEOUS NOTES.

§ 12. **Tests on Polyphase Generators.**—The tests described above may be made on polyphase generators in the same manner as on single-phase machines. The polyphase generator when loaded should ordinarily be given a balanced load, *i. e.*, one that is divided equally between the several circuits. Tests may also be made by loading down one phase only and taking measurements on the unloaded as well as the loaded phases.

In plotting curves, plot voltage and current *per phase* (the more usual way); or, line voltage and *equivalent single-phase* current. See Exp. 6–A, particularly §§ 28–30.

§13. **Power Factors Less than Unity.**—The characteristics of an alternator under load vary with the power factor of the load. With a power factor less than unity and current lagging, the regulation will be poorer, the full-load saturation curve will be lower, the external characteristic lower and the armature characteristic higher than at unity power factor. The reverse is true when the current is leading (instead of lagging), as it may be when there is capacity in the line or in the load, or when the load consists in part of over-excited synchronous motors or converters.

These facts may be fully shown by calculation (Exp. 3–B), or by a complete series of runs made with loads of different* power factors. If such runs are to be made, it will be more profitable to make them after Exp. 3–B. At present, it will suffice to illustrate these facts by a few readings only, as in the next paragraph.

§ 14. *Tests to Compare Effects of Inductive and Non-inductive Loads.*—The difference between inductive and non-inductive loads

composite winding is not, however, being extensively used, for it can not give constant voltage under all conditions—*e. g.*, varying power factor—and the rectifying commutator is liable to spark. The Tirrell regulator (§ 3a, Exp. 1–B), applied to the exciter of an alternator, can maintain constant voltage under all conditions of load.

* (§ 13a). This will require special facilities for adjusting power factor; for an inductive load, this can be done by means of an adjustable resistance and adjustable reactance in parallel. Runs should be made at one high power factor, one medium, and one as low as can be obtained.

can be illustrated by the following tests, or by modifications which may be devised by the experimenter.

1. Load the alternator on inductive load, using for this any one particular load which can be conveniently obtained. An induction motor can be used for a load, as in commercial practice; but a choke coil will serve fully as well.

With the same speed and excitation as were used in taking the external characteristic on non-inductive load, Fig. 4, take readings* of load current and terminal voltage with the inductive load. These readings are plotted,† in Fig. 4, as the point *p*, which is one point on a characteristic for low power factor. (For more complete curves, see Fig. 7, Exp. 3-B.)

Throw off the load and (at the same speed and excitation) read the no-load voltage; the per cent. increase in voltage when the load is thrown off gives the per cent. regulation.

2. With the same speed and excitation, repeat with a non-inductive load, so adjusted as to obtain the same load current as in 1.

Note the terminal voltage under load, the no-load voltage when the load is thrown off, calculate the regulation and compare with the regulation in 1.

3. With the same speed and terminal voltage as were used for obtaining the armature characteristic on non-inductive load, Fig. 5, note the increase in field current required with inductive load to maintain constant terminal voltage and plot the point *q*, Fig. 5.

4. Repeat with a non-inductive load (adjusted for the same load current) and compare results.

§ 15. **Efficiency.**—If the alternator is driven by a direct current motor, the friction and core loss are conveniently determined by the method of § 21, Exp. 2-B. If the driving motor is alternating, a wattmeter is used to measure its input, the increase in motor input

* (§ 14a). If a wattmeter reading is also taken, the power factor can be found by dividing the reading of the wattmeter by the product of current and voltage.

† (§ 14b). Since the same value of exciting current may at different times give different amounts of magnetization (as in the case of the ascending and descending curves), the point *p* thus located—and the point *q* as located later—may not be exact in their positions, as compared with the characteristics previously taken. They will, however, serve to *illustrate* the effects in question.

giving the friction and core loss of the alternator—any changes in motor losses being corrected for, if necessary.

The copper losses of field and armature are calculated from resistance measurements, and the efficiency so determined.

If the armature has large, solid conductors, the loss in them will be greater with alternating than with direct current, this additional loss being a *load loss.* Load losses are losses which occur under load in addition to the losses already accounted for, *i. e.*, in addition to core loss, RI^2, friction and windage. There is no simple and accurate method for determining load losses in alternators. The A. I. E. E. Standardization Rules (116–7) give a method for estimating these losses by assuming them to be—in the absence of more accurate information—equal to one third of the short-circuit core loss.

Experiment 3–B. Predetermination of Alternator Characteristics.*

§ 1. **Introductory.**—It is desirable to be able to predetermine the performance of any machine without loading, and this is particularly true of alternators; for, in the case of large machines, the regulation can not be conveniently found in any other way.

There are two simple methods for predetermining the performance of an alternator *approximately*,—the electromotive force method and the magnetomotive force method. Although other more complex methods are proposed for the more exact determination, no one method has been found which is generally accepted and gives correct results in all cases. It is well to first thoroughly study the electromotive force method, on account of the insight it gives into the general performance of the alternator and into other methods of dealing with the subject. The magnetomotive force method should then follow; after which, other methods (essentially modifications of these two) can be made a special study by those who desire to pursue the subject further. (See Appendices I. and II.)

§ 2. There are primarily two causes for the change in terminal voltage of an alternator with load:

1. The effect of armature resistance, which is small and definite; this causes a drop in electromotive force which is in phase with the armature current and is equal to $R\,I$.

2. The effect of the flux set up by the armature current, a much larger and less definite effect, discussed in the next paragraph.

§ 3. All the flux set up by the armature current encircles the

* To be preceded by Exp. 4–A. See § 9 for a statement of data to be taken. For a short experiment, take §§ 1–18 and 26–30, plotting curves for unity power factor only. The curves used to illustrate this experiment and Exp. 3–A all relate to the same machine.

armature conductors. There are, however, different paths which the flux may follow, causing different inductive effects.

§ 4. (*a*) *True Armature Reaction.*—By one path, flux set up by the armature conductors passes into the pole pieces and through the magnetic circuit of the field magnets (Fig. 10), *linking with* the windings of the field coils. This flux has a *demagnetizing* effect, weakening* the field by a certain *magnetomotive force* produced by the ampere-turns of the armature.

This flux through the field magnets is maintained by successive armature conductors; in a single-phase alternator it is pulsating, but in a polyphase alternator, due to the combined effect of the armature currents in the different phases, it is constant both in position and in magnitude.

§ 5. (*b*) *Local Armature Reactance.*—By a different path, flux set up by the armature current encircles the armature conductors without entering the pole-pieces; this flux (the fine lines in Fig. 9) is entirely in the armature, or partly in the armature and partly in the air gap. The flux surrounding any particular conductor varies periodically and produces a reactance *electromotive force* or reactance drop, XI, in quadrature with the armature current and proportional to it,—as in any alternating current circuit.

§ 6. By another and somewhat similar path, flux encircles the armature conductors by entering into and returning from the poles *without linking* with the windings of the field circuit; this flux is shown by heavy lines, Fig. 9. This is *cross-magnetizing flux* and distorts the field; it does not weaken the field except incidentally to a small extent by saturating the pole pieces. This cross-magnetization is alternating with respect to the armature conductor, as in (*b*); with respect to the pole pieces, it is constant in a polyphase and pulsating in a single-phase alternator, as in (*a*). It may be treated separately; or with (*a*) or (*b*).

* The field is weakened by a lagging current, but strengthened by a leading current, §§ 46–8.

§ 7. It is thus seen that there are two somewhat different effects produced by the armature current: the first (*a*) is a *magnetomotive force,* which reduces the field flux and so reduces the generated voltage; the second (*b*) is an *electromotive force,* which is subtracted from the generated electromotive force (in the proper phase) so as to give a lower terminal voltage.

These two effects operate simultaneously to lower the terminal voltage, the relative amounts of the two varying according to details of design,—saturations, air-gap, shape of slots, etc. To take full and accurate account of the two effects—treating one as a magnetomotive force and the other as an electromotive force—is difficult* and will not be undertaken here.

§ 8. We may, however, instead of treating the two effects separately, treat them combined, following either one of two methods:

(*a*) The *magnetomotive force or ampere-turn method,* which assumes that all the effect is magnetomotive force; or,

(*b*) The *electromotive force or reactance method,* which assumes that all the effect is electromotive force.

If the saturation curve were a straight line, the two methods would be identical;† for, magnetomotive force would produce a proportional electromotive force. With the saturation curve, however, not a straight line, a given increase or decrease in magnetomotive force will cause a less than proportional change in electromotive force.

Hence, if we consider that all the effect of armature flux is a magnetomotive force, we will have a less drop in terminal voltage than if we consider that all the effect is an electromotive force. The magnetomotive force method is, accordingly, *optimistic* (Behrend) and gives the generator a better regulation

* See Appendix II.

† This would be true if the details of the two methods were in all respects the same. Differences in the details of the two methods, as usually applied, cause differences in the results, even though the saturation curve is straight.

than it actually has; the electromotive force method, on the other hand, is *pessimistic,* giving the generator a poorer regulation than the actual.

The two methods, therefore, give the limits between which is the true performance of the machine.

§ 9. **Data.**—For either method, the data required are obtained from the following two* runs, which are made without loading the generator:

1. An open-circuit run, giving the open-circuit voltage E_0, for different field currents,—*i. e.*, the no-load saturation curve, obtained as in § 5, Exp. 3–A. See Curve (1), Fig. 1. To save labor in the many subsequent calculations, it is customary to use only the ascending curve.

2. A short-circuit run, giving the short-circuit current I_S, for different field currents,—called also a synchronous impedance test,—as described in the next paragraph. See Curve (2), Fig. 1.

These data enable us to ascertain the synchronous impedance of the armature and hence to compute the volts impedance drop for the electromotive force method; they also enable us to ascertain the magnetomotive force required to overcome the magnetizing effect of the armature, for the magnetomotive force method.

The hot armature resistance† is to be found by the fall-of-potential method.

§ 10. **Test for Short-circuit Current and Synchronous Impedance.**—With the armature short-circuited through an ammeter,‡

* (§ 9a). Two such runs are common in testing many kinds of apparatus; note, for example, the open-circuit and short-circuit tests for transformers, Exp. 5–B.

† (§ 9b). On account of eddy currents, the resistance will be greater for alternating currents than the value found by direct current. This is of importance as affecting efficiency (§ 15, Exp. 3–A), but is of little consequence so far as regulation is concerned, for *RI drop* has only a small effect at high power factors and is negligible at low power factors, as will be seen later.

‡ (§ 10a). The ammeter leads should be short and heavy; for, by the

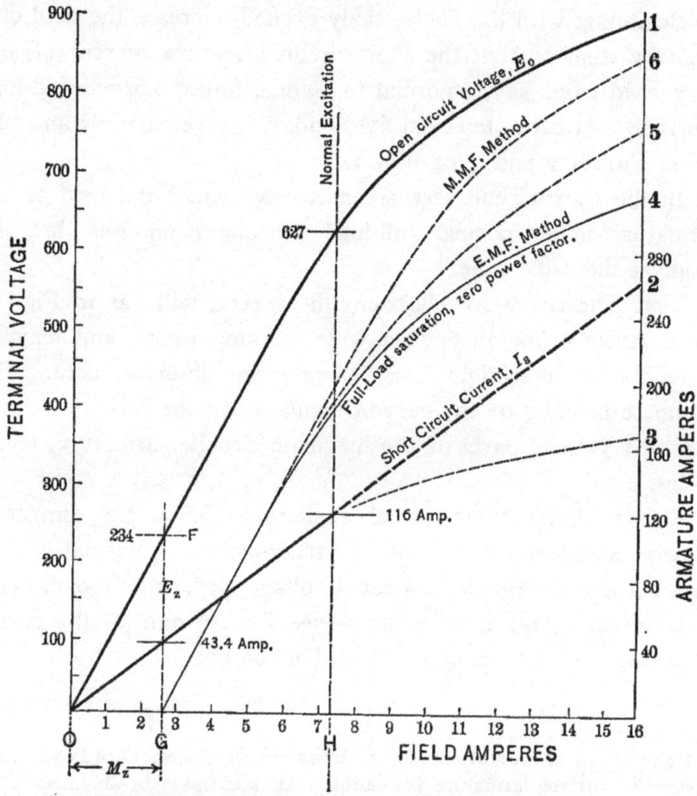

FIG. 1. No-load saturation curve (1) and short-circuit current (2 and 3) for different field excitations. Also full-load saturation curves (4, 5, and 6) for zero power factor, current lagging.

the short-circuit current is found for different values of field current. The ammeter should have a range of about three times full-load current. The speed should be normal, but special care in maintaining constant speed is not necessary.*

methods of computations used later, any drop in them is included in the impedance drop of the armature.

* (§ 10b). If facilities for varying the speed are provided, with constant excitation vary the speed through wide range and note the practical absence of change in the short-circuit current. Note, however, that the open-circuit voltage is proportional to speed. How are these facts explained?

Beginning with the field weakly excited, increase the field current by steps so that the short-circuit armature current (I_S) is increased from, say, $\frac{1}{4}$ normal to $1\frac{1}{2}$ or 2 times* normal full-load current. At each step read field and armature currents and plot as in Curves 2 and 3 of Fig. 1.

In the short-circuit test, we may have either the field or the armature under normal full-load working conditions, but not both at the same time.

§ 11. The curve for short-circuit current, will (as in Fig. 1) be a straight line through a wide working range, and may be extended as a straight line† beyond the observed data. The ultimate bending of the curve depends upon the relative saturations of various parts of the magnetic circuit,—armature, teeth, poles, etc.

Fig. 1 shows that normal excitation, $OH = 7.33$ amperes, gives a short-circuit current of 116 amperes. (Normal excitation is the excitation giving rated voltage, 575, at full load, unity power factor; for this machine—see Figs. 6 and 7—the corresponding no-load voltage is found to be 627.)

* (§ 10c). By taking the run quickly, even higher values of current can be reached.

Running an alternator on short circuit, as described, affords the best means for drying armature insulation. An alternator in shipment may have been unduly exposed to weather or have been allowed to stand in a damp place. The insulation readily takes up moisture and is much impaired thereby. In such a case, as soon as the alternator is installed it should be run for one day with the armature short-circuited, the field excitation being so low that the normal armature current flows; there is no high voltage to break down the insulation. The armature is thus baked and the insulation restored. This precaution, particularly in the case of high voltage machines, may avoid a break-down of insulation upon starting up.

† (§ 11a). Extrapolation as a straight line (2) gives (after saturation is reached) a diminishing value for synchronous impedances $Z = E_0 \div I_S$, as used later. It thus favors the machine by giving a smaller impedance drop; in the electromotive force method this is justifiable because it partially offsets the pessimistic tendency of that method. This justification is empirical.

Curve (3) has been extrapolated by assuming $E_0 \div I_S$ to be constant.

An excitation, $OG = 2.6$ amperes, is required to cause normal full-load current (43.4 amp.) on short circuit. The corresponding impedance voltage is $E_Z = 234$, for on short circuit the whole generated voltage is used in overcoming the internal or armature impedance.

§ 12 *Synchronous Impedance.*—On short circuit, the whole generated voltage is equal to the internal impedance drop in the armature. Impedance is equal to impedance drop divided by current; hence, the synchronous impedance of the armature—*i. e.*, its impedance when running at synchronous speed—is equal

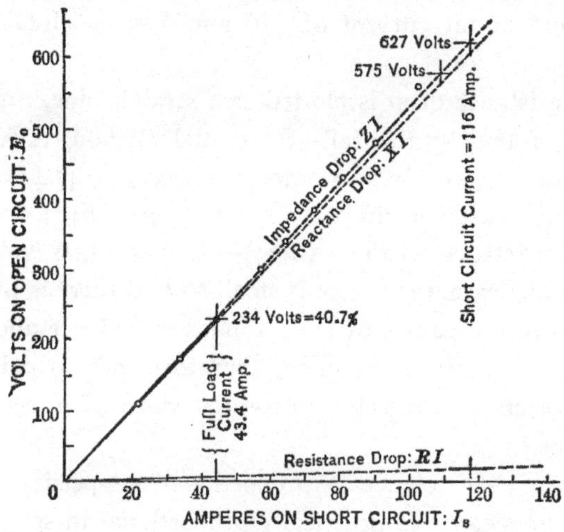

FIG. 2. Impedance, reactance, and resistance drop. (All the curves in Exps. 3–A and 3–B relate to the same machine.)

to the generated voltage E_0, divided by the short-circuit current I_S.

For any field current, the values of E_0 and I_S are obtained from curves (1) and (2), Fig. 1; the corresponding synchronous impedance, $Z = E_0 \div I_S$, should be plotted as a curve (not shown). It will be found nearly constant for a wide range,—diminishing, however, for high values of field current.

§ 13. In Fig. 2, the curve marked *impedance drop* is plotted by taking, from Fig. 1, corresponding values for E_O and I_S. Eventually there is a tendency for the curve to bend, although in this instance there is none within the range for which Fig. 2 is drawn. The ratio of any ordinate to the corresponding abscissa gives the value of the synchronous impedance; thus, in Fig. 2, the impedance drop is 234 volts for a full-load current of 43.4 amperes, and the impedance is, therefore, $234 \div 43.4 = 5.4$ ohms. The normal full-load voltage of this machine is 575; the impedance drop is, accordingly, 40.7 per cent. This is called* the *impedance ratio*. An open-circuit voltage of 627 is seen to give a short-circuit current of 116 amperes, as already seen in Fig. 1.

§ 14. Resistance drop is plotted as a straight line, Fig. 2. The resistance, found by the fall-of-potential method, is 0.17 ohms; the resistance drop, for 43.4 amperes, is $0.17 \times 43.4 = 7.4$ volts.

§ 15. The reactance drop is $E_X = \sqrt{E_Z^2 - E_R^2}$; or, for 43.4 amperes, reactance drop $= \sqrt{234^2 - 7.4^2} = 233.9$ volts. Usually, as in this case, resistance is small so that there is little difference between the values of synchronous impedance and synchronous reactance. It is common, therefore, not to calculate the value of reactance drop, but to use the value of impedance drop in its place.

Synchronous reactance is proportional to speed; hence, synchronous impedance is practically proportional to speed.

Synchronous impedance and synchronous reactance are fictitious quantities, comprising not only the real impedance and reactance of the armature, but also including the effect of armature reactions.

It is instructive to compare the curves of Fig. 2 with similar curves for a transformer; see Fig. 7, Exp. 5–B.

§ 16. **Electromotive Force Method.**—Aside from its usefulness in predetermining the performance of alternators, this method serves as an excellent illustration of the use of vector diagrams

* Standardization Rule, 208.

in solving alternating current problems; it is a practical application* of the elementary principles discussed in detail in Exps. 4-A and 4-B. The electromotive force method is general, applying to all classes of alternating current problems,—transmission lines (§ 56), transformers (Exp. 5-C), etc. For this reason the method will be treated in considerable detail.

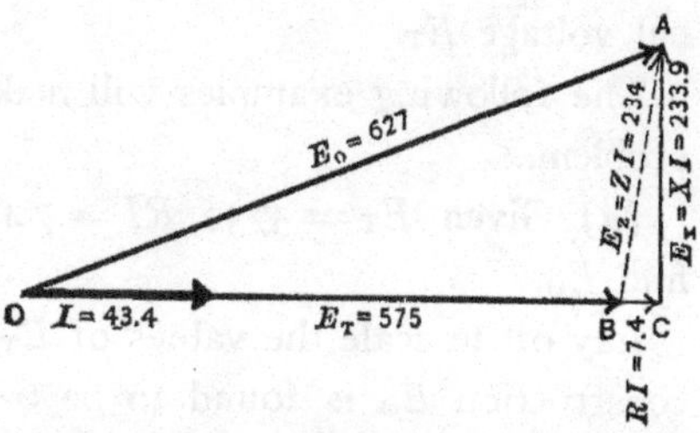

FIG. 3. Electromotive force diagram, at unity power factor; current in phase with terminal voltage.

§ 17. *Unity Power Factor.*—With a non-inductive load, the power factor of the load is unity; the current which flows is, accordingly, in phase with the terminal voltage. This is shown in Fig. 3, in which the terminal voltage E_T, is in phase with the current I. The armature resistance drop, $E_R = RI$, is in the direction of—in phase with—the current I; the reactance drop, $E_X = XI$, is in quadrature with I. The total generated electromotive force E_0, is accordingly the vector sum of the following three electromotive forces: E_T delivered to the load; RI to overcome† armature resistance and XI to overcome armature reactance.

* (§ 16a). This application illustrates the way that general principles can be put to practical purposes; the application was first made independently, and more or less simultaneously, by various engineers. The writer used the method in numerical problems to illustrate the elementary principles of Bedell and Crehore's *Alternating Currents* in the early nineties soon after the issue of that book, and applied it a little later to laboratory data. The data and some of the curves here given are taken from a laboratory outline prepared by the writer for student use and printed in the *Sibley Journal,* 1897-8, p. 215.

† (§ 17a). The arrows show the direction of the vectors in the sense that BC and CA are electromotive forces to *overcome* resistance and reactance, respectively; in the reverse sense, CB and AC are the electromotive forces *produced by* resistance and reactance.

§ 18. Knowing the values of resistance drop RI, and reactance drop XI, we may have either of two problems to solve:

(*a*) Given the terminal voltage E_T, to determine the open-circuit voltage E_O; or,

(*b*) Given the open-circuit voltage E_O, to determine the terminal voltage E_T.

The following examples will make clear the solution of either problem.

(*a*) Given $E_T = 575$; $RI = 7.4$; $XI = 233.9$. Required to find E_O.

Lay off to scale the values of E_T, RI and XI, as in Fig. 3; by construction E_O is found to be 627. Designating the total in-phase voltage by E_P, and the quadrature voltage by E_Q; we have, by computation,

$$E_O = \sqrt{E_P^2 + E_Q^2} = \sqrt{(E_T + RI)^2 + (XI)^2}$$
$$= \sqrt{(575 + 7.4)^2 + 233.9^2} = 627.$$

The regulation is 9 per cent., E_O being 9 per cent. greater than E_T.

(*b*) Given $E_O = 627$; $RI = 7.4$; $XI = 233.9$. Required to find E_T.

Lay off RI and XI to scale, as in Fig. 3. From A as a center and radius $E_O = 627$, strike an arc cutting at O the line OB, drawn as a continuation of BC. By this construction, E_T is found to be 575; by computation

$$E_T = \sqrt{E_O^2 - (XI)^2} - RI = \sqrt{627^2 - 233.9^2} - 7.4 = 575.$$

At unity power factor, it is seen that the terminal voltage is always less than the generated or no-load voltage.

§ 19. *Power Factor Less than Unity, Current Lagging.*—With an inductive load, the power factor of the load is less than unity and the current, accordingly, lags behind the terminal electromotive force. This is shown in Fig. 4 in which the current I lags behind the terminal electromotive force E_T by an angle $\theta = 30°$, the power factor of the load, in this case, being $\cos 30° = 0.866$.

Fig. 4 is drawn by first constructing to scale the triangle *ABC*, with two sides equal to *RI* and *XI*, respectively, and then laying

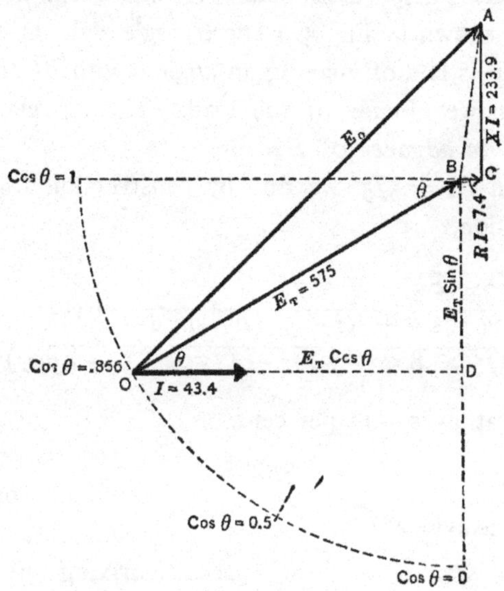

FIG. 4. Electromotive force diagram, at power factor 0.866; current lagging 30° behind terminal voltage.

off *OB* at an angle θ with *BC*, so that $\cos\theta$ equals the power factor of the load.

(*a*) Given $E_T = 575$, we find by construction $E_O = 726$; or, by computation

$$\begin{aligned} E_O &= \sqrt{E_P^2 + E_Q^2} \\ &= \sqrt{(E_T \cos\theta + RI)^2 + (E_T \sin\theta + XI)^2} \\ &= \sqrt{(575 \times .866 + 7.4)^2 + (575 \times .5 + 233.9)^2} = 726. \end{aligned}$$

The regulation is 26.3 per cent. With inductive load, the regulation is always poorer than with non-inductive load. The dotted quadrant indicates the locus of the point *O* for different power factors.

(*b*) Given E_O and power factor; required the terminal voltage E_T. Lay off a line in the direction *BO* making the proper angle θ.

Strike an arc from A as a center, with a radius E_O, cutting the line OB at O, thus giving* $OB = E_T$.

§ 20. *Power Factor Less than Unity, Current in Advance.*—This case is shown in Fig. 5. The triangle ABC is drawn as before, and OB is laid off making an angle θ with BC, so that $\cos\theta$ equals the power factor of the load. The current I, for this case, is 30° *in advance* of E_T.

(*a*) Given $E_T = 575$, we find by construction $E_O = 508$; or, by computation,

$$
\begin{aligned}
E_O &= \sqrt{E_P^2 + E_Q^2}, \\
&= \sqrt{(E_T \cos\theta + RI)^2 + (E_T \sin\theta - XI)^2}, \\
&= \sqrt{(575 \times .866 + 7.4)^2 + (575 \times .5 - 233.9)^2} = 508.
\end{aligned}
$$

The regulation is —12 per cent.

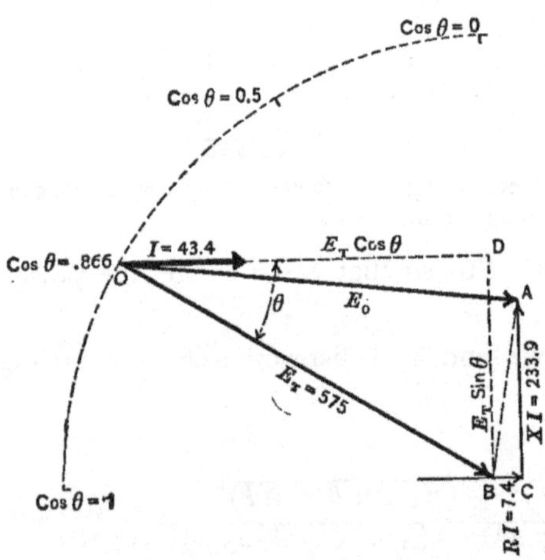

FIG. 5. Electromotive force diagram, at power factor 0.866; current 30° in advance of terminal voltage.

(*b*) Given E_O; the terminal voltage E_T is found, as before, by striking an arc from A as a center, with a radius E_O, cutting the line OB at O.

For a leading current, the terminal voltage is always greater than for a lagging current or for unity power factor, and may even be equal to or greater than the no-load voltage.

§ 21. *Special Case of Zero Power Factor.*—At zero power factor, $\cos\theta = 0$, $\sin\theta = 1$.

* (§ 19a). The graphical construction for this case will usually be preferred; an analytical expression for E_T, derived from the figure, is $E_T = \sqrt{E_O^2 - (XI\cos\theta - RI\sin\theta)^2} - (RI\cos\theta + XI\sin\theta)$.

From Figs. 4 and 5 it is seen that the RI drop becomes ineffective, being at right angles to E_T, and can be neglected. Hence, practically,

$$E_T = E_O - XI, \text{ for lagging current;}$$
$$E_T = E_O + XI, \text{ for leading current.}$$

For this case, the various voltages are combined algebraically. Practically, $XI = ZI = E_Z$, and these expressions become

$$E_T = E_O \pm E_Z.$$

This expression, approximate for $\theta = 90°$, would be exact for a value of θ a little less than $90°$; so that, in Fig. 4, OBA forms a straight line and $\tan \theta = XI \div RI$.

§ 22. *Given the Terminal Voltage at One Power Factor, to Determine it at Any Other Power Factor.*—Given E_T at any power factor, E_O is found by method (*a*) of the preceding paragraphs. With E_O thus known, the value of E_T is readily found for any desired power factor by method (*b*).

In conducting tests, it is often difficult or impossible to determine E_T at unity or high power factors, on account of the power required. The value of E_T can, however, be found by test at a low power factor (§ 52) and then determined by calculation for any desired high power factor. Usually E_O is found by test and resistance drop is known; the reactance drop is not known. In this case the procedure is as follows:

In Fig. 4, lay off resistance drop BC; at right angles draw the indefinite line CA,—the value of reactance drop being unknown. At an angle θ with BC, lay off BO equal to the value of E_T found by test at power factor $\cos \theta$. Draw $OA = E_O$, as found by test, cutting CA at A. The point A being located and E_O known, values of E_T at any power factor are determined by method (*b*) above.

In this manner, if the regulation is known for one power factor, it can be calculated for any power factor. At constant terminal voltage, the locus of the point O will be the arc of a circle with B as a center; at constant excitation, E_0 is constant and the locus of O is the arc of a circle with A as a center.

§ 23. **Application of Electromotive Force Method.**—Knowing the armature resistance and synchronous reactance*—obtained from the short-circuit test,—the electromotive force method can be used for predetermining the regulation, the external characteristic and the full-load saturation curve for any power factor.

§ 24. *Predetermination of Regulation at Different Power Factors.*—By method (*a*) of §§ 17–20, determine the open-circuit voltage E_0, corresponding to rated full-load voltage at rated full-load current, for different power factors. The values of armature RI drop and XI drop corresponding to full-load current will be constant in all the computations, R and X being taken as constant.† Plot the values of E_0, thus obtained, with power factor (or θ) as abscissæ, as in Fig. 6. This is to be done for lagging and for leading currents. Arrange, also, a scale—as on the right of Fig. 6—to show the values of E_0 as per cent. of full-load voltage.

§ 25. The curves show the increase (or decrease) in voltage when full-load current is thrown off at different power factors; in per cent., this gives the regulation. At power factor 1.0, the

* (§ 23a). Synchronous reactance is practically equal to synchronous impedance. In Figs. 1 and 2, synchronous impedance is $Z = E_0 \div I_s$, and is more or less constant; it can be computed for the value of E_0 or for the value of I_s corresponding to working conditions.

Thus, for normal field excitation, corresponding to $E_0 = 627$, we obtain $Z = 627 \div 116 = 5.4$ ohms; the armature current 116 amp. is, however, far *above* normal.

For normal full-load current, 43.4 amp., we obtain $Z = 234 \div 43.4 = 5.4$ ohms; in this case the field excitation is far *below* normal.

It is thus seen that Z can be computed from the short-circuit test either for normal field current or for normal armature current; but field and armature currents can not simultaneously be normal. When Z is constant, the two computations give identical results. When Z is not constant, the two computations give different results; either may be used, but it is justifiable to use the method which gives the smallest value for Z as being least pessimistic. (See §§ 11a and 33.)

† See § 26a.

regulation is 9 per cent.; at power factor 0.5 (lagging current), it is 37 per cent.; at power factor 0.0, it is 40 per cent. At high power factors, it is seen that a small change in power factor causes a marked change in regulation; while at lower power factors the regulation is nearly constant. The reason for this will appear from a consideration of the construction in Figs. 4 and 5. This fact is made use of in § 52.

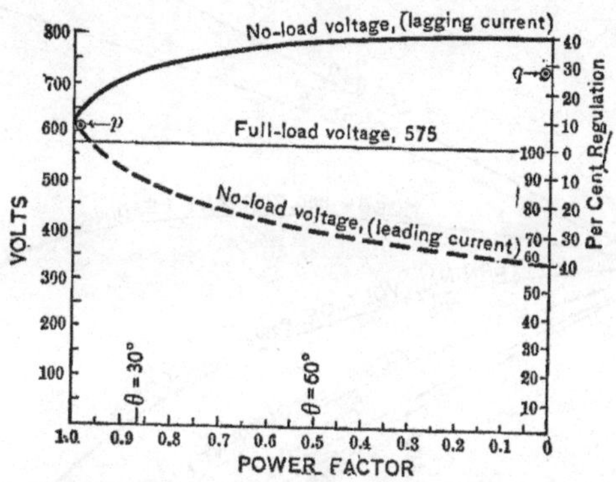

FIG. 6. Curves showing no-load voltage corresponding to a constant full-load voltage (575) for full-load current (43.4 amperes) at different power factors.

§ 26. *Predetermination of External Characteristics.*—For a definite open-circuit voltage E_O and various power factors, compute (by method (*b*) of §§ 17–20) the terminal voltage E_T, for different load currents. Armature XI drop and RI drop are to be taken as proportional to current; *i. e.*, X and R are taken as constant.* Data are thus obtained for plotting the complete external characteristic, from open circuit to short circuit, for different power factors.

* (§ 26a). In §§ 24, 26 and 32, the same constant values of X and Z are to be used. In § 26 it is proper that X and Z be considered constant for the reason that field excitation is constant. In § 24 the armature current is constant, but not the field, and strictly speaking X and Z might not remain constant, although for simplicity and for ease in comparison they are so taken. In the case of § 32, X and Z should only be taken as constant for a certain range, and for very high saturations should be taken as variable as in § 33.

§ 27. Fig. 4, Exp. 3–A, shows the characteristic for unity power factor. Power is zero on open circuit and on short circuit. Maximum power is, in this case, obtained at about twice full-load current; at short circuit, the current is about $2\frac{1}{2}$ times full-load current. A small short-circuit current* is an element

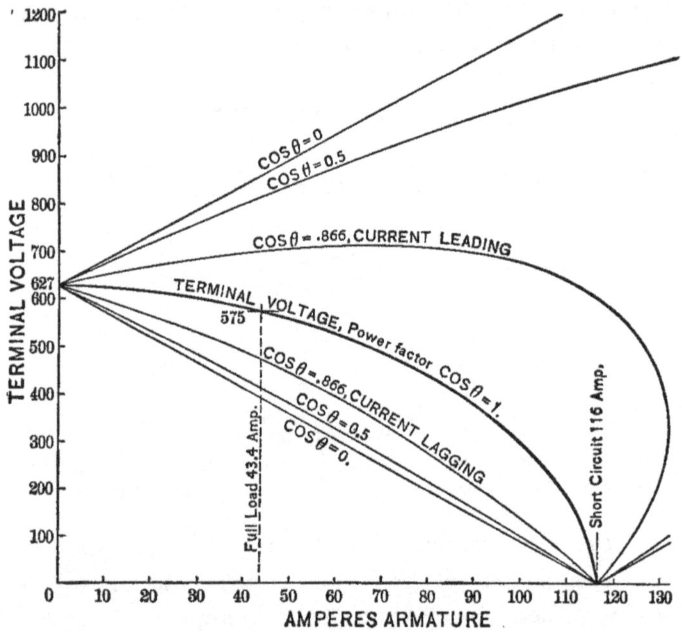

FIG. 7. External characteristics at different power factors.

of safety, obtained, however, by large impedance drop and poor regulation. Compare § 24a, Exp. 5–C.

§ 28. External characteristics for different power factors, with current lagging and leading, should be plotted as in Fig. 7. The lowest possible characteristic is a straight line; it is obtained for a power factor (cos θ) of such a value that tan $\theta =$ (armature reactance-drop) $\div$ (armature resistance-drop). See § 21. The

* (§ 27a). This is the working part of the characteristic for constant current operation, see § 8, Exp. 3–A. The armature should have a high reactance for constant current and low reactance for constant potential.

characteristic for zero power factor is a little higher than the straight line for the limiting case; the difference, however, is inappreciable.

When the scale used is such that the ordinate on open circuit is equal to the abscissa on short circuit, the characteristics are ellipses with a 45° line as axis (Steinmetz, *Alternating Current Phenomena*, 3d ed., p. 304).

In any alternator, armature resistance is small and armature reactance relatively large, so that the armature impedance is practically all reactance; this gives curves as in Fig. 7. If the conditions were reversed, resistance being large and reactance negligible, the curves for $\cos\theta = 1$ and $\cos\theta = 0$ would have to be interchanged. Unity power factor would give the poorest regulation and the straight line characteristic now obtained for zero power factor; for, with reactance zero, $E_T = E_O - RI$, in place of $E_T = E_O - XI$, as in § 21.

§ 29. *Predetermination of Full-load Saturation Curve from No-load Saturation Curve.*—By method (*b*) of §§ 17–20, compute the terminal voltage E_T corresponding to the different open-circuit voltages of the no-load saturation curve; this is to be done* for full-load current at unity power factor and at zero power factor, current lagging. In this manner, full-load saturation curves are plotted for unity power factor (Fig. 2, Exp. 3–A) and for zero power factor (Fig. 1 of this experiment).

§ 30. The interpretation of the full-load saturation curve for unity power factor is given in §§ 10, Exp. 3–A. The curve for zero power factor is capable of similar interpretation. It is seen that, for the same terminal voltage, the excitation must be much greater at zero than at unity power factor; or, for the same excitation, the terminal voltage is much lower.

§ 31. In determining the full-load saturation curves for any power factor, X and Z can be taken as they are (somewhat vari-

* It is unnecessary to construct intermediate curves for part load and for other power factors, unless a special study is to be made.

able, § 33) or they can be assumed constant,* § 32. The computations can be readily made by either method; it is only above saturation that the results differ. This will be discussed in greater detail in the case of zero power factor.

§ 32. For zero power factor, the terminal voltage (§ 21) is $E_T = E_O - E_Z$; that is, the impedance drop, E_Z is subtracted arithmetically from E.

In Fig. 1, if impedance drop E_Z is taken as constant, we obtain Curve (4) differing from Curve (1) by a constant distance (E_Z) vertically.† This is satisfactory below saturation, but above saturation is too pessimistic.

§ 33. If we wish to extend the curve above saturation, it is better to take a variable value, $Z = E_O \div I_S$, computed from Curves (1) and (2), Fig. 1, for each value of E_O,—that is, for each excitation. This gives a decreasing value for Z and results in Curve (5) instead of (4). Instead of subtracting from Curve (1) a constant E_Z, we now subtract

$$E_Z = ZI = \frac{I}{I_S} E_O.$$

Here I is full-load current (43.4 amp.); E_O is taken from Curve (1) and I_S is the corresponding short-circuit current from Curve (2). The formula can be interpreted thus: if a current I_S uses up in the armature a voltage E_O, a current I will use up a proportional voltage, $E_Z = (I \div I_S) E_O$.

* See § 26a.

† By the magnetomotive force method (Appendix I.), Curve (6) differs from Curve (1) by a constant distance (M_Z) horizontally; at high saturations this is too optimistic.

APPENDIX I.

MAGNETOMOTIVE FORCE METHOD.*

§ 34. In the magnetomotive force method, instead of combining vectorially various electromotive forces—as was done in the electromotive force method, Figs. 3, 4 and 5—the corresponding magnetomotive forces are so combined.

§ 35. The magnetomotive force corresponding to any electromotive force is found by reference to the no-load saturation curve, and is commonly expressed in ampere-turns. For a given machine, with constant number of field turns, field amperes are proportional to field ampere-turns and may be used as a measure of magnetomotive force. In Fig. 1 of this experiment and Fig. 2 of Exp. 3–A, it is seen, for example, that 627 volts corresponds to a field excitation of 7.33 field amperes, or 3,401 field ampere-turns, either of which may be taken as a numerical measure of magnetomotive force.

§ 36. It is readily seen that a straight saturation curve gives magnetomotive forces proportional to electromotive forces, so that the same results will be obtained from the use of either, if the procedure is otherwise identical. On the other hand, a saturation curve which is not straight gives values of magnetomotive forces not proportional to electromotive forces, so that different results will be obtained according to whether magnetomotive forces or electromotive forces are used.

§ 37. **Method.**†—The three magnetomotive forces M_O, M_Z and M_T are combined vectorially, as in Fig. 8; $\cos\theta$ is the power factor of the load.

These three quantities M_O, M_Z and M_T may be interpreted by their correspondence‡ to the three electromotive forces E_O, E_Z and E_T,

* No additional data are required; see § 43 for the particular application of the method to be made.

† (§ 37a). This is the common interpretation of the method (see Rushmore, p. 740, Vol. I., St. Louis Elect. Congress, 1904). In Franklin & Esty's *Electrical Engineering*, M_O is obtained as the resultant of two magnetomotive forces which correspond not to E_T and E_Z, but to E_P and E_Q (the in-phase and quadrature components of E_O).

‡ (§ 37b). If the saturation curve were a straight line and magnetomotive forces were proportional to electromotive forces, the triangles for magnetomotive forces and electromotive forces would be similar and each side of one triangle would be perpendicular to the corresponding side of the other.

respectively. A magnetomotive force M_T is required for a terminal voltage E_T, corresponding values being taken from the saturation curve; at no load no other magnetomotive force is required. Under load, an additional magnetomotive force $BA = M_Z$ is required to overcome the magnetizing effect of the armature. In terms of magnetomotive force, M_Z is equal to the ampere-turns of the armature;

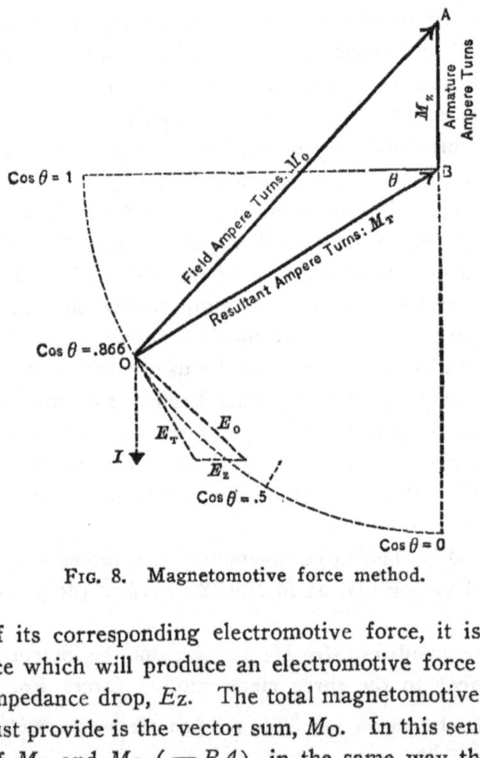

FIG. 8. Magnetomotive force method.

in terms of its corresponding electromotive force, it is a magnetomotive force which will produce an electromotive force equal to the armature impedance drop, E_Z. The total magnetomotive force which the field must provide is the vector sum, M_O. In this sense, M_O is the resultant of M_T and M_Z ($= BA$), in the same way that E_O is the resultant of E_T and E_Z.

Interpreting these quantities further as magnetomotive forces: M_O is the magnetomotive force produced by the field; M_Z ($= AB$, in the direction of armature current, I) is the magnetomotive force *produced* by the armature; M_T is the combined magnetomotive force and produces the electromotive force E_T. In this sense, M_T is the resultant of M_O and M_Z ($= AB$). On open circuit the field ampere-turns (or amperes) give us the value of the magnetomotive force M_T; for, in this case, $M_Z = O$.

On short circuit, the field ampere-turns (or amperes) give us the value of M_Z; for, in this case, $M_T = O$. That is, on short circuit the field and armature ampere-turns are (practically) equal and opposite (compare § 21).

In Fig. 1 it is seen that, on short circuit, full-load current (43.4 amp.) is given by a magnetomotive force $M_Z = OG = 121$ ampere-turns (2.6 amperes); the corresponding impedance voltage, as used in the electromotive force method, is $E_Z = GF = 234$.

§ 38. **Procedure; Any Power Factor.**—The value of M_Z is known, as in the preceding paragraph; also the power factor, $\cos\theta$, of the load.

Given E_T to find E_O. Construct the triangle *OBA*, Fig. 8, from the known values of M_Z and $\cos\theta$, and the value of M_T corresponding to E_T; the value of M_O and the corresponding value of E_O is thus determined.

Given E_O, the converse procedure is followed to obtain E_T.

The most important cases are for unity and zero power factors.

§ 39. **Unity Power Factor.**—For this case, $\cos\theta = 1$, and *OBA* (Fig. 8) becomes a right triangle. The same procedure is followed as in the preceding paragraph.

§ 40. The following procedure, known as the Institute* Method (proposed by a committee but not adopted) differs from the foregoing by taking special account of the armature *RI* drop. Armature *RI* drop is significant at unity power factor; it becomes less so as the power factor decreases and becomes negligible at zero power factor. The Institute Rule is :

"When in synchronous machines the regulation is computed from the terminal voltage and impedance voltage, the exciting ampere-turns corresponding to terminal voltage plus armature resistance-drop, and the ampere-turns at short-circuit corresponding to the armature impedance-drop, should be combined vectorially to obtain the resultant ampere-turns, and the corresponding internal e.m.f. should be taken from the saturation curve."

By the reverse procedure E_T is determined when E_O is known.

* Rule 71, p. 1087, Vol. XIX.

§ 41. **Zero Power Factor.**—When $\cos\theta = 0$, it is seen that, by the construction of Fig. 8, M_Z and M_T are in one straight line; hence

$$M_T = M_O - M_Z; \text{ or, } M_O = M_T + M_Z.$$

At no load $M_O = M_T$. Under load, if M_T (and E_T) is to have the same value as at no load, the field excitation M_O is to be increased by an amount M_Z added *in this case arithmetically.**

§ 42. *Determination of Full-load Saturation Curve.*—Given the no-load saturation curve, Fig. 1; the full-load saturation curve for zero power factor is found by adding the constant magnetomotive force $M_Z = OG$. The two curves (1) and (6) are accordingly a constant distance apart, measured horizontally.

§ 43. **Application.**—To illustrate the use of the magnetomotive force method, it will suffice to apply the method, using observed data, to the following typical cases:

1. Using the Institute Method, § 40, obtain E_O, corresponding to rated voltage, E_T, at full load, unity power factor. Plot this as the point p, Fig. 6. Note that this point is a little lower than E_O obtained by the electromotive force method, *i. e.*, the regulation is better.

2. Also, locate p by the method of § 39.

3. By the method of §§ 38 and 41, locate the point q, Fig. 6, that is E_O corresponding to rated E_T at full load, zero power factor. Note that this is considerably lower than E_O obtained by the electromotive force method.

4. Construct a full-load saturation curve (§ 42) for zero power factor.

§ 44. **Justification of the Magnetomotive Force Method.**—The construction of Fig. 8 shows that the armature ampere-turns are combined with the field ampere-turns in such a way as to have the greatest effect for power factor zero, $\cos\theta = 0$; the least effect for $\cos\theta = 1$; and intermediate effects for intermediate values of $\cos\theta$. This will be shown to be qualitatively correct, although quantitatively it is only correct approximately or under certain assumptions.

§ 45. Fig. 9 shows two conductors of an armature coil, one midway under a north pole, the other midway under a south pole. In this position the electromotive force induced in the armature conductors

* (§ 41a). The corresponding electromotive forces at zero power factor are likewise added arithmetically; $E_O = E_T + E_Z$. (See § 21.)

is a maximum. The armature current will likewise be a maximum, if it is in phase with this electromotive force. In this position, the flux set up by the armature current has a cross-magnetizing effect; the flux passes transversely through the pole piece but does not pass through or link with the field winding and so does not directly oppose the field ampere-turns.

Fig. 10 shows the armature conductors midway between poles; the coil, to which these conductors may be assumed to belong, is exactly opposite a pole. In this position the electromotive force induced in

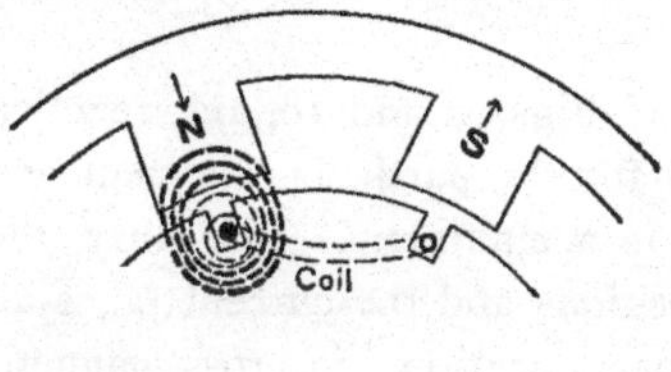

FIG. 9. Distortion of field by transverse magnetization, or cross-magnetizing effect of armature current; produced by an in-phase current, or component of current.

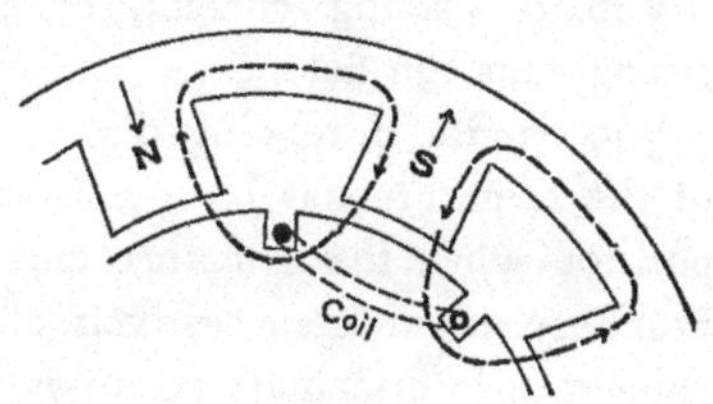

FIG. 10. Weakening of field by demagnetizing effect of armature current; produced by a wattless or quadrature current, or component of current.

the armature conductors is zero; at zero power factor the armature current—lagging 90° behind the electromotive force—is a maximum. It will be seen from the figure that in this position the armature has the greatest demagnetizing effect, the flux produced by the armature passing through the field winding and directly opposing the field ampere-turns.

§ 46. It is seen that when the armature current is in phase with the generated electromotive force it produces distortion and cross-magnetization; when the armature current is in quadrature it produces demagnetization without distortion, the armature ampere-turns being in direct opposition to the field ampere-turns.

When the current has a phase displacement, with respect to the induced electromotive force, between 0° and 90°, it may be considered as composed of two components, an in-phase component producing cross-magnetization and a quadrature component producing demagnetization.

§ 47. On short circuit, the current in the armature lags 90° (or nearly so, on account of high armature reactance). The armature and field ampere-turns on short circuit are, therefore, practically equal and opposite. If they were exactly equal and opposite, there would be no electromotive force generated; as a matter of fact, there is a very small electromotive force equal to the armature RI drop.

That the armature ampere-turns due to a current lagging 90° opposes or weakens (and does not aid or strengthen) the field is verified by this short-circuit test, and its resultant small electromotive force.

§ 48. A leading current, on the other hand, directly aids and strengthens the field.

§ 49. In the foregoing discussion of Figs. 9 and 10, the reaction of the armature has been considered for the particular moment and position when the armature current is a maximum. In reality, the armature assumes successively all positions and the current takes all values; in intermediate positions, demagnetization and cross-magnetization are both present in varying amounts dependent upon the position of the armature and the armature current at any instant. The general nature of the reaction, however, may be considered as defined by its character when the current is a maximum. The real effect is a summation of the effects at each instant through a cycle. A more complete discussion would involve some knowledge or assumption as to flux distribution in the pole pieces, and other design factors.

As a matter of fact, a sinusoidal flux distribution has been assumed in order to make it possible to treat M_O as a vector in Fig. 8; the assumption tacitly made is that the field flux passing through an armature coil varies as a sine function of time, so that the generated electromotive force ($e = -d\phi \div dt$) is also a sine function differing in phase by 90°. This assumption justifies the treatment of M_O and E_O as vectors at 90°.

But distortion, by its very nature, disturbs the flux distribution and makes the assumption necessarily an impossible one. No diagram using plane vectors can exactly represent all the quantities. The justification of the magnetomotive force method is, therefore, partly empirical. It is found to give fairly good result on many modern alternators in which armature reaction is large as compared with armature reactance and in which too high saturation is not reached; it is least accurate in alternators with high saturation and relatively large armature reactance.

APPENDIX II.

OTHER METHODS.

§ 50. There are a number of methods for determining the regulation and characteristics of alternators which are essentially modifications of the electromotive force and magnetomotive force methods, or a combination of the two; these methods are based on test data alone (obtained from open-circuit and short-circuit tests, § 9), on design data alone, or partly on design and partly on test data. Methods based on design data are of particular interest to the designing engineer but cannot be taken up here; they include methods for calculating armature reaction and reactance and for predetermining the behavior of a machine before its construction. (For further discussion, see references, § 55.)

In all methods use is made of the fundamental principles brought out in the electromotive force and magnetomotive force methods, which should therefore be carefully studied before other methods are undertaken. For those whose object is a general understanding of the behavior of alternators, a study of these two methods is sufficient; but those who desire to pursue the subject further should consult the references in § 55. It has been pointed out that, so far as results are concerned, these two methods give the pessimistic and optimistic limits. Other methods give intermediate, and in some cases more correct results; there is, however, no one absolutely correct method. In reference to this, Mr. Behrend says:

"It appears wise to admit the existing dilemma. The question of *accurately* determining the regulation of alternators can not be solved. . . . It seems to the speaker far more dignified and more in accordance with the science that we are working in, to say that this case is so complex, so intricate, there are so many factors to be taken into account, that it can no more be solved than you can state to one thousandth of an inch the distance between two chalk marks drawn on the floor." (A. I. E. E., Vol. XXIII., p. 326.)

§ 51. **Test Methods.**—The aim in various methods is to test the alternator under real or equivalent load conditions with only a small expenditure of power. The machine may be actually loaded and the power returned by some opposition method (§§ 27, 27a, Exp. 2–B), or it may be tested without any load by simulating working load conditions. In the preceding pages this was done by two tests, the open-circuit test at normal voltage and zero current, and the short-circuit test at normal current and zero voltage, in each test the power output being zero. But, inasmuch as power output is the product of current, voltage and power factor, E and I may simultaneously have normal full-load values without involving expenditure of power if the power factor is zero. This leads to the low power factor tests (§ 52) and split field tests (§ 53), concerning which only a brief statement will be made; for fuller information consult references. These tests are used in heat runs and efficiency tests, as well as in test for the determination of regulation.

§ 52. **Tests at Low Power Factor.**—When operated at low power factor, an alternator may have full-load current and normal voltage with only a small expenditure of energy. If E_O and E_T are thus determined for one power factor, their values and the regulation can be calculated (§ 22) for unity or any other power factor. This calculation is usually made either for the same terminal voltage or for the same excitation (same E_O). The load may consist of reactances, unloaded induction motors or a synchronous motor with low or no-field excitation. The power factor is known from readings of ammeter, voltmeter and wattmeter. Any power factor less than 0.20 or 0.25 may be considered as zero, for between these limits (see Fig. 6) there is practically no change in regulation.

When a synchronous motor is used, the generator voltage is adjusted by the field rheostat of the generator; the armature current by the field rheostat of the motor. In this way a full-load saturation curve for low power factor can be obtained (Fig. 1) and compared with the no-load curve; or points can be plotted for an external characteristic, as in Fig. 7.

§ 53. **Split Field Method.**—When an alternator is operated at low power factor with a synchronous motor load, as in the preceding paragraph, electric energy is given out by the alternator to the motor one quarter-cycle and is practically all returned the next quarter-cycle; power circulates between the two machines. Circulation of power in one machine was first proposed by Mordey*; this was accomplished by dividing the armature coils in two parts, one opposed to the other. In this way part of the armature acted as a generator and part as a motor. This, however, proved open to objection.

* W. M. Mordey, *Journal Brit. Inst. of Elect. Eng'rs,* Vol. II., 1893.

Behrend (see his St. Louis paper, § 55) has developed a method for circulating power in one machine by dividing not the armature but the field and reversing the excitation of one part of the field. The armature acts as a generator with respect to one part of the field and as a synchronous motor with respect to the other part. Each part of the field has its own rheostat, one controlling the generator and the other the motor action. Tests are made in much the same way as though two machines were used, § 52. For a later modification of this method, see paper by S. P. Smith, § 55.

§ 54. **Arguments for and Against Specifying Regulation at Zero Power Factor.**—The opinion is growing among engineers that regulation should be specified at zero power factor. Tests at unity power factor are objectionable, not only on account of the use of much power which may be prohibitive, but also on account of errors in the results. In Fig. 7 it is seen that the difference in regulation for a small change in power factor is very small near zero power factor, but is considerable near unity power factor.

At unity power factor, therefore, any inductance or capacity in the load introduces a large error. The use of a water rheostat as a load causes an error for this reason, for it possesses a capacity which, though small, is sufficient to give an alternator a better regulation than it would have if the power factor were unity.

Tests at zero power factor, on the other hand, have the advantage that such errors are insignificant; furthermore, the tests are less difficult to make on account of the small amount of power required. They can often be made when tests at unity power factor are not possible.

For these reasons, specification of regulation at zero power factor (rather than unity power factor) has been advocated; such specification can be checked by experiment and, furthermore, it gives the regulation under the worst conditions. On the other hand, this is objected to because, by itself, the regulation at zero power factor is no positive indication of the behavior of the machine at unity power factor; two machines with the same regulation at zero power factor may have very different regulations at unity power factor. This is largely due to resistance drop, which is of importance at unity power factor, but has practically no effect at zero power factor. Specification of regulation at zero power factor is, therefore, insufficient—

unless, in addition, the resistance drop is separately stated. Tests at zero power factor are also objected to because such tests are made when the distorting influence of cross-magnetization is absent. (See Vol. I., p. 761, *Int. Elec. Cong.*, 1904.)

§ 55. **References.**—References are given below to a few leading articles on the subject of alternator regulation. A complete list would be a long one, but the references here given are the best ones to consult first; they contain references to practically all that has been written on the subject. Rushmore's paper, with twenty-four references, summarizes the work of others and is one of the best papers to read first, particularly in connection with variations of the magnetomotive force and electromotive force methods. The discussion, found at the close of some of these papers, will be found very valuable.

Transactions International Elect. Congress, St. Louis, 1904:

The Regulation of Alternators, by D. B. Rushmore, Vol. I., p. 729;

The Testing of Alternating Current Generators, by B. A. Behrend, Vol. I., p. 528;

Methods of Testing Alternators According to the Theory of Two Reactions, by A. Blondel, Vol. I., p. 620;

Methods of Calculation of Armature Reactions of Alternators, by A. Blondel, Vol. I., p. 635.

Transactions American Inst. of Electrical Engineers:

The Determination of Alternator Characteristics, by L. A. Herdt, Vol. XIX., p. 1093, 1902;

The Experimental Basis for the Theory of the Regulation of Alternators, by B. A. Behrend, Vol. XXI., p. 497, 1903;

A Contribution to the Theory of the Regulation of Alternators, by Hobart and Punga, Vol. XXIII., p. 291, 1904.

Journal British Inst. of Electrical Engineers:

Henderson and Nicholson, p. 465, 1905; S. P. Smith, paper read November 12, 1908 (also *Lond. Electrician,* November 13).

See also Guilbert, *Elect. World,* 1902–3; Torda-Heymann, *Lond. Electrician,* Vol. LIII., p. 6, 1904; C. A. Adams, *Harvard Eng. Journal,* 1902–3.

Parts of the subject will be found treated in various text-books: S. P. Thompson's *Dynamo Electric Machinery,* Karapetoff's *Exp. Elect. Eng.,* Franklin and Esty's *Elect. Eng.,* etc.

APPENDIX III.

MISCELLANEOUS NOTES.

§ 56. **Transmission Line Regulation.**—In the electromotive force method, §§ 16–22, a complete treatment is given of the effect upon delivered voltage of resistance drop and reactance drop in the armature of an alternator. The treatment, however, is general and is not limited to alternators. The same treatment will apply to *any* resistance and reactance drop, wherever located, and may accordingly be applied to the case of a transmission line. In the geometrical treatment of any problem, resistance drop is always in phase with the current, reactance drop in quadrature.

Example 1.—Given a transmission line in which RI drop $= 7.4$; XI drop $= 233.9$. What must be the voltage E_O applied at the sending end of the line to maintain a voltage of 575 at the receiver for a load of 43.4 amperes, at unity power factor, at power factor 0.866 (current lagging 30°), and at power factor 0.866 (current leading 30°)? Figs. 3, 4 and 5 show that 627, 726 and 508 volts, respectively, are required at the sending end in the three cases, the corresponding line regulation being 9, 26.3 and — 12 per cent. In this example the same numerical values have been used for a transmission line as were used in Figs. 3, 4 and 5 for an alternator. Practical values for a transmission line would give a relatively greater resistance drop and smaller reactance drop, as in example 2.

Example 2.—A transmission line gives 1,000 volts at the receiver. The resistance drop is 100 volts, reactance drop is 200 volts; what is the regulation for different power factors?

Curves as shown in Figs. 6 and 7 can be drawn for a transmission line. These curves have been discussed for an alternator; the discussion can, however, be applied to a transmission line.

In calculating the regulation of a transmission line, the values of resistance and reactance can be taken from tables in various handbooks and elsewhere.

In testing a transmission line, the reactance drop can be found by an open-circuit test and a short-circuit test, as in the case of alternators. With a low voltage, short-circuit the line and measure I_S; open-circuit the line and measure E_O. The line impedance is $Z = E_O \div I_S$; the line reactance is $X = \sqrt{Z^2 - R^2}$.

In the laboratory a line with resistance and reactance can be tested in this way as a transmission line; the regulation for loads of different power factors can be predetermined (Figs. 6 and 7) and compared with actual load tests.

CHAPTER IV.

SINGLE-PHASE CURRENTS.

EXPERIMENT 4–A. Study of Series and Parallel Circuits Containing Resistance and Reactance.

§ 1. **Introductory.**—The object of this experiment is to acquaint one with the fundamental relations between currents and electromotive forces in alternating current circuits. These relations will be brought out by a study of series and parallel circuits containing resistance and inductance, the clear understanding of which is essential for one undertaking any study of alternating currents. Practically every problem in alternating currents involves—or can be reduced to—a problem of series and parallel circuits. A study of alternator characteristics (see Figs. 3–5, Exp. 3–B) is a study of series circuits; the transformer (see Figs. 6–9, Exp. 5–C) can be reduced to equivalent series and parallel circuits, and so, too, the induction motor. This is true of nearly all types and kinds of alternating current apparatus. It will be found that the study of series and parallel circuits brings out the general principles that are common to *all* alternating current problems. Such circuits are studied, therefore, not merely as *leading up* to the subject proper, but as actually *being* the subject matter of all alternating current testing.

Part I. contains an outline of the underlying principles of the subject, which will be found discussed in detail in Bedell and Crehore's *Alternating Currents* and in other treatises. Part II. describes the tests to be made and Part III. describes the results derived from them. For the convenience of the reader, some paragraphs on theory are included in Part III.

PART I. ELEMENTARY PRINCIPLES.

§ 2. **Defining Relations.**—In a direct current circuit, the current which flows is $I = E \div R$, irrespective of whether the circuit is inductive or not; the power expended is the product of electromotive force and current.

§ 3. In a *non-inductive** alternating current circuit, this is also true; the current is determined by the resistance, as in a direct current circuit, and the power is the product of electromotive force and current; thus,

$$I = E \div R; \qquad W = EI.$$

The impedance, defined below, consists in this case of the resistance R only.

§ 4. In an *inductive* alternating current circuit, the current is *less* than $E \div R$ and the power is *less* than EI; thus,

$$I = E \div Z; \qquad W = EI \times \text{power factor}.$$

The impedance Z, defined as the *volts per ampere*, is greater than the resistance R on account of the reactance X; thus,

$$Z = E \div I = \sqrt{R^2 + X^2} = \sqrt{R^2 + L^2\omega^2}.$$

The reactance (defined in § 40) for an inductive circuit has a value $X = L\omega$, where L is the inductance, or coefficient of self-induction of the circuit, and ω is $2\pi \times$ frequency in cycles per second (§ 1, Exp. 3–A). Impedance and reactance are expressed in ohms. It is seen that inductive reactance† depends not only

* (§ 3a). A circuit is inductive when a current in it sets up a magnetic field (§ 14); it is non-inductive when a current in it produces no magnetic field. A circuit is never entirely non-inductive, but may be made nearly so. This is practically accomplished when the outgoing and return conductors are placed so close together that the magnetic effects of the currents in the two conductors neutralize each other. In a solenoid this is accomplished by using a double winding, the currents in the two halves of which flow in opposite directions.

† (§ 4a). In a circuit with capacity C, the reactance is $1/C\omega$. When L and C are both present, the total reactance is the *difference* between the capacity reactance and inductive reactance; $X = L\omega - 1/C\omega$. See § 57.

upon L, which is a constant of the circuit depending upon its form and dimensions, but also upon the frequency of the alternating current supply.

§ 5. The preceding equations can be written

$$I = \frac{E}{Z} = \frac{E}{\sqrt{R^2 + X^2}} = \frac{E}{\sqrt{R^2 + L^2\omega^2}} = E \times Y.$$

The admittance Y of an alternating current circuit, defined as the *amperes per volt*, is the reciprocal of impedance; $Y = I \div E$. The unit of admittance is commonly called the *mho*.

§ 6. Power factor, defined as the ratio of true power W to apparent power or volt-amperes EI, is always less than (or equal to) unity. Power factor $= \cos\theta$, where θ is the phase difference between E and I; see Figs. 2 and 7 discussed later. In a circuit with resistance R and reactance X,

$$\tan\theta = X \div R.$$

The subject will be most readily understood by considering: first, circuits with R, only; second, circuits with X, only; and finally circuits with both R and X.

§ 7. **Series Circuit with Resistance Only.**—In an alternating current circuit containing only a resistance R, the electromotive force required to make flow a current I, is

$$E_R = RI,$$

as in a direct current circuit.

The current is *in phase* with the electromotive force. As the electromotive force rises from zero to a maximum and falls again to zero, the current i at each instant is proportional to the electromotive force e at that instant; $e = Ri$. The current is zero when the electromotive force is zero, and is a maximum when the electromotive force is a maximum.

§ 8. If E is represented as a vector, Fig. 5, the current I is represented as a vector in the same direction or phase as E;

that is, to cause a current I to flow through a resistance R, an *in-phase* electromotive force equal to RI is required.

§ 9. **Significance of Vectors.**—In developing the theory of vector diagrams for alternating current quantities, the vectors represent the *maximum* values of quantities which vary according to a sine law. In applying these diagrams, however, the vectors are usually drawn to represent the *effective* (or virtual) values, as measured by ammeter and voltmeter,—the effective value of a sine wave being $\frac{1}{2}\sqrt{2}$ times its maximum value.* Furthermore, vectors are used for currents and electromotive forces which do not vary exactly as a sine law, although the results in these cases are not, in general, theoretically correct.† In drawing vector diagrams, it is implied, therefore, that the currents and electromotive forces have wave forms which are sine waves or may be represented by *equivalent* sine waves of the same effective values. The phase difference θ, between equivalent sine waves for current and electromotive force, is determined by the relation: $\cos\theta = \text{power factor} = W \div EI$.

§ 10. *Direction of Rotation.*—Counter-clockwise rotation is usually taken as the direction of rotation of alternating current vector diagrams, and this convention will be here followed.

By considering a diagram as making one complete revolution (360°) in one cycle, the projections, from instant to instant, of the various lines of the diagram upon any fixed line of reference will be proportional to the instantaneous values of the quantities represented by those lines. By reversing all diagrams as in a mirror, the corresponding diagrams for clock-wise rotation will be obtained.

§ 11. *Electrical Degrees.*—In alternating current vector diagrams, "angle" is a measure of time, 360° indicating the time

* See Bedell and Crehore's *Alternating Currents*, p. 38, and other textbooks.

† (§ 9a). Compare §§ 60–64; for further discussion, see references given in § 9b, Exp. 5–C.

of one complete period or cycle, 90° indicating $\frac{1}{4}$ period, etc. A degree is, therefore, a unit of time, being sometimes designâted a "time-degree" or "electrical degree." This designation is, however, unnecessary except in discussions where "space-degrees" are also used.

§ 12. **Series Circuit with Reactance Only.**—In an alternating current circuit containing only a reactance of X ohms, the electromotive force required to make flow a current I, is

$$E_X = XI; \text{ and } I = E_X \div X,$$

as shown in §§ 14–17.

When the reactance X, is due to inductance, the electromotive force to overcome reactance is

$$E_X = XI = L\omega I.$$

Reactance is the same as resistance in that an electromotive force proportional to it is required to cause a current to flow, the electromotive force being XI for reactance and RI for resistance. Reactance is, however, different from resistance in that it consumes no energy; when the current is increasing, energy is stored* in the magnetic field (as in a fly-wheel), this energy being returned to the circuit when the current is decreasing. In a reactance, the current and electromotive force are not in phase but are *in quadrature* with each other, *i. e.*, the current and electromotive force differ in phase by a quarter of a cycle or 90°, and when one is a maximum the other is zero.

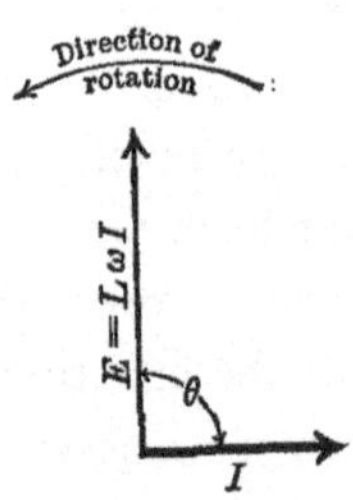

FIG. 1. Vector diagram for circuit with inductive reactance.

* (§ 12a). The energy of the magnetic field is equal to $\frac{1}{2}LI^2$, corresponding to the energy of a moving body, $\frac{1}{2}MV^2$.

§13. For inductive reactance,* the electromotive force to overcome reactance is *in advance* of the current by 90°, as in Fig. 1, and is not in phase as in Fig. 5. The current *lags behind* the electromotive force by 90°, that is, the current reaches a positive maximum $\frac{1}{4}$ cycle later than the electromotive force reaches its positive maximum. When $R=0$, $\tan\theta=X\div R=\infty$; $\theta=90°$; power $=EI\cos\theta=0$. A current and electromotive force in quadrature represent no power and are said to be "wattless."

§14. *Theory.*—When a current flows in an inductive circuit, the current sets up magnetic flux which is linked with the circuit. When the current changes, this flux changes and a counter-electromotive force is induced in the circuit tending to oppose any change in the current,—the current seemingly possessing inertia.

The electromotive force *produced by* self-induction depends upon the rate of change of current,† and is

$$e \propto -di/dt;\ \text{or,}\ e=-L\,di/dt.$$

The negative sign indicates that the electromotive force is counter to the impressed electromotive force.

The equal and opposite impressed electromotive force *to overcome self-induction is*

$$e=L\,di/dt.$$

* (§ 13a). For capacity reactance, the electromotive force to overcome reactance $X=1/C\omega$ is $XI=I\div C\omega$ and is 90° *behind* the current; the current is 90° in advance of the electromotive force; see § 55.

† (§ 14a). The electromotive force produced by self-induction, expressed in terms of rate of change of flux, is $e=-S\,d\phi/dt$. (Compare §§ 33, 33a, Exp. 5–A.) In the absence of iron, i and ϕ are proportional to each other and L is constant. In this case $Li=S\phi$, and $L=S\phi\div i$; or, the inductance of a coil is equal to the flux-linkages or flux-turns $S\phi$ for unit current. Since $\phi\propto Si$, it follows that $L\propto S^2$, other things (including dimensions of coil and leakage) being equal; the inductance of a coil is *approximately* proportional to the square of the number of turns. In the presence of iron, i and ϕ are not proportional, and L is not constant but varies with saturation.

§ 15. The inductance L of a circuit is defined by the foregoing equations. When e is in volts and i is in amperes, L is in *henries.* A circuit has an inductance of one henry when a change of current at the rate of one ampere per second induces an electromotive force of one volt.

§ 16. When the current varies according to a sine law,

$$i = I_{\max} \sin \omega t.$$

The impressed electromotive force is, accordingly,

$$e = L\, di/dt = L\omega I_{\max} \cos \omega t = L\omega I_{\max} \sin (\omega t + 90^\circ).$$

The impressed electromotive force to overcome self-induction is, therefore, 90° in advance of the current; the current, on the other hand, lags 90° behind the electromotive force.

§ 17. The maximum value of this electromotive force is seen to be $L\omega$ times the maximum value of the current; hence, the *effective* value of this electromotive force is $L\omega$ times the effective value of the current, that is, $E_X = L\omega I = XI$. Fig. 1 and the statements in §§ 12, 13 are thus established.

§ 18. **Series Circuit with Resistance and Inductive Reactance.**—In a circuit with both R and X, the electromotive force required to cause a current I to flow consists of two components, which have been separately discussed in the preceding paragraphs:

RI, in phase with I, to overcome resistance;

XI, 90° ahead of I, to overcome reactance.

Thus in Fig. 2, if OD is current, OC is the electromotive force to overcome resistance and CA is the electromotive force to overcome* reactance, OA being the total impressed electromotive force. These electromotive force relations are fundamental and

* (§ 18a). These electromotive forces, CA and OC are components of the impressed electromotive force. In the opposite sense, as counter-electromotive forces, we have the counter-electromotive force AC, lagging 90° behind the current, *produced by* inductive reactance; and, the counter-electromotive force CO, opposite in phase to the current, *produced by* resistance. Compare §15, Exp. 6–A.

are shown by the electromotive force triangle, Fig. 2, and by the following equations:

$$E = \sqrt{E_R^2 + E_X^2} = \sqrt{\overline{RI}^2 + \overline{L\omega I}^2} = I\sqrt{R^2 + L^2\omega^2}.$$

The impedance triangle, Fig. 3, is *derived* by dividing the electromotive forces, Fig. 2, by I.

§ 19. It is seen that the electromotive forces XI and RI are added as vectors. If, instead of a single X and R, there were

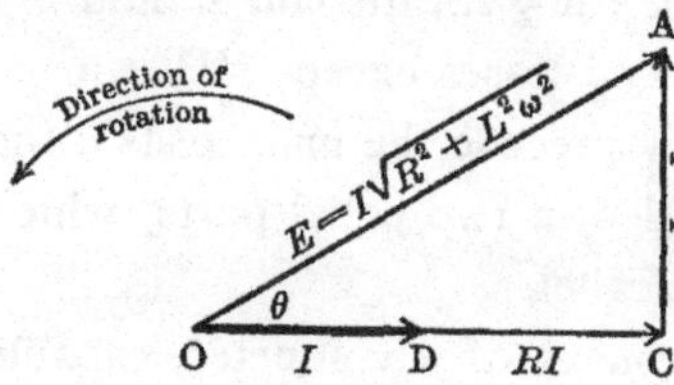

FIG. 2. Electromotive force triangle.

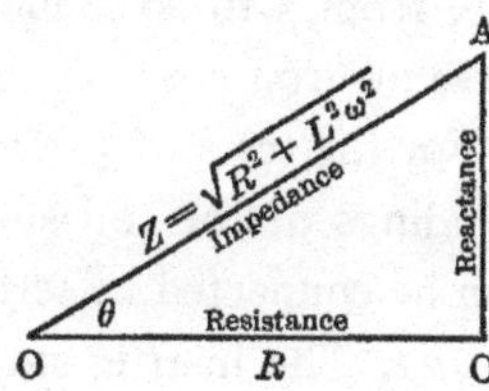

FIG. 3. Impedance triangle.

several, the same procedure could be followed: R_1I, R_2I, R_3I, etc., would be laid off in phase with I; and X_1I, X_2I, X_3I, etc., in quadrature with I.

Electromotive forces in a series circuit are added as vectors. Impedances, resistances and reactances in a series circuit are added as vectors.

§ 20. The total drop in phase with I is ΣRI; the total drop in quadrature with I is ΣXI. Hence, for any series circuit,

$$E = \sqrt{(\Sigma RI)^2 + (\Sigma XI)^2}, \text{ and } Z = \sqrt{(\Sigma R)^2 + (\Sigma X)^2}.$$

The total resistance of a series circuit is seen to be the arithmetical sum of the separate resistances; the total reactance is the arithmetical sum of the separate reactances.

For further discussion of series circuits, see §§ 38–50; for parallel circuits see §§ 51–53.

PART II. MEASUREMENTS.

§ 21. The following tests require a *resistance,* which is non-inductive and is designated R_1; and a *coil,* which is inductive and is designated R_2L_2. It is desirable to have the resistance and the coil take currents which are comparable in value with each other, for the frequency at which the tests are made; thus, if at 110 volts, 60 cycles, the coil takes a current of 10 amperes, the resistance should be so selected that at 110 volts it takes a current of, say, from 5 to 20 amperes. Except for § 28, the coil should not have an iron core, so that there are no losses except RI^2.

For the tests of § 26a (which may precede the main tests), the windings of the coil should be divided in two equal parts, which can be connected in series and in parallel.

§ 22. The instruments required consist of a voltmeter, capable of reading the supply voltage and lower voltages; an ammeter capable of measuring the combined currents of the coil and resistance; and a wattmeter having a voltage range corresponding to the range of the voltmeter and a current range corresponding to the range of the ammeter.

A voltmeter switch will be found convenient for the series tests (§§ 29–31) and an ammeter switch for the parallel tests (§§ 32–34).

On all tests the frequency should be known.

§ 23. (*a*) **Resistance Alone.**—With an adjusting resistance in series, as in Fig. 4, connect the resistance R_1 to the supply circuit (say 110 volts, 60 cycles) and measure the current I, the voltage E at the terminals of R_1, and the watts W consumed by R_1. The current coil of the wattmeter is connected in series as an ammeter and the potential coil in shunt as a voltmeter, the arrangement* of instruments being shown in Fig. 1, Exp. 5–B.

* (§ 23a). In these tests no account is ordinarily to be taken of the fact that the instruments themselves consume a certain small amount of power, as fully discussed in Appendix III., Exp. 5–A; this fact, however, should not be neglected in accurate testing, as for example in the accurate determination of L by the impedance method, § 47.

§ 24. Vary the adjusting resistance,* and in this way take several sets of readings.

If there is any question as to the accuracy of the instruments, assume the ammeter and voltmeter to be correct and determine a correction for the wattmeter, so that in (*a*) the watts as read by the wattmeter are equal to the product of volts and amperes, as read by the voltmeter and ammeter. This serves as a calibration of the wattmeter, to be used in this and subsequent tests.

§ 25. Take readings, in a like manner, at a second frequency.

§ 26. (*b*) **Coil Alone.**†—Repeat (*a*) using the coil R_2L_2 alone, as in Fig. 6, instead of the resistance R_1.

§ 27. Take readings at a second frequency.

§ 28. *Effect of Iron.*—Gradually introduce an iron core and watch the ammeter; or, introduce iron wires, a few at a time, thus gradually increasing the amount of iron. At present, only the general effect of iron is to be noted and explained; a more complete study of iron in the form of a closed magnetic circuit is made in the subsequent experiments on the transformer.

§ 29. (*c*) **Resistance and Coil in Series.**—Connect the resistance R_1 and the coil R_2L_2 in series, and, together with an adjusting resistance, connect to the supply, as in Fig. 8. For a certain current, take readings of the voltage drop, the current and the watts consumed as follows: first, for the resistance; second, for

* (§ 24a). This adjustment should be so made that the readings of the various instruments are taken at open parts of the scales.

† (§ 26a). *Series and Parallel Connections.*—It is instructive to use a coil with two equal windings. In this case, the regular tests should be made with the two windings either in parallel or in series and *additive,*—*i. e.,* setting up magnetic flux in the same direction. If one winding is reversed, it will *oppose* the other so that the resultant flux (and hence the impedance) is small. *A few volts may cause a very large current.*

Preliminary Test.—With the resistance R_1 in series as a safeguard, to avoid excessive current, measure the current and voltage and determine the impedance of each winding alone and of the two windings connected in series and in parallel, *additively* and *differentially.* The additive winding is *inductive;* the differential winding is *non-inductive,*—except so far as there is magnetic leakage.

the coil; and third, for the resistance and coil combined. Vary the current, by means of the adjusting resistance, and take several sets of readings, the current being kept constant for each set; see § 24a.

§ 30. The ammeter and current coil of the wattmeter are in series with the circuit for all readings and their location is unchanged. The voltmeter and the voltage coil of the wattmeter are in parallel with each other and are connected: first, across the terminals of the resistance; second, across the terminals of the coil; and third, across the terminals of the resistance and coil combined. These changes can be most readily made by means of a voltmeter switch, the current being maintained constant during one set of readings by means of the adjusting resistance. Some error is here introduced on account of the power consumed in the instruments.

§ 31. Repeat at a second frequency.

§ 32. (*d*) **Resistance and Coil in Parallel.**—Connect the resistance R_1 and the coil R_2L_2 in parallel, and, together with the adjusting resistance, connect to the supply as in Fig. 10. For a certain constant voltage E, take readings of current, voltage and watts: first, for the resistance alone; second, for the coil alone; and third, for the resistance and coil together in parallel.

Vary the voltage by means of the adjusting resistance, and take several sets of readings, the voltage being kept constant for each set; see § 24a.

§ 33. The voltmeter and potential coil of the wattmeter are not changed during the readings. The ammeter and the current coil of the wattmeter are shifted from one circuit to another, being: first, in series with the resistance; second, in series with the coil; and third, in the main circuit. Since, during one set of readings, the voltage is maintained constant, the readings thus obtained*

* This would be true if the instruments themselves took no power; § 23a.

will be the same as readings obtained simultaneously with three ammeters and three wattmeters.

§ 34. Repeat at a second frequency.

§ 35. (*e*) **Measurement of Resistance.**—Measure the resistances R_1 and R_2 by direct current, § 17, Exp. 1-A.

PART III. RESULTS.

§ 36. In each test, (*a*), (*b*), (*c*), and (*d*), select say two sets of readings at each frequency and construct vector diagrams showing the magnitude and relative phase positions of the various currents and voltages. Compute for the various circuits, and parts of circuits, the power factor and the phase difference between current and voltage. The prime object is to obtain a clear understanding of the relations between the various quantities, rather than to obtain exact numerical values.

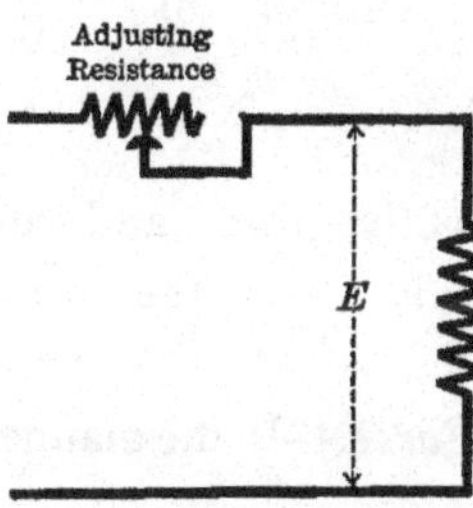

FIG. 4.

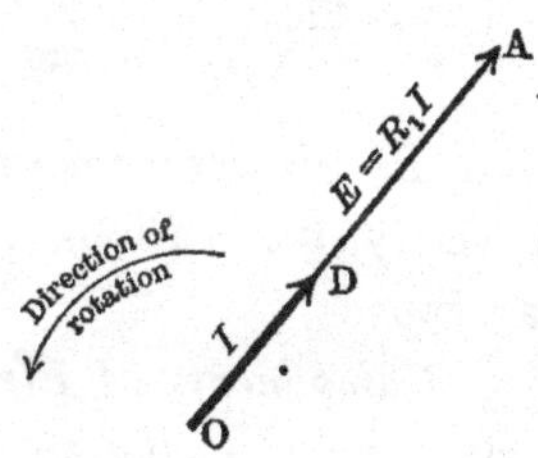

FIG. 5.

Circuit containing resistance R_1.

§ 37. (*a*) **Resistance Alone.**—For this case, the current and electromotive force are in phase, and true power is equal to the product, volts × amperes. Power factor $= W \div EI = 1$; $\cos \theta = 1$; $\theta = 0$. See Fig. 5.

§ 38. (*b*) **Coil Alone.**—The current I lags behind the electromotive force E by an angle θ, as in Fig. 7. The *true power W*, indicated by the wattmeter, is less than the volt-amperes or *apparent power, EI*; thus

$$W = EI \times \text{power factor} = EI \cos\theta.$$

Hence

$$\cos\theta = \text{power factor} = W \div EI.$$

The angle θ is thus computed from the readings of the wattmeter, voltmeter and ammeter.

In constructing Fig. 7, lay off $OA = E$; then lay off $OD = I$, at an angle θ determined as above, and construct the right triangle of electromotive forces, OCA.

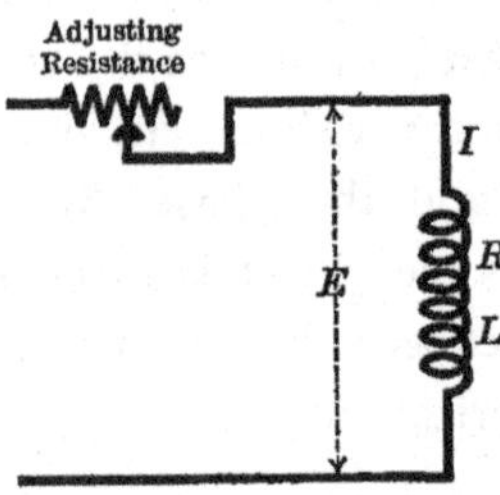

Fig. 6.

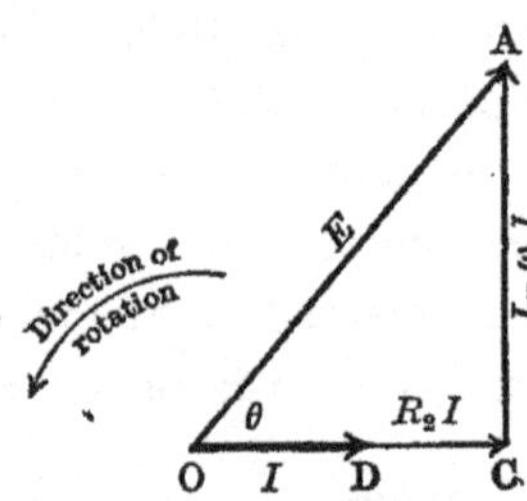

Fig. 7.

Circuit containing coil R_2L_2.

Compute the components of electromotive force and current, and verify the various relations discussed in the following paragraphs.

§ 39. *Components of Electromotive Force.*—In the manner just described, the electromotive force is resolved into the power component, $E_P = OC$, in phase with I, and the wattless component, $E_Q = CA$, in quadrature with I. These components are

$$E_P = E \cos\theta = E \times \text{power factor};$$

$$E_Q = E \sin\theta = E \times \text{reactive factor}.*$$

The vector sum of these two components gives the total impressed electromotive force.

$$E = \sqrt{E_P^2 + E_Q^2}.$$

* (§ 39a). Designating power factor by p and reactive factor by q, it is seen that $p^2 + q^2 = 1$. Compare Standardization Rule 56.

§ 40. From these electromotive forces, we have the definitions:

Impedance is total electromotive force divided by current; $Z = E \div I$.

Resistance is power or in-phase component of electromotive force divided by current; $R = E \cos \theta \div I$. (In general, when motors, transformers, etc., are included in the circuit, this gives *apparent* resistance.)

Reactance* is the wattless or quadrature component of electromotive force divided by current; $X = E \sin \theta \div I$.

§ 41. *Components of Current.*†—In a similar manner, the current may be resolved into a power component, $I_P = I \cos \theta$, in phase with E, and a wattless component $I_Q = I \sin \theta$, in quadrature with E; the total current is $I = \sqrt{I_P^2 + I_Q^2}$.

§ 42. From these currents, we have the definitions:

Admittance Y is total current divided by electromotive force; $Y = I \div E$.

Conductance g is the power or in-phase component of current divided by electromotive force; $g = I \cos \theta \div E$.

Susceptance b is the wattless or quadrature component of current divided by electromotive force; $b = I \sin \theta \div E$.

We have, then, the following relations;

$$\text{Total current} = I = E \times Y.$$
$$\text{Power current} = I \cos \theta = E \times g.$$
$$\text{Wattless current} = I \sin \theta = E \times b.$$
$$g = Y \cos \theta.$$
$$b = Y \sin \theta.$$
$$\text{Admittance} = \sqrt{g^2 + b^2}.$$

---

* This is the general definition, $L\omega$, $1/C\omega$, etc., being merely particular values; see paper on *Reactance*, by Steinmetz and Bedell, p. 640, Vol. XI., *Transactions A. I. E. E.*, 1894.

† (§ 41a). As an illustration of the resolution of current, see Fig. 2 and other figures in Exp. 5-C. It is usual to resolve electromotive force into components for series circuits and current into components for parallel circuits.

Admittance is the reciprocal of impedance; but conductance is not the reciprocal of resistance (as with direct currents), nor is susceptance the reciprocal of reactance.

§ 43. *Power.*—It is seen that the expression for true power, $EI\cos\theta$, may be written in two ways:

$$W = E\cos\theta \times I \text{ (resolving electromotive force);}$$

or,

$$W = I\cos\theta \times E \text{ (resolving current).}$$

§ 44. Resolving the electromotive force into components, we have: *True power is equal to the product of current* (I) *and the component of electromotive force* ($E\cos\theta$) *which is in phase with the current.*

§ 45. Resolving the current into components, we have: *True power is equal to the product of the impressed electromotive force* (E) *and the component of current* ($I\cos\theta$) *which is in phase with the electromotive force.*

§ 46. *Calculation of L and X by Wattmeter Method.*—Reactance is by definition (§ 40) equal to the quadrature electromotive force, E_X, divided by current.

Referring to Fig. 7, the reactance and inductance of the coil R_2L_2 are computed as follows:

$$L_2\omega = X_2 = CA \div I, \text{ ohms;}$$
$$L_2 = X_2 \div \omega = X_2 \div 2\pi n, \text{ henries.}$$

By this method, X_2 and L_2 are determined by measurements of E, I and W, and are independent of the measured value of R_2. (See §§ 47 and 49.) Note also that $R_2 = OC \div I = W \div I^2$, and that $\tan\theta = X_2 \div R_2$.

§ 47. *Calculation of L and X by Impedance Method.*—By the impedance method, L_2 depends upon E, I and the measured value of R_2, and is independent of the wattmeter reading. The calculations are made as follows:

$$\text{Impedance (ohms)} = Z_2 = E \div I.$$
$$\text{Reactance (ohms)} = X_2 = \sqrt{Z_2^2 - R_2^2}.$$

Here R_2 is the resistance of the coil, as measured by direct current. The inductance, in henries, is $L_2 = X_2 \div 2\pi n$.

For the accurate determination of L by either of these methods, the wave form of electromotive force should be sinusoidal and the losses in instruments should be taken into consideration, § 23a.

§ 48. (c) **Resistance and Coil in Series.**—In a series circuit there is one current which is the same in all parts of the circuit; electromotive forces are added vectorially, *i. e.*, the voltage drops around the separate parts of the circuit, when added as vectors, give the total impressed electromotive force of the circuit.

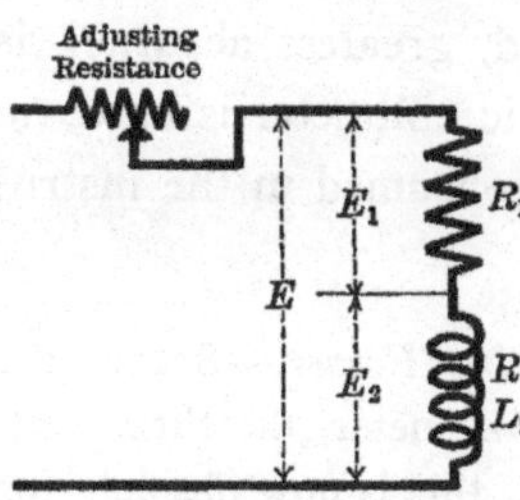

FIG. 8.

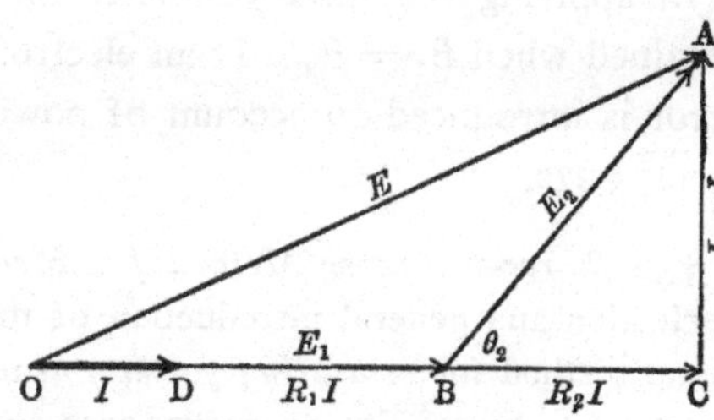

FIG. 9.

Resistance and coil in series.

The three readings of the voltmeter, E, E_1 and E_2, are, accordingly, drawn to scale so as to form the triangle OAB, in Fig. 9. The current I is laid off in phase with E_1, since the current and electromotive force in the non-inductive resistance are in the same phase. OCA is then drawn as a right triangle.

We have then the in-phase electromotive forces, $OB = R_1I$ to overcome the resistance R_1, and $BC = R_2I$ to overcome the resistance R_2; and the quadrature electromotive force, $CA = L_2\omega I = X_2I$, to overcome the reactance X_2. It will be seen that Fig. 9 is the same as Figs. 5 and 7 combined in one diagram so drawn that the current in both parts of the circuit is the same in magnitude and in phase.

§ 49. *Three-voltmeter Method.*—The foregoing construction, known as the three-voltmeter method, enables us to calculate L_2

and X_2, the results being dependent upon three voltmeter readings and current, and not dependent upon the wattmeter (as in the wattmeter method, § 46), nor upon the measurement of resistance (as in the impedance method, § 47).

Referring to Fig. 9, we have

$$CA = X_2 I = L_2 \omega I;$$

hence

$$X_2 = L_2 \omega = CA \div I,$$

and

$$L_2 = X_2 \div \omega = X_2 \div 2\pi n.$$

In applying the three-voltmeter method, greatest accuracy is obtained when $E_1 = E_2$. If an electrostatic voltmeter is used, no error is introduced on account of power consumed in the instrument, § 23a.

§ 50. *Three-voltmeter Method for Measuring Power.*—Before the perfection and general introduction of the wattmeter, the three-voltmeter method for *measuring power* was used; this is now obsolete for practical testing. The procedure was as follows:

Given a device R_2L_2 (which might be, for example, a transformer) the power in which is to be measured. Connect in series a non-inductive resistance R_1, as in Fig. 8, and read E, E_1, E_2 and I. The power in R_2L_2 (see Fig. 9) is

$$W_2 = E_2 I \cos \theta_2 = (I \div 2E_1)(E^2 - E_1^2 - E_2^2).$$

See Bedell and Crehore's *Alternating Currents,* p. 232. The weak point in the method is that small errors in observation make large errors in the result. The *three-ammeter* method, with a non-inductive resistance in parallel with the apparatus under test as in Fig. 10, is open to the same objection.

§ 51. (*d*) **Resistance and Coil in Parallel.**—In parallel circuits,* currents combine vectorially, the main current being the vector sum of the branch currents.

* (§ 51a). Currents are proportional to admittances; hence admittances may be added as vectors. The admittance of several circuits in parallel

The main current I is laid off, in Fig. 11, as the diagonal of a parallelogram with sides equal to the branch currents I_1 and I_2. The electromotive force E is laid off in the direction of I_1, since

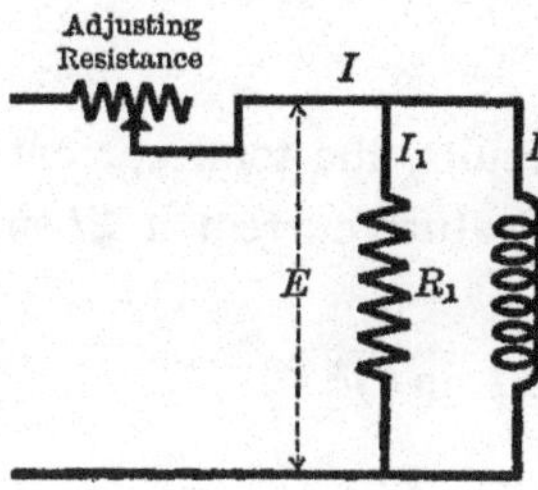

FIG. 10.

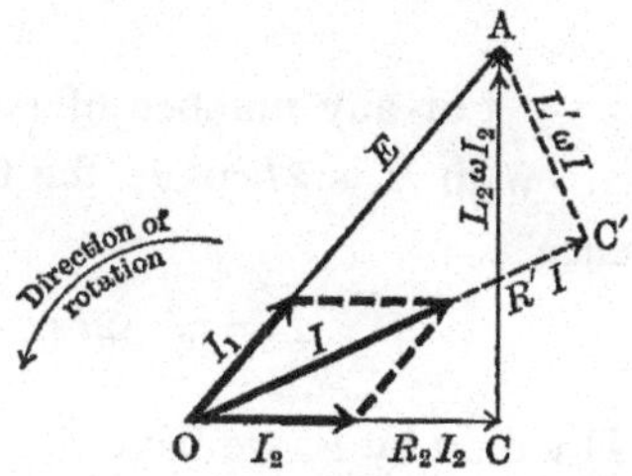

FIG. 11.

Resistance and coil in parallel.

the current and electromotive force in the non-inductive branch are in phase. E is the common terminal electromotive force and is the same for both branches.

For the inductive branch, the electromotive force triangle OCA is constructed, as in (b). For this branch, the power electromotive force is OC, in phase with I_2; the wattless electromotive force is CA in quadrature with I_2. Fig. 11 is seen to be the same as Figs. 5 and 7, drawn with a common E and combined. In Fig. 9, these figures were combined with a common I.

§ 52. The right triangle $OC'A$ is the electromotive force triangle for an equivalent* single circuit, $R'L'$, which could be substituted for the two parallel circuits. Since $OC' = R'I$, and

is the vector sum of the admittance of the separate branches. In parallel circuits we may add as vectors either currents or admittances; while in series circuits we may add as vectors either electromotive forces or impedances, §§ 19 and 20.

* (§ 52a). For a more complete discussion of equivalent resistance and inductance, see Bedell and Crehore's *Alternating Currents*, pp. 238–41. Both R' and L' depend upon frequency and are not constants of the circuits; the equivalent resistance of parallel circuits is not the same for alternating as for direct current.

$C'A = L'\omega I$, the resistance and reactance of this equivalent circuit are computed as follows:

$$R' = OC' \div I;$$
$$L'\omega = C'A \div I.$$

§ 53. For any number of parallel circuits, the total current in phase with E is $\Sigma I \cos\theta$; the total quadrature current is $\Sigma I \sin\theta$. Hence

$$I = \sqrt{(\Sigma I \cos\theta)^2 + (\Sigma I \sin\theta)^2}.$$

Dividing by E, we have

$$Y = \sqrt{(\Sigma g)^2 + (\Sigma b)^2}.$$

The total conductance of a number of parallel circuits is the arithmetical sum of the separate conductances; the total susceptance is the arithmetical sum of the separate susceptances. (Compare with § 20 for series circuits.)

APPENDIX I.

CIRCUITS WITH CAPACITY.

§ 54. It is not intended in this experiment that tests with capacity be included, the following summarized statements concerning capacity being made for reference and for comparison with the relations already discussed concerning inductance.

§ 55. **Circuits with Resistance and Capacity.**—In theory, circuits containing capacity (C) can be treated exactly the same as circuits containing inductance, if the following differences are noted:

Inductive reactance $= L\omega$; current *lags* behind impressed electromotive force.

Capacity reactance $= 1 \div C\omega$; current is *in advance* of impressed electromotive force.

In either circuit, $\tan\theta = X \div R$.

All diagrams for inductive circuits can be applied to capacity circuits by writing $1 \div C\omega$ in place of $L\omega$, and reversing the diagrams (as in a mirror) so that current is leading instead of lagging.

Inductance produces effects similar to mass in a moving mechanism; capacity produces effects similar to a spring. Inductance and capacity store, but do not consume,* energy; the stored energy being $\frac{1}{2}LI^2$ in inductance and $\frac{1}{2}CE^2$ in capacity.

§ 56. As frequency is increased, the impedance of an inductive circuit becomes greater; the impedance of a capacity circuit becomes less. Furthermore, as frequency is increased, θ in an inductive circuit becomes greater; θ in a capacity circuit becomes less.

§ 57. **Circuits with Inductance and Capacity.**—When a circuit contains both inductance and capacity, the total reactance of the circuit is the *difference* between the inductive reactance and the capacity reactance; $X = L\omega - 1/C\omega$. Inductance and capacity tend to neutralize each other. When the inductive reactance is greater than the capacity reactance, the current lags behind the electromotive force, as in an inductive circuit; when, on the other hand, the capacity reactance is the greater, the current is in advance of the electromotive force, as in a capacity circuit. In either case, $\tan \theta = X \div R$.

§ 58. **Voltage Resonance.**—In a series circuit, the total impedance may, therefore, be less than the impedance of part of the circuit only, and the total impressed voltage may, accordingly, be less than the voltage drop around part of the circuit only. The voltage around part of the circuit is thus increased by resonance so as to be greater than the impressed electromotive force.

§ 59. **Current Resonance.**—In parallel circuits with two branches, one with inductance and the other with capacity, the current in the inductance branch is lagging while the current in the capacity branch is leading. The two branch currents are to a certain extent in phase opposition so that the total or main line current, which is the vector sum of the two, may be less than the current in either branch. Due to resonance, a local circulating current is obtained which is greater than the current from the generator.

§ 60. **Non-Sine Waves.**—When the impressed electromotive force does not follow a sine law, there are present—in addition to the

* In an inductance with iron, some energy is lost in magnetic hysteresis; similarly, in a condenser, a small amount of energy is lost in dielectric hysteresis.

fundamental — harmonics with frequencies higher than the fundamental.

§ 61. In inductive circuits, reactance increases with frequency. Inductive reactance is, accordingly, greater for these harmonics than for the fundamental; harmonics in the current are *choked out* and the current wave is more nearly a sine wave than is the wave of impressed electromotive force. For this reason, vector diagrams can be used for representing experimental results, obtained from measurements on inductive circuits, without much error—even when the impressed electromotive force is not a true sine wave.

§ 62. Capacity reactance, on the other hand, decreases with frequency and harmonics in the current are, accordingly, *augmented*; hence, a small distortion in the electromotive force wave may make a large distortion in the current wave.

§ 63. For this reason, vector diagrams are less accurate for representing experimental results for capacity circuits than for inductive circuits, when the electromotive force is not a true sine wave. For example, in the laboratory without special precautions, capacity reactance can not be measured by alternating current methods as accurately as inductive reactance.

§ 64. When capacity and inductance are both present, a small distortion in the electromotive force wave may be much exaggerated by the resonance of a particular harmonic, with corresponding error in any vector representation.

EXPERIMENT 4–B. Circle Diagram for a Circuit with Resistance and Reactance.

§ 1. **Introductory.**—If a circuit with resistance and reactance is supplied with a constant impressed electromotive force, the current will have a certain value and a certain phase position with reference to the electromotive force, as discussed in Exp. 4–A.

$$I = E \div \sqrt{R^2 + X^2}; \quad \tan\theta = X \div R.$$

These values of current and phase angle will be changed, if either the resistance or the reactance of the circuit is changed.

§ 2. In a circuit in which the reactance X is constant, and the resistance R is varied, the value of I and θ will increase when R is decreased; as resistance is cut out of circuit, the current will, accordingly, not only be larger but will be more out of phase with reference to the electromotive force. In the limiting cases: when $R = 0$, the current is $E \div X$ and (in the case of inductive reactance) lags 90° behind the electromotive force; when $R = \infty$, $I = 0$, and $\theta = 0$.

§ 3. The object of this experiment is to show the change of current in magnitude and phase, in a circuit with constant inductive* reactance, when the resistance is varied and the impressed electromotive force is maintained constant. It will be found that the locus† of the current vector is the arc of a semicircle, as in Fig. 2; this is true of any constant potential circuit, in which the reactance is constant and the power consumption is variable—as in a transformer (Exp. 5–C) or in an induction motor.

* (§ 3a). A similar experiment may be performed with capacity reactance; see Appendix I., Exp. 4–A.

A converse experiment may also be made with constant resistance and variable reactance, in which case the diameter of the semi-circle locus is in the direction of E, instead of at right angles to it; see reference, § 3b.

† (§ 3b). Established by Bedell and Crehore, *Alternating Currents*, pp. 223 and 275.

§ 4. **Data.**—Let the circuit be as shown in Fig. 8, Exp. 4–A, in which R_1 is a non-inductive resistance and R_2L_2 is a coil (without iron) with resistance R_2, inductance L_2 and reactance X_2. The impressed electromotive force E should be constant; in case E varies, all readings should be reduced by direct proportion to correspond to some constant value of E; an adjusting resistance, as shown in Fig. 8, Exp. 4–A, is unnecessary. The frequency should be constant.

With an ammeter, measure the current I. With a voltmeter, measure the various falls of potential as follows: E, the impressed electromotive force; E_1, the fall of potential around the non-inductive resistance R_1; E_2, the fall around the coil R_2L_2.

The error due to the current taken by the voltmeter, although negligible for a circuit in which the current is large, becomes appreciable when the current is small; this error may be avoided by using an electrostatic voltmeter, which takes only sufficient current to charge the instrument.

§ 5. Take a series of readings for decreasing values of R_1 throughout the range that it is possible to read E_1 and E_2.

§ 6. Repeat at a second frequency.

§ 7. Repeat at one frequency with an iron core in the coil.

§ 8. Measure the resistance of the coil, R_2, by direct current, fall-of-potential method, § 17, Exp. 1–A.

§ 9. **Results.**—For one set of readings, draw a triangle, OAB, Fig. 1, with the observed values of E, E_1 and E_2 as the three sides. Lay off OD in the direction of OB, equal to the current I, in any convenient scale. Produce OB to C by an amount $BC = R_2I$, the electromotive force to overcome the resistance and supply the RI^2 losses in the coil. OC is the electromotive force to overcome the resistance of the entire circuit. The current and electromotive forces are now represented in magnitude and direction for one value of the resistance. Fig. 1 is the typical diagram for a series circuit, being the same as Fig. 9, Exp. 4-A, and Fig. 3, Exp. 3-B, for the alternator; compare, also, Fig. 9, Exp. 5-B and the transformer diagrams, Exp. 5-C. As explained in § 11, OCA is not an exact right angle.

§ 10. For a second set of readings, locate the points B', C', D', in the same way as the points B, C, D were located.

Locate points in this manner for all the readings, thus defining the curves in Fig. 2, which are the loci of the points B, C and D.

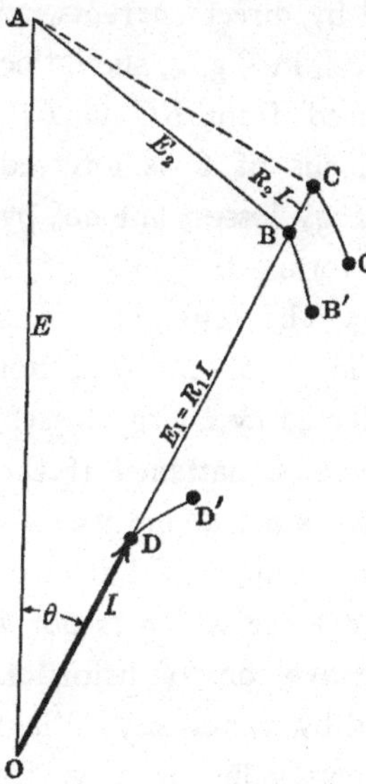

FIG. 1. Method of plotting results.

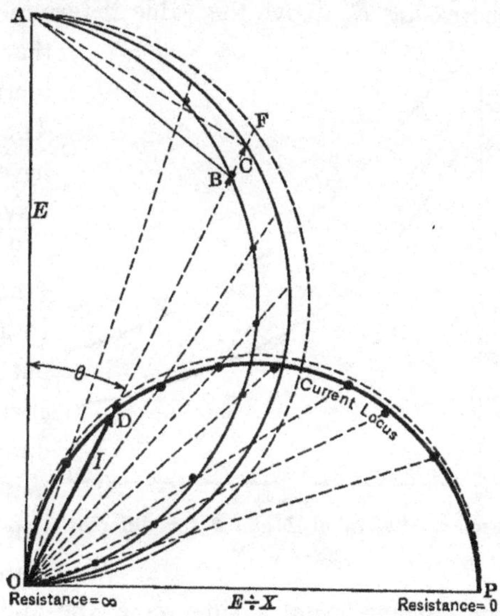

FIG. 2. Circle diagram for a constant potential circuit with constant reactance, when resistance is varied.

§ 11. It is seen that, as the resistance is decreased, the current increases and lags more and more behind the electromotive force. If the impressed electromotive force, and hence the current which flows, are sinusoidal, and if there is no power lost in the reactance coil R_2L_2, except R_2I^2 (supplied by the power electromotive force $BC = R_2I$), OCA will be a right triangle. In this case, the locus of C will be a semicircle with diameter $OA = E$, and the locus of D (the current locus) will be a semicircle* with diameter $E \div X$, at right angles to E. The theoretical semicircles are shown by dotted lines in Fig. 2.

§ 12. If, however, the power consumption in the reactance coil is more than R_2I^2, the locus of C will be flattened so as to lie inside of a semicircle. This would be the case in a reactance with iron, and is likely to be the case even in a coil without iron on account of eddy currents in the copper. Eddy currents have the effect of increasing R_2 above the value determined by direct currents, so that R_2I, in Fig. 2, should be increased from BC to BF. The locus of C is affected by energy losses, but not by wave form.

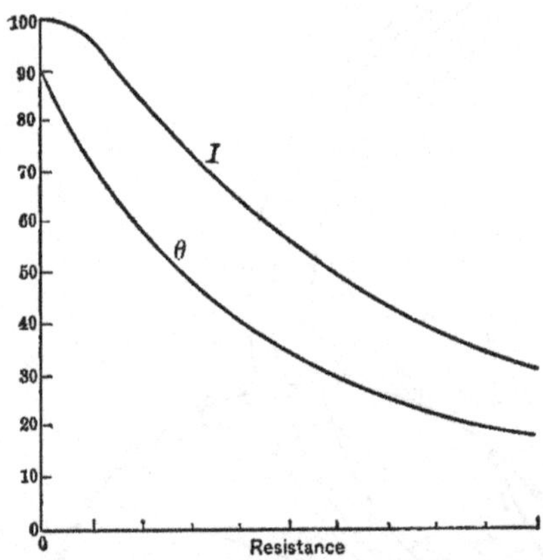

FIG. 3. Values of I and θ for different values of R.

§ 13. The current locus (the locus of D) will not be affected by energy losses, but will be flattened if the current is not a sine wave—due to an impressed electromotive force which is not a sine wave, or to distortion caused by hysteresis. When the current is not a sine wave, the apparent value of X varies somewhat with R; since X is not constant, the locus of D is not an exact semicircle.

§ 14. *Constant Current Operation.*—It is seen that when θ is large, *i. e.*, when $X \div R$ is large, the current remains nearly constant, irrespective of any variation of R and θ; between $\theta = 70°$ and $\theta = 90°$ the value of the current varies only 6 per cent. This is the condition for constant current operation and is obtained

* (§ 11a). If this is a semi-circle, $I = (E \div X) \sin \theta$ and $XI = E \sin \theta$; this accords with facts (see Fig. 2, Exp. 4–A) and the proposition is accordingly proved.

in any apparatus by means of high reactance within, or external to, the apparatus. Constant current generators (§ 8, Exp. 3-A, §§ 27, 27a, Exp. 3-B) and constant current transformers (§§ 24, 24a, Exp. 5-C) are so constructed.

§ 15. *Rectangular Coördinates.*—The results shown in polar coördinates, in Fig. 2, should also be shown in rectangular coördinates, as in Fig. 3, the values of I and θ being plotted for different values of the total resistance of the circuit. For small values of R it will be seen that the current is nearly constant.

CHAPTER V.

TRANSFORMERS.

Experiment 5–A. Preliminary Study and Operation of a Transformer.

PART I. INTRODUCTORY.

§ 1. A transformer consists of three elements: a *core* of laminated iron; and a *primary* and a *secondary* winding upon this core. The two windings are insulated from each other and usually from the core; they are in close proximity to each other or are so inter-spaced that practically all the flux which passes through one must pass through the other—*i. e.*, there is the least possible magnetic leakage.

The transformer is used on alternating current circuits to increase or *step-up* the voltage, or to decrease or *step-down* the voltage, in the ratio of the number of primary to secondary turns ($S_1 : S_2$); there is a corresponding opposite change in the current in the ratio $S_2 : S_1$, an increase in voltage being accompanied by a decrease in current, and *vice versa*. It is chiefly the transformer which makes alternating current superior to direct current for power transmisssion, for it makes possible a high potential on the transmission line, with consequent copper economy (§ 50, Exp. 6–A), and any desired lower potential at the generating and at the receiving apparatus.

§ 2. In operation, the primary winding is connected to an alternating current supply (see Fig. 1). A current flows in the primary which magnetizes the core, *i. e.*, it sets up an alternating magnetic flux which induces an electromotive force in the secondary winding and, when the secondary is closed through a resistance or other load, this electromotive force causes a current to flow.

The condition very nearly attained in the operation of a transformer is the transference of power from the primary to the secondary without loss, the current and voltage being one in-

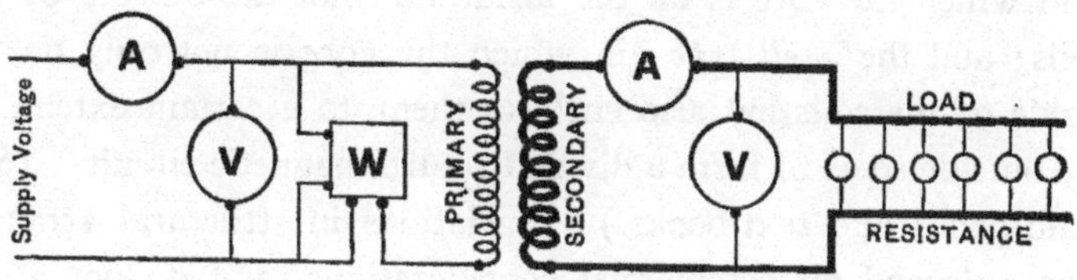

FIG. 1. Connections for loading a transformer.

creased and the other decreased in the ratio of turns. Using subscripts 1 and 2 to refer to the primary and secondary, respectively, the product E_2I_2 is accordingly nearly equal to E_1I_1, being in fact only a few per cent. less.

As an example, let $S_1 = 10S_2$ in a 20 K.W. transformer. The condition very nearly attained is

Primary watts	= 20,000	Secondary watts	= 20,000
Primary volts	= 1,000	Secondary volts	= 100
Primary amperes	= 20	Secondary amperes	= 200

On account of losses, however, if the secondary is to have its full rated watts, volts and amperes, the corresponding primary quantities must be slightly more than the amounts shown above.

There are, accordingly, the following losses: lost watts; lost volts; lost amperes.

The lost watts determine the efficiency and are due to *core losses* (hysteresis and eddy currents) and *copper losses* in both the primary and secondary windings.

The lost volts determine the regulation, and are due to *resistance drop* or copper drop (which, for a given load, is proportional to copper loss, see § 30, Exp. 5-B) and *reactance drop* due to magnetic leakage.

The lost amperes are due to the fact that, even on no load, a transformer takes a certain exciting current to maintain the flux and to supply the core losses. For a more detailed discussion, see Exps. 5–B and 5–C.

§ 3. Structurally, transformers are of two types, the *core type* —in which the core is on the inside and not the outside of the coils; and the *shell type*—in which the core is not only on the inside of the coils but also encloses them, to a certain extent, on the outside so as to form a divided return magnetic circuit. (See hand-books and text-books.) Variations in structural arrangements depend on commercial considerations, and do not affect at all the principle of operation.

Transformer losses eventually appear as heat and a transformer is so designed that this heat can be radiated or disposed of without exceeding a limiting safe rise in temperature.* The magnetic circuit is laminated to minimize the eddy current loss. For all usual purposes, the magnetic circuit is closed. A transformer with an open magnetic circuit takes excessive magnetizing current and—while it might be used for some special purpose—it is never used for power and lighting. (The "Hedgehog" transformer was of this type.)

* (§ 3a). *Heating of Transformers.*—This necessitates, on the part of the designer, a consideration of radiating surface, etc., or the provision of some special means of cooling. The radiating surface usually found necessary varies between 2 and 4 sq. in. per watt. For the allowable rise of temperature, see *Standardization Rules* which at present allow a rise of 50° C. above the air. Run at higher temperatures, transformer iron *ages, i. e.,* the core losses increase in the course of time. While this has been true of the iron ordinarily used for years in transformer construction, it is less true of the improved alloy steels which are being introduced. Hence, aging ceases to be a factor and the allowable temperature rise might be increased as much as the insulating material will stand. Good insulation will stand continuously a temperature of 90° C. This will increase the rating of a given size transformer, or will reduce the size and cost of a transformer of a given rating. In rating new iron the allowable magnetising current, and not temperature, may become the limiting consideration.

§ 4. In the majority of cases transformers are used on *constant potential* systems, the primary and the secondary potentials being substantially constant. The secondary current, accordingly, varies with the load. The primary current varies nearly in proportion to the secondary current and to the load. Transformers connected in different parts of a constant potential system are in parallel, *i. e.*, the primary of each transformer is connected directly across the line so as to receive the full line voltage. It will be seen (compare Appendix II.) that a constant potential transformer is essentially a *constant flux* transformer. Other uses† of the transformer may be considered special.

Commonly, transformers are made for single-phase currents, there being a single primary and a single secondary winding. On polyphase circuits, several single-phase transformers are used (see Exp. 6–A), one on each phase. A special 3-phase transformer, with three primary and three secondary coils is frequently used (see § 26, Exp. 6–A).

§ 5. **Object and Apparatus.**—The object of this experiment is to familiarize one with the structure and general behavior of a transformer and with some of the more important relations between the different quantities involved in its operation. Subsequent experiments go more fully into test methods (Exp. 5–B) and theory (Exp. 5–C), parts of which can be read to advantage in connection with the present experiment.

A transformer with several coils having the same number of turns is well suited for the purposes of this experiment. The

† (§ 4a). Here may be mentioned the series or current transformer, with primary in series with the line and secondary supplying current for an ammeter or wattmeter; the constant current transformer for supplying (constant current) arc lights from a constant current series circuit, one time of importance; arc-light transformers for supplying (constant current) arc lights from a constant potential primary circuit, depending for their operation on magnetic leakage, the present form being the "tub" type with movable secondary which is repelled by the primary and counterbalanced by weights.

following outline is written specifically for such a transformer having four equal coils, each for 55 volts, but the experiment may be modified so as to apply to a transformer wound in some other manner. See Appendix I. for polarity and ratio tests on a commercial transformer.

PART II. TESTS.

§6 **Polarity Test; Series and Parallel Connection of Coils.**— Use one coil as a primary and connect it (with a resistance in series as a safeguard) to a 55-volt alternating current supply. Use the other three coils as a secondary, connecting them in series in such a manner that the three electromotive forces are additive and do not oppose one another. To prove this, measure the electromotive force of each coil and the electromotive force across the three; the latter value should equal the sum of the three other values. This establishes the polarity of the coils.

§7. With the primary circuit unchanged, join the three secondary coils in parallel. In doing this it is necessary to make sure that terminals about to be connected together are of *like polarity,* as in connecting batteries. When the polarity has been already established, as in the preceding section, the proper parallel connection can be readily made. The following procedure, however, insures the right connection independent of previous knowledge of polarity, two coils being first joined in parallel and the third coil being then joined in parallel with these two. To determine which terminals should be connected together to connect two coils in parallel, join a terminal of one coil to a terminal of the other coil and connect a voltmeter to the two remaining terminals. If the voltmeter reads zero, the two terminals connected to the voltmeter may be joined together and the two coils will be in parallel. If the voltmeter does not read zero or very nearly zero, the terminals connected to the voltmeter cannot be joined together without causing a short circuit of the two coils which would give rise to excessive current and burn out the transformer.

§ 8. *Marking Polarity.*—It is convenient and common to designate the polarity of several coils by some systematic marking; thus, all terminals of one polarity may be marked prime (′) and those of opposite polarity be unmarked. To connect coils in parallel, marked terminals are connected to one line and unmarked terminals to the other; to connect in series, the marked terminal of one coil is connected to the unmarked terminal of the next coil, as in Figs. 2, 3 and 5.

In marking polarity it is always best to have the marked secondary terminals of the same polarity as the marked primary terminals. Some positive convention* of this kind becomes important whenever proper polarity is essential, as in the case of transformers supplying the same secondary main, transformers on polyphase circuits (Exps. 6-A and 7-A), transformers used for reducing the current or voltage supplied to wattmeters, etc.

For polarity and ratio tests on commercial transformers, see Appendix I.

§ 9. **Ratio Test.**†—Compute and verify, experimentally, the different ratios of voltage transformation, $E_1 \div E_2$, which are possible with the transformer. At any instant the electromotive force of a coil is, by Faraday's Law,

$$e = S\frac{d\phi}{dt},$$

where S is number of turns and ϕ is flux. The instantaneous value of the voltage, and hence the effective or virtual value, is accordingly proportional to the number of turns, and the ratio of voltages in any two coils is the ratio of the number of turns in the coils. (See Appendix I.)

* In connecting together transformers of different makes, care must be taken, for their polarities may be indicated by different systems.

† For current ratio and tests on commercial transformers, see Appendix I.

If any combination of coils gives a voltage which is beyond the range of the voltmeter, these tests can be made by using a lower supply voltage; it may be found convenient to connect the high potential side of the transformer to the line, thus stepping the voltage down to a lower voltage in the secondary.

§ 10. Prove that the voltage of the secondary is either in phase, or 180° out of phase,† with the primary voltage. To do this, join together one terminal of the primary and one terminal of the secondary, so that the two windings are in series; the supply voltage is connected to the terminals of the primary. Measure the voltage across the primary, the voltage across the secondary, and the voltage across the two, measured between the terminal of the primary and the terminal of the secondary which are not joined together. Either the sum or the difference of the first two readings will equal the third reading; whether it is the sum or the difference will depend on which terminals of the two windings are connected together. If the two voltages were of different phase, the total would be found to be not the algebraic sum but the vector sum.

§ 11. **Use as an Auto-transformer.**—As ordinarily used, a transformer has two independent circuits, a primary and a secondary, and any particular winding is used as part of one of

† (§ 10a). The secondary eletromotive force is in the same phase as the primary *counter* electromotive force, being induced by (substantially) the same flux; hence it is opposite in phase to the primary *impressed* or line electromotive force. It follows that the secondary current, when the transformer is loaded, is nearly opposite in phase to the primary current, this being discussed more fully in Exp. 5–C. This opposition of currents is verified by the auto-transformer test, § 11.

That primary and secondary currents are opposite to each other in phase may be further illustrated by the following experiment. Take a straight upright core surrounded by a primary circuit. Place around it (loosely) a closed ring forming a secondary circuit. Connect the primary to an alternating current supply. When the primary circuit is closed, the secondary will be thrown off violently, showing that the currents in the two circuits are in opposite directions. The secondary ring may be held down by threads, so as to float as a halo.

these only. In the auto-transformer,* or single-coil transformer, part of the windings is common to both primary and secondary.

Connect the transformer coils as an auto-transformer, and verify the different values of voltage transformation. To do this connect all coils in series and consider any one or more of the coils, as may be desired, to be primary or secondary. Some of the coils will at the same time form part of both primary and secondary; these coils will carry, therefore, both the primary and the secondary currents, which are opposite in phase (§ 10a) and so give a resultant current approximately equal to the arithmetical *difference* of the two.

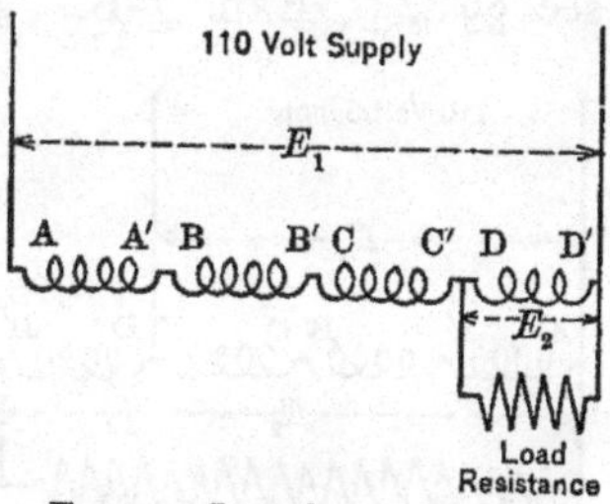

FIG. 2. Step-down auto-transformer, using coils *A B C D* as primary; coil *D* is also used as secondary.

§12. Connect the coils as a *step-down* auto-transformer (Fig. 2) and as a *step-up* auto-transformer or "booster" (Fig. 3). Using suitable resistances as a load, determine the current† in each coil, in the resistance and in the supply line and explain their relative values. The currents and voltages for other combinations of coils can be computed and compared, or determined experimentally. Suppose a 3:2 ratio is desired; with *A, B, C* as primary, how would the use of *C, D* as secondary compare with the use of *B, C*?

§ 13. *Advantages of the Auto-transformer.*—It will be found that the auto-transformer requires less copper than a transformer with separate primary and secondary coils; it has, therefore, not only lower first cost but less copper loss and copper drop, giving better efficiency and regulation. The saving in space on account

* Also called "balance coil" or "compensator"; the term auto-converter should be discarded.

† In making measurement of current, it will be found convenient to use one ammeter and a 3-way ammeter switch.

of less copper makes it possible to reduce also the iron and iron loss.

This advantage of an auto-transformer will be seen to be greater the nearer the ratio of transformation is 1:1. For a comparison of output of transformers and auto-transformers, see §§ 8, 9, Exp. 7–B. An auto-transformer cannot be used when complete insulation of the primary from the secondary is necessary, as in house lighting from high potential lines.

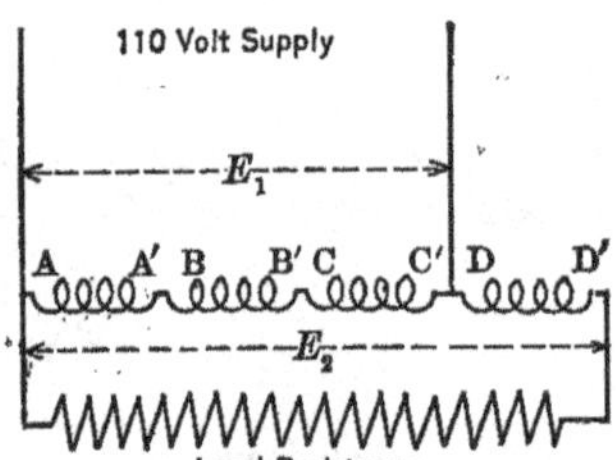

FIG. 3. Step-up auto-transformer or booster, using coils *A B C* as primary; coils *A B C D* are used as secondary.

The step-down auto-transformer of Fig. 2 is in common use as a starting device for induction motors, giving a lower voltage than full line voltage while the motor is coming up to speed; see Fig. 6, Exp. 7–A.

A common use of the step-up arrangement of Fig. 3 is as a booster to raise the voltage on remote parts of a distribution system, say from 2,000 to 2,200 volts. For this a standard 2,000/200 volt transformer can be used, with the low-potential coil in series with the primary to boost the voltage, as in Fig. 2, Exp. 7–B. This becomes a "negative booster" if the connections of the low-potential coil (coil *D* in Fig. 3) are reversed. (If a standard transformer is to be tried in the laboratory, a 100-volt circuit may be boosted to 110 volts, or reduced to 90 volts.)

§ 14. **Constant Potential Operation.**—Transformers are usually operated from a constant potential circuit, so as to transform—either step-up or step-down—from a constant primary potential to a constant secondary potential.

§ 15. *Open Circuit.*—Connect a 110-volt alternating current supply circuit across two of the transformer coils *in series*, as a primary. Measure the no-load primary current, I_0, called the *exciting current.* Predict, and then measure, the value of I_0 when the two primary coils are *in parallel* and connected to a 55-volt supply—*i. e.,* half the preceding voltage. Compare the relative values, for the two cases, of primary turns, ampere turns, volts, volts per turn and flux density.

Measure I_0 when the two primary coils are *in series,* and connected to a 55-volt supply; and interpret the results (see Fig. 2, Exp. 5–B).

Commercial transformers are commonly built with two primaries for connection in series (for, say, 2,200 volts) or parallel (for 1,100 volts); and two secondaries for connection in series (for, say, 220 volts) or parallel (for 110 volts).

§ 16. *Operation Under Load.**—Join two† of the coils in series to form a primary and join the other two coils in series to form a secondary—or make such other arrangement of coils as may be desired. Connect the primary with an alternating current supply—say 110 volts, 60 cycles—appropriate to the arrangement of coils adopted. A voltmeter, ammeter and wattmeter are connected‡ in the primary, as in Fig. 1.

§ 17. With the secondary on open circuit, measure the primary voltage, the primary current (in this case, the no-load current, I_0) and the primary power (in this case, the no-load or core losses, W_0).

* Time should not be spent in an attempt to get very accurate results in this test, particularly if it is to be followed by the more accurate test by the method of losses, Exp. 5–B.

† (§ 16a). Where there is a choice of coils, select an arrangement which avoids great magnetic leakage. If each coil forms one layer or section, to take the first two for primary and the other two for secondary would not be a good arrangement. In a commercial transformer, the primary and secondary windings are so placed as to reduce magnetic leakage; to secure this end, however, all the windings should be used, that is, no coil should be left idle. An arrangement of coils commonly used is as follows: low, high, high, low, potential.

‡ With instruments arranged as in Fig. 1, no corrections need be made. (See Appendix III.)

§ 18. Load the secondary by means of suitable non-inductive resistance. Change this resistance by steps so as to vary the secondary current between no load and 25 per cent. overload. At each step measure the primary voltage E_1, current I_1, and power, W_1; also the secondary* voltage E_2, and secondary current I_2. The product of the secondary voltage and current will give the secondary power W_2, the secondary load being non-inductive. In practice, a load of incandescent lamps is non-inductive, but not so a motor load.

§ 19. Measure the resistance of primary and secondary. (See § 15, Exp. 5-B.)

§ 20. For each load, compute the power factor ($W_1 \div E_1I_1$); also the angle θ by which the primary current lags behind the electromotive force. (Power factor$=\cos \theta$.)

Plot I_1, W_1, power factor, θ, E_2 and W_2 for different values of I_2, as in Fig. 4. Plot, also, the copper loss for primary ($R_1I_1^2$) and for secondary ($R_2I_2^2$) and the core loss W_0 (the value of W_1 on open circuit) which is constant at all loads, as in Fig. 8, Exp. 5-B.

Note that E_2 decreases with load. Determine the per cent. regulation—the per cent. increase in E_2 in going from full load to no load.

Note the current ratio for different loads. It will be seen that as the transformer becomes loaded (by decreasing resistance in the secondary) the secondary current becomes more nearly equal to the primary current (multiplied by $S_1 \div S_2$). In a loaded transformer, primary and secondary ampere-turns are practically (but not exactly) equal.

It is seen that in a transformer there is a loss in volts, a loss in amperes and a loss in watts, this last determining the efficiency. While best for illustrating the operation of a transformer, the

* By means of suitable transfer switches one voltmeter and one ammeter may be used for both primary and secondary.

loading method is not so good for the accurate determination of efficiency and regulation. These can be computed much more

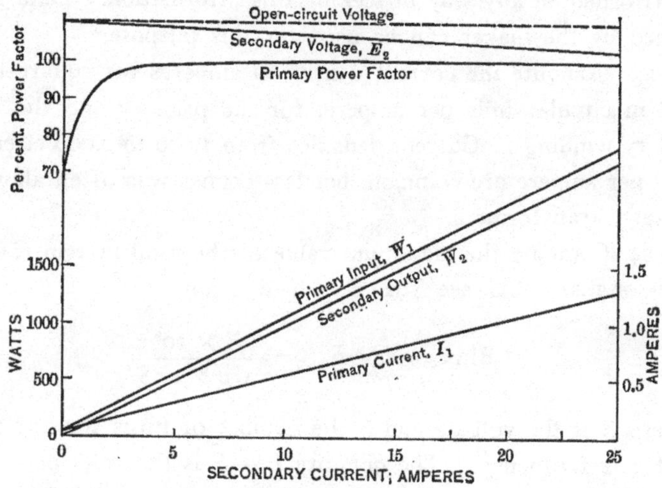

FIG. 4. Curves for 2,000/100 volt, 2 K.W. transformer; see also Fig. 8, Exp. 5–B.

accurately from the losses, determined without loading, as in Exp. 5–B.

§ 21. Load the transformer with an inductive load and take one reading of the instruments. It will be seen that the secondary voltage is somewhat less than it was with non-inductive load—that is, the regulation is poorer.* This happens when induction motors are operated from transformers. In this case the secondary current is lagging. If the secondary current were leading, the secondary voltage in some cases would increase, instead of decrease, with the load. The results are similar to those obtained for an alternator; see Exp. 3–B, particularly Fig. 7.

§ 22. **Design Data and Computation of Flux Density.**—Note the construction and essential dimensions of the transformer,

* (§ 21a). If the leakage reactance of a transformer is small, compared with its resistance, the regulation may be better at low than at high power factor; compare § 28, Exp. 3–B.

including the cross section of the magnetic circuit and size of wire, but do not remove parts, destroy insulation or damage the transformer in any way in seeking this information. Data furnished by the maker can be used for this purpose.

§ 23. Compute the current density in amperes per square inch and in circular mils per ampere, for the primary and the secondary windings. Current densities from 1,000 to 2,000 circular mils per ampere are common, but less copper was often allowed in early transformers.

§ 24. Compute the maximum value of the total flux in C.G.S. lines or maxwells (see § 9a, Exp. 1–B); thus

$$\text{Flux} = A \times B_{\text{max.}} = \frac{E \times 10^8}{\sqrt{2} \times \pi n S},$$

where E is the voltage and S the number of turns for any coil, and n is frequency. The quantity $E \div S$ is the volts per turn. For proof of formulæ, see Appendix II.

Compute the maximum value of the flux density in gausses (flux per sq. cm.); thus

$$\text{Flux density} = B_{\text{max.}} = \frac{E \times 10^8}{\sqrt{2} \times \pi A n S},$$

where A is the cross section* of the core in sq. cms. If A is in square inches, $B_{\text{max.}}$ is the flux density in lines per square inch. If A, in square inches, is multiplied by 6.45, the formula gives $B_{\text{max.}}$ in gausses—for, unfortunately, this mixture of C.G.S. and English units is in common use.

§ 25. The computations for B should be made for standard frequency (60 cycles), and two other frequencies (30 and 120) with the same value of E. If values of A and S are not obtainable, assumed values may be assigned for practice computations. If the cross section of the core is not uniform, B will have dif-

* (§ 24a). The net cross section is, say, 15 per cent. less than the gross cross section on account of lamination.

ferent values for different parts of the magnetic circuit. From these computations it can be seen whether B will be more or less, if a transformer is operated at a higher or lower frequency than rated and at the same voltage. (Note in what manner E should be changed to maintain B the same at different frequencies.) Practically, transformers are run at different frequencies without changing E, if the frequency is not too far below the frequency for which the transformer is designed. For a discussion of the effect of frequency upon core loss, see §§ 8–14, Exp. 5–B.

In transformer design,* B is given a wide range (4,000–14,000 gausses at 60 cycles), being sometimes greater in small than in large transformers and greater in transformers designed for low than in those for high frequency. In design, E and n being given, B may be assumed and the product $A \times S$ determined. This product being fixed, the designer may adjust the values of A and S to suit his purpose, increasing A and decreasing S to use more iron and less copper, or *vice versa.*

§ 26. From the formula for flux density, it will be seen that the electromotive force of any coil of a transformer is proportional to the number of turns in the coil, a fact already noted. The *volts-per-turn* should be computed as a constant for the transformer. For small transformers this may be one third or one half, being greater for large transformers, perhaps 2 to 4 for transformers above 30 K.W. The reciprocal gives the *turns-per-volt.* The volts-per-turn, when known for a certain type and size of transformer, may be used as a design constant.

§ 27. Other data of interest to the designer (which may be determined when worth while) are the weight of copper and of iron, total and per K.W. This may range from 5 to 25 lbs. per K.W. for either copper or iron. The space factor for copper is the

* (§ 25a). For more complete design data, see handbooks, etc. As magnetic material is improved, higher magnetic densities are possible for the same loss. While densities of 4,000–8,000 were used with ordinary grades of iron, densities of 6,000–12,000 are now common with alloy steel.

ratio of the cross section of copper to the total cross section of the windings, *i. e.*, to the cross section of copper plus insulation and air space. Similarly the space factor for the iron is the ratio of its net to gross section.

APPENDIX I.

POLARITY AND RATIO OF COMMERCIAL TRANSFORMER.

§ 28. **Polarity; Alternating Current Method.**—The coils are connected in series, two at a time, and notice is taken whether the voltage around the two is the sum or the difference of the separate voltages. There are several ways in which this can be carried out.

As an example, let us take a transformer with two primaries for 1,000 volts each and two secondaries for 50 volts each. Connect the two 1,000-volt primaries in series and connect the terminals of one* of the primaries to a low potential supply circuit, say 50 volts, as in Fig. 5. If a voltmeter across the two coils together reads zero, reverse the connections of one of the coils. The voltmeter should then read 100 volts across the two coils together, and 50 volts across each one separately. Terminals *A* and *B* are now of one polarity; terminals *A′* and *B′* are of the opposite polarity, to be marked with a prime (′) or ×.

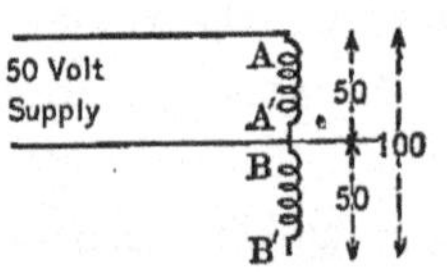

FIG. 5. Polarity test by alternating current method.

Each secondary is then connected in series with one primary, the primary being connected to the 50-volt supply circuit; the secondary in series with it is so connected that the voltmeter reading around the two coils in series is greater (52.5 volts) than the potential from the mains (50 volts). If the reading is less (47.5 volts), reverse the secondary. Secondary terminals are marked with a prime (′) or × to correspond with the primary.

Small transformers are commonly so wound that, when the primary and secondary leads on one side of the transformer are connected

* If the two coils in series were connected to the supply circuit, a burn-out might result if the coils were opposed to each other.

together, the voltage measured across the two primary and secondary leads on the opposite side will be the sum of the voltages of the two windings.

§ 29. **Polarity; Direct Current Method.**—The alternating current method is usually preferred, but sometimes the following method will be found convenient. The primary is supplied with a direct current sufficient to give a reading on a direct current voltmeter connected to the primary terminals. The voltmeter terminals are then connected to what are supposed to be corresponding terminals of the secondary. If, when the primary circuit is closed,* the voltmeter needle is thrown in the same direction as the preceding reading, the voltmeter has been connected to the secondary terminals corresponding to primary terminals; *i. e.*, the voltmeter lead from the primary terminal (′) or × is connected to the secondary terminal to be marked (′) or ×. If the voltmeter needle is thrown in the opposite direction, the reverse is true.

§ 30. **Potential Ratio.**—Where one transformer alone is to be tested, the transformer should be supplied with any convenient voltage and the voltage of each circuit measured either by two voltmeters, one of which has been calibrated in terms of the other, or by one voltmeter reading direct on the low potential side and with a multiplier on the high potential side.†

When one transformer has been tested in this manner, or a small potential transformer of accurate ratio is available, two transformers can be run in parallel from the same circuit and their secondary voltages on open circuit compared by readings taken with one voltmeter or with two voltmeters whose relative calibration is known.

If the secondaries of two similar transformers are connected in series and in opposition, any difference will be shown by a voltmeter connected across the two.

§ 31. **Current Ratio.**—For commercial testing of the ratio of a transformer, test may be made by comparison of primary and secondary currents instead of voltages. The secondary circuit is short-circuited through an ammeter of low resistance and the

* The current should be small so as not to injure the voltmeter by slamming the needle when the circuit is made and broken.

† It is not necessary to run the transformer at full rated potential. When high potentials are used, due caution should be observed.

primary and secondary currents measured when a proper voltage (a few per cent. of normal primary voltage) is applied to the primary, so that about the normal current flows.

§ 32. **Circulating Current Test.**—As a shop test, after one standard transformer has been tested, other transformers designed for the same ratio may be operated from the same primary mains and tested one at a time by connecting each secondary to be tested in parallel with the secondary of the standard, terminals of the same polarity being connected together. If an ammeter shows a circulation of current through the secondaries, the two transformers are not of the same ratio.

Commercially a small difference in ratio is allowable as shown by the circulating current, which, however, should never exceed one per cent. of the rated full-load current. Instead of an ammeter a suitable fuse may be conveniently used, and more safely where much difference in ratio may exist.

APPENDIX II.

RELATION BETWEEN FLUX AND ELECTROMOTIVE FORCE.

§ 33. The fundamental relation between flux and electromotive force is expressed by Faraday's law; that is, in a closed circuit* of S turns embracing a varying flux ϕ, the induced electromotive force is $-S \cdot d\phi/dt$. In a transformer, this applies alike to primary or secondary. In the case of a primary coil this induced electromotive force is a *counter* electromotive force and requires to overcome it an equal and opposite impressed† electromotive force

$$e = S \cdot d\phi/dt.$$

§ 34. **Sine Assumption.**—Assuming the wave of electromotive force to be a sine wave, we have

$$e = E_{\text{max.}} \sin \omega t;$$

* Not limited to a transformer.

† (§ 33a). The actual terminal voltage includes also resistance drop, thus

$$e = Ri + S \cdot d\phi/dt.$$

The resistance drop, however, is practically negligible in the primary of a transformer on open circuit.

$$S d\phi = E_{max.} \sin \omega t \, dt;$$

$$\phi = -\frac{E_{max.}}{\omega S} \cos \omega t = \frac{E_{max.}}{\omega S} \sin (\omega t - 90°).$$

The maximum value of the flux is

$$\phi_{max.} = \frac{E_{max.}}{\omega S}.$$

and the flux per unit area is

$$B_{max.} = \phi_{max.} \div A = \frac{E_{max.}}{\omega S A}.$$

To express in terms of effective voltage, substitute $\sqrt{2}E$ for $E_{max.}$. Multiplying by 10^8 to change from C.G.S. units to volts, and remembering that ω is 2π times the frequency (n), we have as a working formula for flux density

$$B_{max.} = \frac{\sqrt{2}E \times 10^8}{2\pi n S A} = \frac{E \times 10^8}{4.45 n S A}.$$

It follows from this formula that a constant potential transformer is a constant flux transformer. It also follows that, if a certain flux is maintained in the transformer, the voltage in any coil is proportional to the number of turns in that coil. For further interpretation, see §§ 24–26.

§ 35. **Without Sine Assumption.**—We have the fundamental relation

$$d\phi = e dt/S.$$

Integrating for half of a period T, during which time the flux changes from a minus to a plus maximum,

$$\int_{-\phi}^{+\phi} d\phi = \frac{1}{S}\int_0^{\frac{T}{2}} e dt,$$

$$2\phi_{max.} = E_{av.}/S \times T/2.$$

Writing $1/n$ for T, and multiplying by 10^8 to reduce to volts, we have

$$B_{max.} = \phi_{max.} \div A = \frac{E_{av.} \times 10^8}{4 n S A}.$$

This is true of any shaped electromotive force wave, of which $E_{av.}$ is the average value. It is seen that the maximum value of the flux depends upon the *average* and not the *effective* value of electromotive force. If we let the form factor f designate the ratio of effective to average value, $E_{eff.} = fE_{av.}$ and

$$B_{max.} = \frac{E_{eff.} \times 10^8}{4fnSA}.$$

For a sine* wave $f = 1.1$, substituting which gives the formula of § 34.

APPENDIX III.

USE OF A WATTMETER, VOLTMETER AND AMMETER; ARRANGEMENT OF INSTRUMENTS AND CORRECTIONS TO BE APPLIED.

§ 36. The ordinary measuring instruments consume, in themselves, a small amount of power,—usually only a few watts. In many cases this can be neglected, particularly in testing apparatus requiring considerable power, but in precise measurements of small quantities the effect of these losses should be considered. So far as the load run in the present experiment is concerned, the losses in instruments can be neglected; the errors and the methods of correcting for them should, however, be noted for use whenever necessary.

Some arrangements of instruments introduce larger errors than others. Furthermore, the errors can be readily corrected for in some cases and not in others.

In selecting a method for arranging instruments, choose one in which the errors (even if large) can be best corrected for, or else choose one in which the errors are as small as possible and no correction is necessary. So far as convenience is concerned, the latter is to be preferred.

§ 37. **The Wattmeter.**—A wattmeter has two coils: a series or *current coil,* connected in series with one line of the circuit as an ammeter, and a shunt or *potential coil,* connected in shunt, from one line to the other, as a voltmeter.

* (§ 35a). For a sine wave, $E_{av.} = \frac{2}{\pi} E_{max.}$; and $E_{eff.} = \frac{1}{\sqrt{2}} E_{max.}$ See page 37, Bedell and Crehore's *Alternating Currents.* Form factor is $f = E_{eff.} \div E_{av.} = 1.1$ for a sine wave. (Form factor was first used by Roessler as $E_{av.} \div E_{eff.}$, which for a sine wave is .9.)

A wattmeter* may be connected in two ways, as follows.

§ 38. In the first and usual method, Fig. 6, the potential coil is connected between the line wires on the *supply* side. In this method,

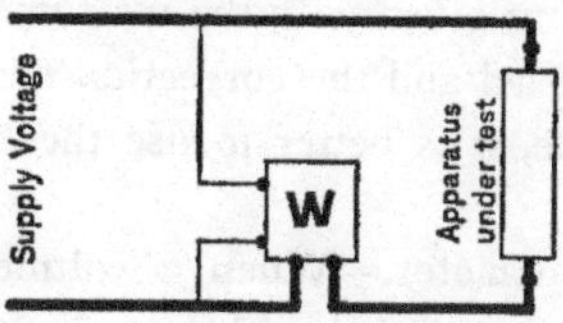

FIG. 6. First method.

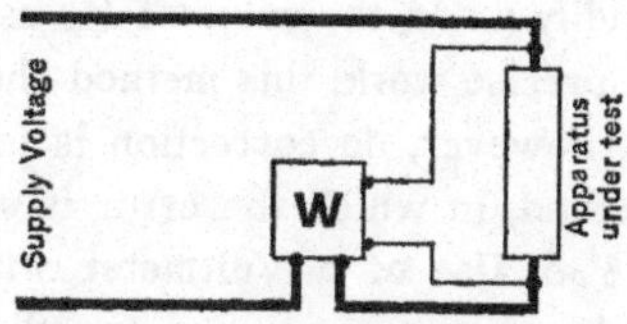

FIG. 7. Second method.

Methods for connecting a wattmeter.

the wattmeter reading is too large, including not only the true watts of the load but also the RI^2 loss in the current coil of the wattmeter. This error, which may amount to several watts, can frequently be neglected; correction for it is not easily made. In measuring small amounts of power, in order that the error may be neglected, the current should not exceed one half the rating of the current coil of the wattmeter. With current greater than this, the loss in the series coil, which increases with the square of the current, may be too large to neglect.

§ 39. In the second method, shown in Fig. 7, the potential coil is connected across the terminals of the *load*. In this method the wattmeter reading is also too large, since it includes the RI^2 loss in the potential coil of the wattmeter. This error is larger than the error in the first method and should be corrected for by subtracting $E^2 \div R_w$ from the wattmeter reading, E being the line voltage and R_w the

* (§ 37a). *Lag Error.*—A wattmeter measures $EI \times$ power factor. For accuracy, a wattmeter must have the resistance of the potential circuit so large, compared with its reactance, that the circuit is practically non-inductive. The current in the potential circuit is then practically in phase with the electromotive force; in reality it lags by a small angle, $\theta = \tan^{-1}(L\omega \div R)$. The error due to this lag angle is different for different power factors, $\cos\phi$, of the load. The true watts are equal to the wattmeter reading multiplied by

$$\frac{\cos\phi}{\cos\theta\cos(\phi-\theta)}.$$

In commercial wattmeters, at commercial frequencies, this correction can be neglected. It becomes appreciable on higher frequencies, particularly on loads of low power factor and at low voltages—*i. e.,* when the resistance of the potential circuit is small.

resistance of the potential coil of the wattmeter. This correction might be, for example, 5 watts in a 100-volt wattmeter, 10 watts in a 200-volt wattmeter, etc. The correction, however, is exact and is readily made, the value of R_w usually being given with the instrument. In precise work, this method should be used and the correction made. If, however, no correction is to be made, it is better to use the first method, in which the error is smaller.

§ 40. **Use of a Voltmeter with a Wattmeter.**—When a voltmeter and wattmeter are used together, the voltmeter should (usually) be connected between the same points as the potential coil of the wattmeter. There are, therefore, the same two methods of connection as with a wattmeter alone.

§ 41. *First Method.*—The voltmeter is connected between the lines on the *supply* side of the wattmeter. The reading of the voltmeter includes the RI drop in the current coil of the wattmeter; the error is small and may often be neglected.

§ 42. *Second Method.*—The voltmeter is connected on the *load* side of the wattmeter, directly to the terminals of the load. The voltmeter reading is now correct. The wattmeter, however, includes the watts consumed in the voltmeter. The reading of the wattmeter should, accordingly, be corrected by subtracting $E^2 \div R_v$, where R_v is the resistance of the voltmeter. The whole correction for the wattmeter is now $E^2(1/R_w + 1/R_v)$, which is to be subtracted from the wattmeter reading.

§ 43. **Use of an Ammeter with a Wattmeter.**—If an ammeter is connected in circuit on the load side of a wattmeter, as the current coil in Fig. 6, the ammeter reads the true load current. The wattmeter reading, however, includes the watts loss in the ammeter—a small error which is neglected.

If the ammeter is connected on the supply side of the wattmeter no error is introduced in the wattmeter reading; the ammeter reading, however, is too large,* since it includes the current in the potential

* (§ 43a). *Ammeter Correction.*—The ammeter reading can be corrected by subtracting $(1/R_v + 1/R_w)W/I$. The current which flows in the potential circuits of the voltmeter and wattmeter is $E/R_v + E/R_w$. This is in phase with the electromotive force and not with the current, and must be multiplied by the power factor of the load W/EI to get its component in phase with the current.

coil of the wattmeter. This error can be neglected, when the load current is large. The ammeter reading would be correct if the potential coil of the wattmeter (and voltmeter, if one is used) were opened when the ammeter is read; sometimes this is allowable under steady conditions, but simultaneous readings of all instruments are usually more accurate.

§ 44. **Combinations of Instruments.**—In the combined use of ammeter, wattmeter and voltmeter, the best method to use depends somewhat upon the conditions of the test. The arrangement of Fig. 1 is, for most purposes, as good as any; no corrections are made. In the short-circuit test of a transformer, the reading of the current is most important; hence, Fig. 6, Exp. 5-B, the ammeter for this test can best be connected on the load side of the other instruments. For the open-circuit test, voltage is important and not current; the ammeter is, therefore, in the supply line and the instruments arranged as in Fig. 1, Exp. 5-B (requiring a correction) or as Fig. 1 of this experiment (requiring no correction and hence simpler to use). See § 3a, Exp. 5-B.

§ 45. **Multipliers.**—To extend the range of a voltmeter, either a series resistance (called a multiplier) or a potential transformer can be used. The potential range of a wattmeter is extended in the same way.

To extend the range of an ammeter, a current transformer is used; the primary of the transformer is connected in series with the line, the secondary being short-circuited through the ammeter. The current range of a wattmeter is extended in the same way.

The ratio of transformation of any potential or current transformer must be accurately known, and, for a current transformer, this ratio must be known in connection with the particular instrument and secondary leads with which it is to be used. Any small phase shifting, due to the fact that primary and secondary quantities are not exactly in phase opposition, introduces no error in the use of instrument transformers with ammeters or voltmeters, but with wattmeters such phase shifting may introduce considerable error and needs to be taken into consideration for accurate work. For a complete discussion, see *Electric Measurements on Circuits Requiring Current and Potential Transformers,* a paper by L. T. Robinson, read at the June, 1909, meeting of the A. I. E. E.

Experiment 5–B. Transformer Test by the Method of Losses.

§ 1. **Introductory.**—The losses in a transformer are the *core loss,* which is dependent upon and varies with voltage, and the *copper* (and *load*) losses, which are dependent upon and vary with current. The most accurate* and the most convenient method for testing a transformer is to measure these losses separately, without loading the transformer, and compute the efficiency and regulation.

This requires two simple tests, each employing† a voltmeter, ammeter and wattmeter: an *open-circuit* or no-load test for determining the no-load or core loss and the exciting current at various voltages, particularly at normal voltage; and a *short-circuit* test at a low voltage (a few per cent. of normal) for determining the copper and load losses and impedance drop for various currents, particularly for normal full-load current. The latter test gives, also, the equivalent resistance and leakage reactance of the transformer.

Measurements are also made of primary and secondary resistance.

§2. This method may be employed in testing any transformer, whether it is intended for constant potential, constant current or other service; the method will be described in detail with reference to its application to a constant potential transformer.

* (§ 1a). This is most accurate for the reasons explained in § 1b, Exp. 2–B. It is not practicable to determine efficiencies accurately by loading a transformer (§ 16, Exp. 5–A) and measuring the input and output directly—unless exceeding care be taken—the two quantities measured being so nearly equal. The indirect method of losses is, furthermore, most convenient because no load is required and no high-potential measurements are necessary.

† In many cases the same instruments can be used in the two tests; compare § 44. Two similar tests are made in testing alternators; see § 9, Exp. 3–B.

In a constant potential transformer the magnetization and hence the core loss and exciting current are (substantially) the same at all loads, being dependent upon voltage and not upon current. The copper and load losses, on the other hand, depend upon current and vary with the load. (In a constant current transformer, the conditions are reversed; copper losses are constant and core loss varies with the load.)

PART I. OPEN-CIRCUIT TEST.

§ 3. With the secondary open, measurements are made on the primary with ammeter, voltmeter and wattmeter, one method for making the connections* being shown in Fig. 1. Although any coil or combination of coils could be used as a primary in this test, it is most convenient to use a low potential coil (50, 100 or 200 volts) as primary to suit the instruments and supply voltage available; furthermore, there is less danger in working on the low potential side.

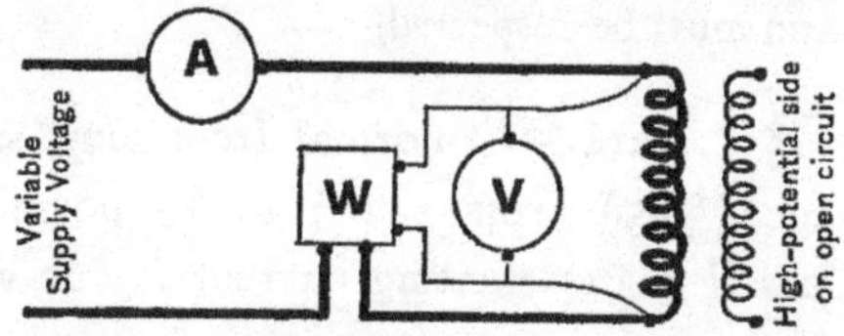

FIG. 1. One method of connection for open-circuit test for core loss and exciting current. See § 3a for method of connecting instruments requiring no corrections.

For the same degree of magnetization, the exciting current in a 100-volt coil is ten times as large as in a 1,000-volt coil, the ampere turns being the same. It is, accordingly, a simple matter

* (§ 3a). For selection of instruments, see § 44. The arrangement of instruments shown in Fig. 1 should be followed when the highest accuracy is desired; the wattmeter reading is to be corrected by subtracting $E^2(1/R_w + 1/R_v)$, which is the power consumed in the potential coils of the wattmeter and voltmeter.

It is, however, much simpler—and in many cases sufficiently accurate—to arrange the instruments as in Fig. 1 of Exp. 5-A, and to make no correction. See Appendix III., Exp. 5-A.

to reduce the exciting current measured on a coil of one voltage to its value for a coil of another voltage. The watts core loss is the same measured on one coil as on another, for the same magnetization.

CAUTION. If two coils are to be connected in parallel or series, to avoid a burn-out it is necessary to first make sure of their polarity, as described in §§ 6 and 28, Exp. 5–A.

Be careful of the high-potential terminals in this test. It should be made impossible for loose wires, or for persons making measurements, to come in contact with these terminals. Although the testing current and instrument are all of low voltage and although the high-potential coil is open and has no current in it, *the potential is there* and must be respected.

§ 4. **Data.**—At normal frequency, say 60 cycles, vary the voltage (§ 45) from say $\frac{1}{4}$ to $1\frac{1}{4}$ normal and determine the core loss W_0, and exciting current I_0 for various voltages. Note the frequency at which the test is made. It is desirable that the frequency be maintained constant, and that the voltage be of sine wave-form.

Take very accurate readings at two points (within, say, 5 or 10 per cent. of half and full voltage) by taking at each of these points a series of five readings and averaging. This two-voltage method is very convenient, since normal and half voltage (as 55/110 or 110/220) are often available—or their equivalent can be obtained by series and parallel connections, as described in § 46. As will be seen later, Figs. 2 and 5, it is very accurate for transformers built of ordinary iron, at normal and higher frequencies, but not at frequencies far below normal. For transformers with improved iron, the two-voltage method is not correct unless one observed point is taken very near full voltage, little error being then introduced by obtaining values for full voltage from the curves.

§ 5. If possible, repeat the data at a second frequency. If

the frequency is higher than normal, complete data can be taken as before. If the frequency is lower than that for which the transformer is intended, the core loss and exciting current will be greater and the voltage should not be raised so that they become excessive for the transformer or the instruments;* W_0 should not exceed say twice and I_0 four or five times their respective values at normal frequency. It will be understood, however, that these limits are only arbitrary.

§ 6. **Curves for Exciting Current.**—For each frequency, plot a curve showing the exciting current for different voltages, as in Fig. 2. Locate by heavy black dots the two points accurately

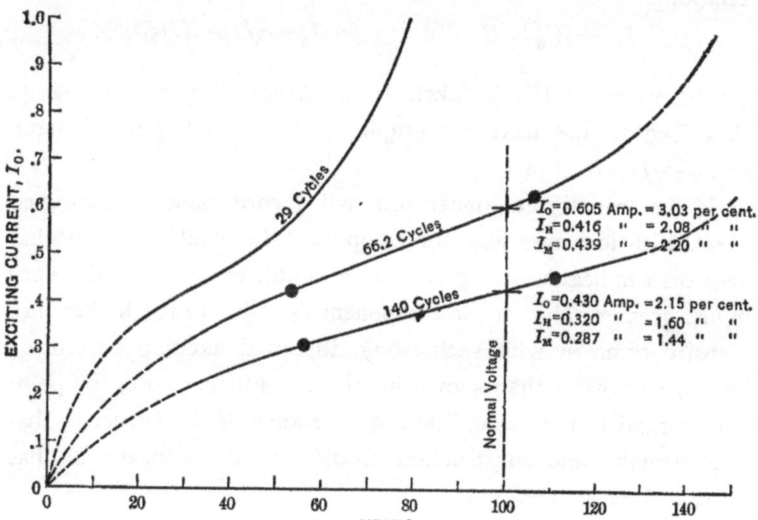

FIG. 2. Observed exciting current for varying voltage at different frequencies; 2 K.W. transformer, 100-volt coil. To obtain the exciting current for the 2,000-volt coil, divide these values by 20.

* (§ 5a). The current coil of the wattmeter has a certain rated current carrying capacity which should not be exceeded for any length of time. It may, however, be exceeded *for a few moments only* by 30 or even 40 per cent.; readings are taken quickly and the wattmeter is then cut out.

determined at about half and full voltage and note how closely a straight line drawn through them coincides with the curve through the working range. (It will be found that this straight line construction, based on two readings, can not be used when the transformer is worked at a very high flux density, as in the 29-cycle curve of Fig. 2.)

§ 7. Take from each curve the value of I_0 for normal voltage (§ 48). Resolve the exciting current, I_0, into two components: the in-phase power component I_H (which supplies the core losses due to hysteresis and eddy currents); and the quadrature magnetizing* component, I_M. These are determined by the following relations:

$$I_H = W_0 \div E. \qquad I_0 = \sqrt{I_H^2 + I_M^2}.$$

The value of W_0 is taken from curves, Fig. 3 and Fig. 5, described in the next paragraph. At no load, power factor $= I_H \div I_0$.

If the transformer under test had a core made of improved steel, with less core loss, the component I_H would be somewhat less than indicated in Fig. 2. On account, however, of the very much greater value of the component I_M (due to the higher flux density common with such iron), the total exciting current I_0 would be greater than shown in Fig. 2. Furthermore, the point for normal voltage would be near the knee of the curve, so that the straight line construction would not be accurate, as has already been pointed out.

Compute, as in Fig. 2, the values of I_0, I_H and I_M as per cent. of the normal full-load current of the coil on which the test is made; thus, the full-load current for a 100-volt coil of a 2 K.W. transformer, Fig. 2, is 20 amperes. Expressed as per cent., the results will apply to any coil.

* Usage is not fixed in regard to the terms "exciting" and "magnetizing" currents, these terms being not infrequently interchanged.

§ 8. **Curves for Core Loss.**—The readings of the wattmeter in the open-circuit test (after corrections are made, if there are any, § 3a) gives the core loss plus a small RI_0^2 loss due to the heating effect of the exciting current. The latter loss can be computed and deducted from the wattmeter reading; it will generally be found negligible.*

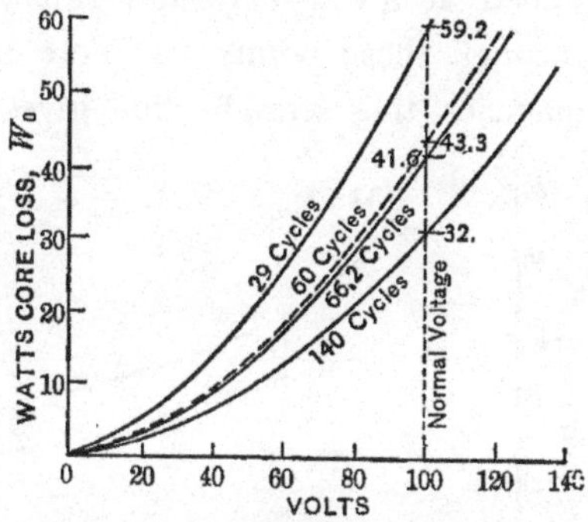

FIG. 3. Watts core loss for varying voltage at different frequencies; 2 K.W. transformer, 100-volt coil.

The curves showing the change of core loss with voltage can be plotted on ordinary cross-section paper as in Fig. 3; a derived curve, showing the variation of core loss with frequency, being plotted as in Fig. 4. It is much better, however, to use a logarithmic scale for ordinates and abscissæ, in which case the curves become (within limits) practically straight lines. For this purpose, it is convenient to use logarithmic cross-section paper.†

Above normal voltage, as higher densities are reached, the curve tends to bend upwards, due to the fact that the hysteretic exponent (which has a value of about 1.6 up to 10,000 gausses) becomes greater. Transformers with improved iron are run at higher densities, so that at normal frequency this bend may be reached at normal, or even below normal, voltage.

§ 9. For each frequency plot a curve on logarithmic paper

* (§ 8a.) Although any RIo^2 loss should be deducted for obtaining true iron loss, for the calculation of efficiency it is better not to make such a deduction but to include the RIo^2 loss with the iron loss Wo.

† This paper can be obtained from the Cornell Coöperative Society, or Andrus and Church, Ithaca, N. Y.

The same results can be obtained on plain paper by plotting the logarithms of the observed quantities—a laborious process—or by using a slide rule as a scale.

showing the core loss, W_0, for different voltages, as in Fig. 5. Locate by heavy black dots the two points accurately determined at about half and full voltage. Draw a straight line through these points and note that at normal and higher frequencies this straight line gives the curve accurately through the working range, so that W_0 for normal voltage can be readily obtained from it. At frequencies much below normal, and at high flux-densities, this straight line relation may not hold.

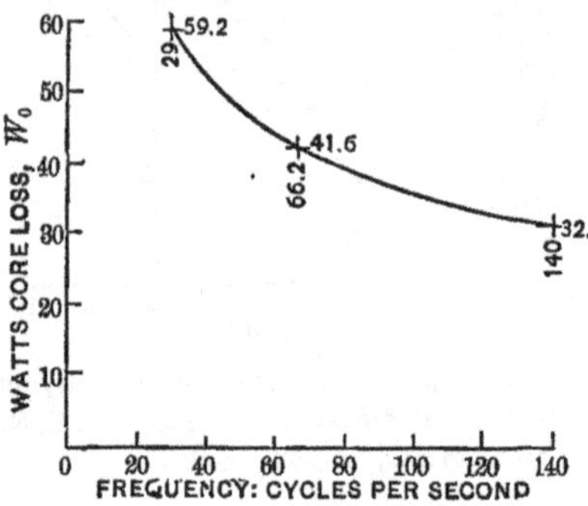

FIG. 4. Watts core loss for different frequencies at normal voltage; 2 K. W. transformer.

If it was impossible to get data for any curve up to normal valtage, extend the curve that far as a dotted line. This extension is quite accurate at frequencies near normal or higher, but can not be depended upon at frequencies way below normal. The curves, however, can be more readily and more accurately extended on logarithmic paper than on ordinary coördinate paper.

§ 10. The slope of these curves (the actual tangent with the horizontal) is the exponent (a) of E or B in the formula showing the law of core-loss variation for different voltages and flux densities at a constant frequency;

$$W \propto E^a \propto B^a.$$

This exponent (a) should be determined and interpreted; see §§ 49–51.

§ 11. *Variation of Core Loss with Frequency.*—On the same logarithmic sheet, see Fig. 5, plot a derived curve showing the core loss, W_0, at normal voltage for different frequencies. If

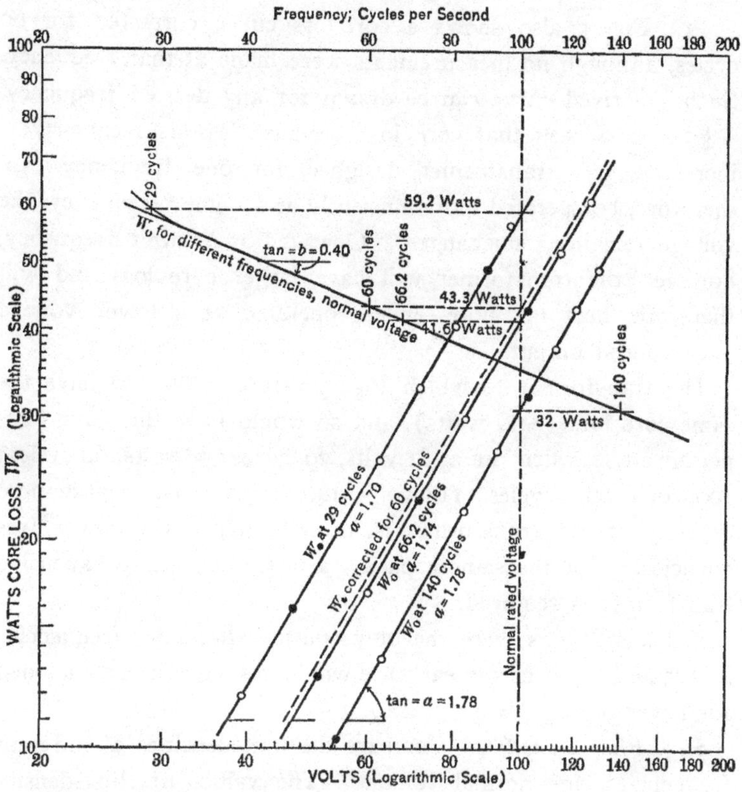

FIG. 5. Curves for core loss plotted with logarithmic scale.

there are data for only two frequencies and hence only two points on this curve, plot it as a straight line—as it will be practically a straight line through a considerable range. (When the test is made at only one frequency, see § 52.)

The slope of this line gives the exponent (b), showing the law of core-loss variation for different frequencies at constant voltage:

$$W_0 \propto n^b.$$

The exponent (b) should be determined; see § 51. Plotted on ordinary coördinate paper, this curve appears as in Fig. 4.

§ 12. Fig. 5 also shows a core-loss curve corrected for 60 cycles, although no measurements were made at that frequency. Such a derived curve can be drawn for any desired frequency.

§ 13. It is seen that core loss becomes less as frequency is increased. A transformer designed for one frequency can, therefore, be operated more efficiently at a higher frequency, the voltage remaining unchanged. Operated at a lower frequency, however, the transformer will have larger core loss and will therefore, heat up more—unless operated at a lower voltage and reduced output.

The transformer to which Fig. 5 refers is seen to have the same core loss (41.6 watts), and so would have the same temperature rise, when run at 81 volts, 29 cycles; 98 volts, 60 cycles; 100 volts, 66.2 cycles; 118 volts, 140 cycles. The volt-ampere capacity of any transformer is, accordingly, less at lower frequencies. For the same capacity, a larger and more expensive transformer is required.

If transformers were the only consideration, the frequencies of 125 and 133 cycles in early use would not have been abandoned for lower ones. (See § 3, Exp. 3–A.)

§ 14. *Flux Densities.*—Compare the flux densities at different frequencies, for normal voltage. The values of flux density can be computed, as in §§ 33–35, Exp. 5–A, if the number of turns and iron cross-section are known. Without calculating the actual values and without knowing the construction data of the transformer, the *relative* values can be found by the relation $B \propto E \div n$. Thus, if at 60 cycles B is taken as 1.0, the flux density at 30 cycles is 2.0, at 120 cycles 0.5, etc.

PART II. RESISTANCE MEASUREMENTS.

§15. **Data.**—The primary and secondary resistances are measured by means of a bridge or by direct current, fall-of-potential method (§ 17, Exp. 1–A). Disconnect the voltmeter before the current is thrown off, to avoid damage by inductive kick. Avoid heating the coils and so causing their resistance to increase; the testing current should not exceed 25 per cent. of the full-load current for the coil, or should not be long continued. The range of the ammeter to be used is thus determined from the known value of full-load current.

The range of voltmeter is found by assuming an approximate value for resistance drop; thus, if the resistance drop were one per cent. in primary or secondary for full-load current, this would be 10 volts in a 1,000-volt coil and 1 volt in a 100-volt coil. If only one fourth of full-load current were used for testing, the voltage readings would be 2.5 and 0.25 volts, respectively.

The resistance measurements by direct current are to be used as a check and for comparison with the results obtained in the short-circuit test.

Temperature conditions should be taken account of (§ 22).

§ 16. **Equivalent Primary Resistance.**—From the measured values of R_1 and R_2, compute the equivalent resistance* R,

* (§ 16a). The equivalent resistance R must have such a value that

$$RI_1^2 = R_1I_1^2 + R_2I_2^2.$$

Dividing by I_1^2 and writing the ratio of turns $(S_1 \div S_2)$ in place of the ratio of currents $(I_2 \div I_1)$, we have

$$R = R_1 + (S_1 \div S_2)^2 R_2.$$

It is obvious, also, that $R =$ (copper loss) $\div I_1^2$.

Any resistance in the secondary, either within the transformer or in the external circuit, has the same effect as though it were multiplied by the square of the ratio of turns and placed in the primary circuit. It may be noted here that the same is true of reactance.

which is the joint resistance of the primary and secondary in terms of the primary:

$$R = R_1 + (S_1 \div S_2)^2 R_2.$$

The value thus determined will be used for comparison with the value ($R = W_C \div I^2$) determined from the copper loss in the short-circuit test.

PART III. SHORT-CIRCUIT TEST.

§ 17. **Method of Test.**—For the short-circuit test, the secondary (low-potential side) of the transformer is short circuited and the primary (high-potential side) is supplied with a small difference of potential, just sufficient to cause the full-load or desired current to flow. This is rarely more than 5 or 6 per cent. of normal voltage. Instruments* are connected in the primary, as shown in Fig. 6. The current might be measured by an ammeter in either circuit, but it is better† to have the ammeter on the primary side with the voltmeter and wattmeter.

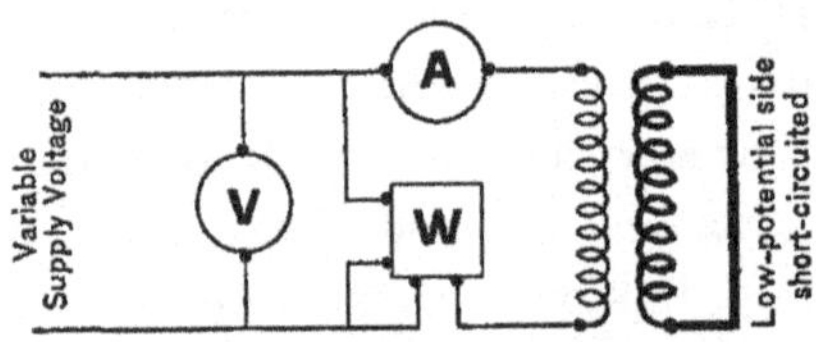

Fig. 6. Connections for short-circuit test for copper loss and impedance voltage.

* (§ 17a). The most important reading to have correct is that of the ammeter, since the wattmeter reading varies as I^2 and the voltmeter reading varies as I, and all results are calculated for values of current. For this reason it is well to place the ammeter directly in the primary circuit, as in Fig. 6, in which case no correction is necessary. If the ammeter is connected in the supply line, as in Fig. 1, and the instruments are read simultaneously, a small error is introduced (tending, in this case, to favor the transformer) unless a correction is applied. See Appendix III., Exp. 5-A. Connected as in Fig. 6, the wattmeter usually needs no correction; but for the accurate measurement of small power the method of connection shown in Fig. 7, Exp. 5-A, should be used and a correction applied.

For selection of instruments, see § 44.

† (§ 17b). It is important for accuracy to have the short circuit of the secondary as "short" as possible, *i. e.*, with practically zero impedance

In this test, when full-load current flows in the primary, full-load current flows in the secondary; when half-load current flows in the primary, half-load current flows in the secondary, etc. The flux density is very low, so that there is practically no core loss. The wattmeter reading gives, therefore, the total copper losses—both primary and secondary—for any particular current. Included with the copper losses are the load losses.

§ 18 *Load Losses.*—Load losses are due chiefly to eddy currents in the copper and are greatest, therefore, in large solid conductors. They have the effect of causing a greater loss in a conductor when traversed by alternating current than when traversed by direct current, the resistance being apparently increased. The term load losses includes all losses* which increase with load and depend upon current, over and above the copper losses as determined by direct current. Evidently such losses should be taken into consideration in calculating efficiency,† and the Standardization Rules of the A. I. E. E. so specify.

outside of the transformer. An ammeter and its leads in the secondary, sometimes used, tends to give the transformer a poorer regulation and efficiency. It is instructive, however, before taking readings, to insert an ammeter in the secondary, as well as the primary, and to note that the ratio of currents is practically equal to the ratio of turns.

If there are two secondaries, it makes no difference whether they be put in parallel or in series; two primaries should be put in parallel or series to suit the range of instruments. No coil should be left idle.

* For example, eddy-current loss in the core due to local flux set up by the current in a loaded transformer in addition to the normal core loss.

† (§ 18a). Very commonly, however, this is not done, copper losses (neglecting load losses) being determined by direct current measurement of resistance. This tends to favor the transformer. In justification of this, it may be said that it has not been fully established that the load losses under actual load conditions are the same as those obtained on short circuit—it being held that they may be less. The two methods serve as a check. If the losses by the wattmeter are only slightly greater than by direct current, the result is satisfactory for the transformer. Any considerable difference, however, shows the existence of load losses. For an accurate comparison, great care is necessary in regard to temperature conditions and the calibration of instruments.

Although these losses may be considerable in large transformers, in small well built transformers they are usually insignificant.

§ 19. *Impedance Voltage.*—In an ideal transformer on short circuit, with zero secondary resistance and no magnetic leakage, the only voltage necessary to cause a given current to flow would be R_1I_1, to overcome the resistance of the primary. The effect of secondary resistance is to apparently increase the resistance of the primary to

$$R = R_1 + (S_1 \div S_2)^2 R_2,$$

and the resistance drop is, accordingly, RI_1.

On account of the magnetic leakage, the transformer apparently has a reactance X, called leakage reactance;* this causes (in terms of the primary) a reactance drop XI_1, in addition to the resistance drop, RI_1. For a given frequency, this reactance due to leakage is a constant of the transformer, the same as resistance. It is the same on open circuit or short circuit, and is the same at no load or full load.

The total impedance, which limits the flow of current in the short circuit test, is a combination of the equivalent resistance and leakage reactance, being

$$Z = \sqrt{R^2 + X^2}.$$

The voltmeter reading gives the total *impedance voltage*,

$$E_Z = \sqrt{R^2I^2_1 + X^2I^2_1},$$

necessary to overcome both resistance and leakage reactance.

§ 20. **Data.**—At rated frequency, take, say, five readings of the impedance voltage (E_Z) and the copper loss (W_C) for various currents from about ¼ to 1¼ full-load current.

§ 21. Readings at various currents are chiefly for illustration and are not essential. When facilities for varying the current are lacking, one accurate reading (or better the mean of five

* Discussed more fully in Exp. 5-C.

readings) at any convenient value of current is sufficient for all results. Slight changes in wave form are immaterial and a series resistance or any other means may be used for adjusting the current (§ 45).

§ 22. These readings vary with temperature, being dependent upon resistance, and should be taken at some definite temperature or under some definite temperature conditions to be specified in the report, as hot after a heat run of a certain duration, or cold before the transformer is heated up. In this latter case readings must be taken quickly to avoid rise in temperature due to the testing current. Commercial tests should be under specified service conditions, commonly after a three hour heat run at full load or the equivalent, the room temperature being 25° C.; in this the A. I. E. E. Standardization Rules should be consulted and followed.

§ 23. At a second frequency repeat the readings. (In a commercial test, readings would be taken at rated frequency only.) It will be found that the copper loss and apparent resistance vary but slightly with frequency and that the leakage reactance is proportional to frequency. Known for one frequency, it can be computed for any other.

§ 24. **Results.**—The impedance voltage, E_Z, and the copper losses, W_C, can be plotted directly from the voltmeter and wattmeter readings, with primary current as abscissæ, as in Fig. 7 and Fig. 8. It is better, however, to proceed as follows:

§ 25. From the readings of the ammeter, voltmeter and wattmeter in the short-circuit test, compute, for each observation, the values of Z, R and X, as given below, and determine an average value for all the observations.

Impedance: $Z = E_Z \div I_1$.

Resistance: $R = W_C \div I_1^2$.

Reactance: $X = \sqrt{Z^2 - R^2}$.

These are equivalent or apparent values, in terms of the primary, and include the effect of both primary and secondary. The effect of load losses is included in the values of W_C and R.

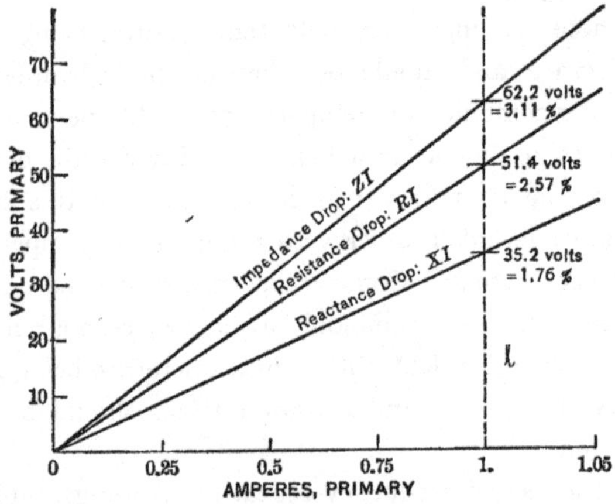

FIG. 7. Short-circuit test, 2 K.W. transformer, 2,000-volt coil. Voltage drop due to impedance resistance and leakage reactance.

§ 26. Compare this value of R with the value found from resistance measurements of R_1 and R_2 in § 16.

§ 27. Where tests are made at different frequencies, compute the mean value of $X \div n$ for each frequency; the value should be about the same for all frequencies.

§ 28. The values of X affect regulation but not efficiency; while the values of R affect not only regulation (on account of RI drop) but also efficiency (on account of RI^2 loss).

§ 29. *Curves for Voltage Drop.*—Using the values of Z, R and X thus determined (§ 25), plot curves for ZI_1, RI_1 and XI_1 drops for different values of primary current, as in Fig. 7, these curves being straight lines. Compare these curves with the corresponding curves for an alternator, Fig. 2, Exp. 3–B.

For normal full-load current, mark the value of each drop in volts and as per cent. of normal full-load voltage. The per cent. impedance drop is also called *impedance ratio* (§ 13, Exp. 3–B).

§ 30. *Curves for Copper Losses.*—Calculate RI_1^2 for $\frac{1}{10}$, $\frac{1}{4}$, $\frac{1}{2}$, $\frac{3}{4}$, 1 and $1\frac{1}{4}$ load. Plotted as in Fig. 8, these give the values of

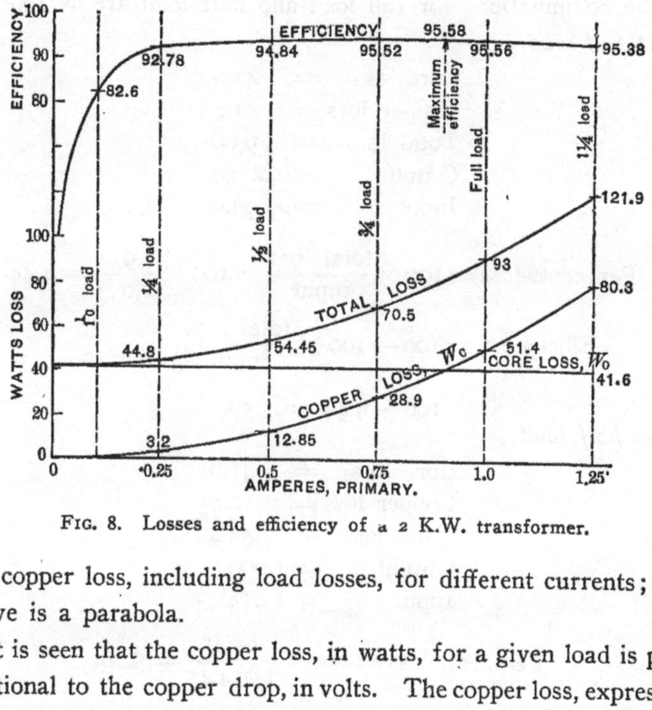

FIG. 8. Losses and efficiency of a 2 K.W. transformer.

the copper loss, including load losses, for different currents; the curve is a parabola.

It is seen that the copper loss, in watts, for a given load is proportional to the copper drop, in volts. The copper loss, expressed as a per cent. of rated volt-amperes, is equal to the copper drop, expressed as a per cent. of rated volts.

Per cent. copper loss $= RI^2/EI$. Per cent. copper drop $= RI/E$.

PART IV. RESULTS. EFFICIENCY AND REGULATION.

§ 31. **Efficiency.**—Efficiency is equal to output divided by input and is readily determined when the losses are known. For a particular frequency and normal voltage, take the value of core loss from the curves already determined, Figs. 3, 4 and 5.

Thus, let $W_0 = 41.6$ watts. The total losses are found by adding this constant core loss to the copper losses for each load, as in Fig. 8. The efficiency* should be computed for $\frac{1}{10}$, $\frac{1}{4}$, $\frac{1}{2}$, $\frac{3}{4}$, 1 and $1\frac{1}{4}$ load.

The computations for full load and half load are as follows:

At full load.

Core loss	=	41.6
Copper loss	=	51.4
Total loss	=	93.0
Output	=	2,000.0
Input	=	2,093.0

$$\text{Per cent. loss} = 100 \times \frac{\text{total loss}}{\text{input}} = 100 \times \frac{93}{2,093} = 4.44.$$

$$\text{Efficiency}\dagger = 100 - 100 \times \frac{\text{total loss}}{\text{input}},$$

$$= 100 - 4.44 = 95.56.$$

At half load.

Core loss	=	41.6
Copper loss	=	12.85
Total loss	=	54.45
Output	=	1,000.
Input	=	1,054.45

$$\text{Per cent. loss} = 100 \times \frac{54.45}{1,054.45} = 5.16.$$

$$\text{Efficiency} = 100 - 5.16 = 94.84.$$

* (§ 31a). The efficiency will be different for different frequencies and for different rating of voltage and current; see § 48. See § 57 for a more exact method of determining W_0 for full-load voltage.

Referring to Fig. 5, the core loss at 100 volts is 41.6 for the frequency (66.2 cycles) used in the test. Corrected for 60 cycles, $W_0 = 43.3$, it being possible to thus determine the efficiency for a frequency not used in the test. This is useful in comparing guarantees.

† (§ 31b). This formula will be found much better for making computations than the equivalent and more usual form,

$$\text{Efficiency} = \text{Output} \div \text{Input}.$$

§ 32. Maximum efficiency* occurs at such a load that the copper loss is equal to the core loss; in Fig. 8, this is at 0.9 full load.

Note the similarity between the curves for a transformer, shown in Fig. 8, and the corresponding curves for a shunt motor, Fig. 3, Exp. 2–B.

§ 33. *All-day efficiency* is computed on some assumption, as 5 hrs. full load and 19 hrs. no load. Other assumptions can be made to suit specific service conditions. Except under special conditions, the term "all-day efficiency" has no useful significance.

§ 34. **Regulation.**—The regulation of a constant potential transformer is the per cent. increase in secondary voltage in going from full load to no load. See Appendix I., Exp. 5–C. There are various graphical methods for determining regulation, which are necessarily unsatisfactory on account of the small values of some of the quantities and the consequent difficulty in making an accurate drawing to scale. There are also various analytical methods, many of which are equally unsatisfactory on account of their involved character and the unnecessary labor required in using them.

The regulation of a transformer can be determined for all power factors—current lagging or leading—by the same method as is used in determining the regulation of an alternator by the electromotive force method (§§ 16–22, Exp. 3–B), either graphically or analytically.

A modified method, however, is easier to apply to a transformer on account of the fact that the resistance and reactance drops in a transformer are comparatively small.

§ 35. What the writer believes to be the simplest and most practicable method† for determining the regulation of a trans-

* See § 28, Exp. 2–B.

† (§ 35a). From a paper "Transformer Regulation," by F. Bedell, *Elec. World*, Oct. 8, 1898; the term with i_M is now dropped on account of difference of definition (see Appendix I., Exp. 5–C).

former is given below, any errors being less than the usual errors of observation.

Regulation is to be computed for non-inductive load and for loads of various power factors, with current lagging and leading. (It is suggested that the reader compares the results obtained by this method and by other methods with which he may be familiar, and that he also compares the labor required in applying the different methods.)

Let r be the per cent. resistance drop and x the per cent. reactance drop, as determined by the short-circuit test. Thus, in Fig. 7, $r=2.57$ and $x=1.76$ (not .0257 and .0176).

§ 36. *Non-inductive Load.*—The regulation on non-inductive load is computed as follows:

$$\text{Per cent. regulation} = r + \frac{x^2}{2(100+r)}.$$

For all practical purposes, as a glance at the numerical example will show, this may be written

$$\text{Per cent. regulation} = r + \frac{x^2}{200}.$$

For example, when $r=2.57$ and $x=1.76$;

$$\text{In-phase drop} = r = 2.57 \text{ per cent.}$$

$$\text{Effective quadrature drop} = \frac{x^2}{200} = 0.015 \text{ per cent.}$$

$$\text{Regulation} = 2.585 \text{ per cent.}$$

It is seen that the regulation is practically determined by the resistance drop; the effect of reactance drop on non-inductive load is nearly negligible. This is seen in Fig. 9 which is discussed later. In computing the regulation, therefore, the accuracy of the results depends directly upon the accuracy with which the resistance is determined. Regulation varies with temperature and to be definite must be for a specified temperature.

§ 37. *For Lagging Current.*—When the load has a power

factor ($\cos\theta$) less than unity and the current is lagging, the regulation is practically* as follows:

$$\text{Per cent. regulation} = r\cos\theta + x\sin\theta.$$

For example, let $\cos\theta = 0.866$; $\sin\theta = 0.5$;

In-phase resistance drop $= r\cos\theta = 2.57 \times 0.866 = 2.23$ per cent.
In-phase reactance drop $= x\sin\theta = 1.76 \times 0.500 = \underline{0.88}$ per cent.
Regulation $= 3.11$ per cent.

§ 38. *For Leading Current.*—When the load has a power factor ($\cos\theta$) less than unity and the current is leading, the regulation is practically†

$$\text{Per cent. regulation} = r\cos\theta - x\sin\theta.$$

For example, let $\cos\theta = 0.866$; $\sin\theta = 0.5$;

In-phase resistance drop $= r\cos\theta = 2.57 \times 0.866 = 2.23$ per cent.
In-phase reactance drop $= x\sin\theta = 1.76 \times 0.500 = \underline{0.88}$ per cent.
Regulation $= 1.35$ per cent.

§ 39. **Proof.**—Fig. 9 shows a simple graphical method for obtaining regulation at non-inductive load. (Compare also Fig. 11, Exp. 5-C, in which the same lettering is used, and Fig. 3, Exp. 3-B.)

Referring to Fig. 9, lay off AL equal to the secondary full-load voltage $E_2 = 100$ per cent. (A scale of volts could be used, if

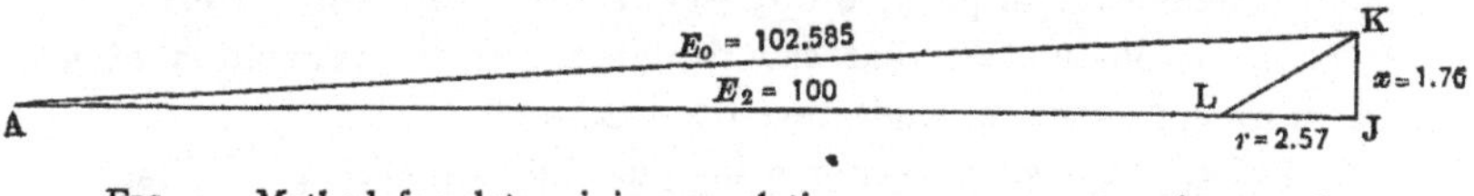

FIG. 9. Method for determining regulation; r = per cent. resistance drop; x = per cent. reactance drop; regulation $= E_0 - E_2 = 2.585$. (This Fig. is not drawn to scale.)

* (§ 37a). For greater accuracy, a term for effective quadrature drop ($q^2 \div 200$), should be added, § 42. In the present example this term is only .0003, making the regulation 3.1103. In any ordinary case, on lagging current, this term can be neglected.

† (§ 38a). For greater accuracy, a term $q^2 \div 200$ should be added, § 43; in the present example, this term equals .039.

desired, instead of per cent.) Lay off $LJ=r$ and, $JK=x$. Then $AK=E_0$, the secondary terminal voltage at no load. Per cent. regulation $=E_0-E_2=2.585$.

As already stated, the graphical method can not be accurately applied on account of the small values of r and x. From the graphical method, an analytical method is derived as follows.

§ 40. Analytically, we have from Fig. 9,

$$E_0=\sqrt{(100+r)^2+x^2};$$

or, more simply (see § 41),

$$E_0=100+r+\frac{x^2}{2(100+r)}.$$

Transposing, we have

$$\text{Regulation}=E_0-100=r+\frac{x^2}{2(100+r)}.$$

§ 41. Expressed more generally, let

$p=$ per cent. in-phase voltage drop.

$q=$ per cent. quadrature voltage drop.

$$E_0=\sqrt{(100+p)^2+q^2}=100+p+\frac{q^2}{2(100+p)}.$$

This practical identity* can be seen by squaring, or by solving a numerical example. Transposing, we have

$$\text{Regulation}=E_0-100=p+\frac{q^2}{2(100+p)},$$

or, for practical purposes, $=p+(q^2\div 200)$.

Regulation = (in-phase drop) + (effective quadrature drop).

It is the in-phase drop that chiefly determines the regulation of a transformer; effective quadrature drop is small.

§ 42. *Lagging Current.*—For a lagging current, with power factor $=\cos\theta$, the resistance drop r makes an angle θ with the terminal

* (§ 41a). *Theorem.*—In a right triangle in which the height is small compared with the base, the hypotenuse = base + [(height)² ÷ 2(base)]. This is convenient in solving many alternating current problems.

For example, let base = 100; when height = 5, hypotenuse = 100.125 (true value = 100.124922); when height = 10, hypotenuse = 100.5 (true value = 100.498756).

voltage E_2. Fig. 9 would then be drawn as Fig. 4, Exp. 3-B, in which the line BC makes an angle θ with the line OB.

Resolve r into an in-phase component, $r \cos \theta$, and a quadrature component, $r \sin \theta$; resolve x into an in-phase component, $x \sin \theta$, and a quadrature component, $x \cos \theta$.

$$\text{In-phase drop} = p = r \cos \theta + x \sin \theta.$$
$$\text{Quadrature drop} = q = x \cos \theta - r \sin \theta.$$

$$\text{Regulation} = p + \frac{q^2}{2(100 + p)}.$$

$$= p + (q^2 \div 200), \text{ approximately.}$$

For practical purposes

$$\text{Regulation} = p = r \cos \theta + x \sin \theta.$$

§ 43. *Leading Current.*—Similarly, for a leading current,

$$p = r \cos \theta - x \sin \theta.$$
$$q = x \cos \theta + r \sin \theta.$$

$$\text{Regulation} = p + \frac{q^2}{2(100 + p)}.$$

$$= p + (q^2 \div 200), \text{ approximately.}$$

APPENDIX I.

MISCELLANEOUS NOTES.

§ 44. **Selection of Instruments.**—In selecting instruments, the necessary range can be approximately told by assuming some reasonable value for efficiency or losses. Thus, in a 2 K.W. transformer, let us assume that the efficiency is 95 per cent. and that the iron losses and copper losses are equal. Assume the power factor in the open-circuit and short-circuit tests to be 0.66.

In the open-circuit test the core loss will be 50 watts; the ammeter and current coil of the wattmeter must, accordingly, carry a current of 1.5 amperes, if the test is made on a 50-volt coil; 0.75 amp. for a 100-volt coil; 0.375 amp. for a 200-volt coil, etc.

In the short-circuit test, the copper loss will be 50 watts; if the test is made on a 2,000-volt coil, full-load current will be 1 ampere and the impedance voltage will be 75 volts; for a 1,000-volt coil, the values become 2 amp. and 37.5 volts, etc.

In many cases, by a proper connection of coils in series or parallel, one set of instruments may be selected which will be suitable for both open-circuit and short-circuit tests. It is to be understood that this method of selection will give only the approximate range. For frequencies much below normal, it is to be borne in mind that the current will be much greater than at normal frequency and instruments with say four or five times the current-carrying capacity will be required.

§ 45. **Adjustment of Supply Voltage.**—The best way to get various voltages for the open-circuit test is by means of a transformer or an auto-transformer with a number of taps. An adjustment of voltage by means of a series resistance distorts the wave form and hence introduces error in the readings of core loss and exciting current. This error is small if the reduction in voltage caused by the resistance is small, as from 110 to 104 volts, and in this case the use of series resistance is permissible. It should not be used, however, for large reduction in voltage or in any case when high accuracy is wanted. (In a particular case,* reducing the voltage from 220 to 110 volts by a series resistance caused a decrease in core loss and in magnetizing current of about 6 per cent.) If a resistance is to be used, less error is introduced when the resistance is bridged across the line and the transformer supply shunted off of part of the resistance than is introduced with the resistance in series. It is best, however, to avoid resistance control entirely.

Another way to vary the voltage is to vary the field excitation of the supply alternator. No other load should be on the alternator, nor should the transformer under test form an appreciable load on the alternator; otherwise change in wave-form may materially change the core loss. (A change of 20 per cent. can be thus produced.)

§ 46. The two-voltage method (§ 9) obviates the necessity of voltage adjustment. If normal and half voltage are not available, the method may often be used with one voltage only by connecting coils

* This was for an old transformer. With the new transformer iron and higher densities, the errors due to wave distortion become greater.

first in parallel on the appropriate normal voltage and then in series on the same voltage (which will then be half normal); thus, two 55-volt coils in parallel on a 55-volt circuit give the reading for normal voltage, while the two coils in series on the 55-volt circuit give the reading for half normal voltage. For the same degree of magnetization, the wattmeter will indicate the same core loss whatever coil is used; the ammeter will read twice as much for the parallel as for the series arrangement (the ampere turns being the same) and the ammeter reading must, accordingly, be divided (or multiplied) by two to reduce all readings to common terms. The voltmeter reading must be multiplied (or divided) by two.

§ 47. In the short-circuit test, the matter of wave form is practically of no consequence; the only result affected is the reactance drop and only a very large change in wave form could materially affect its value. For the short-circuit test, therefore, any means of adjustment may be used which is found convenient.

§ 48. **Normal Voltage and Current.**—In determining normal full-load values of current and electromotive forces, the assumption is commonly made that the efficiency is 100 per cent. and that currents and voltages are transformed exactly in the ratio of turns. Thus in a 2,000/100 volt, 2 K.W. transformer, the secondary current is taken as 20 amperes and the primary current as 1 ampere (whereas strictly the latter should be a trifle more); the secondary voltage is taken as 100 volts and the primary 2,000 volts, neglecting the fact that these no-load values of voltages do not strictly hold at full load. Any change of rating changes the results of a test. If the voltages are rated as 2,200 and 110, the corresponding primary and secondary currents are 0.909 and 18.2 amperes; the copper loss is less and the core loss greater. In comparing transformers and their guarantees, each transformer should be tested at its rating. In comparing transformers for a specific service, the tests of all should be made at a common voltage according to the conditions of the service. From the curve sheets it is very easy to pick results for different ratings from one set of data. In the laboratory, to facilitate comparison of data, certain voltages should be adopted as standard, as 100–200/1,000–2,000, 110–220/1,100–2,200, etc.

§ 49. **Core Losses and Their Variation.**—Eddy currents flow in local short-circuited secondary circuits which are practically non-inductive.

If e and r represent the electromotive force and resistance of one of these circuits, the watts loss is $e^2 \div r$. But $e \propto E$. It accordingly follows that eddy current loss* varies as the square of the voltage and is independent of frequency and wave form. As the temperature of a transformer increases the eddy current loss decreases.†

Referring to Fig. 5, if all the core loss were due to eddy currents, we would have $b = 0$ and $a = 2$. The curve for losses at different frequencies and constant voltage would be a horizontal line; the curves for losses at different voltages and constant frequency would slope at an angle with a tangent 2.

§ 50. Hysteresis loss in watts per cu. cm. is practically equal to $\eta n B^{1.6} 10^{-7}$. Here η is a coefficient of hysteresis, equal‡ to about .002 for what was formerly good iron, but is now little more than half that value for the best alloy steel. The hysteretic exponent 1.6, first determined by Steinmetz, is only approximate, being less than this for low magnetic densities and considerably more than this for high densities.

Referring to Fig. 5, if all the core loss were due to hysteresis, we would have (taking 1.6 as the hysteretic exponent) $a = 1.6$ and $b = 1 - a = -0.6$.

§ 51. In an actual transformer both hysteresis and eddy current loss are present, so that a has a value between 1.6 (hysteresis) and 2 (eddy currents), and b has a value between -0.6 (hysteresis) and 0

* (§ 49a). In terms of B, eddy current loss in watts per cu. cm. is $\gamma (dnB)^2 10^{-21}$, where d is the thickness of lamination expressed in mils and γ is the conductance (the reciprocal of resistance in ohms per cu. cm.) of the material; for iron, γ is about 10^5. There is a *very slight* change of eddy currents with frequency and wave form due to local inductance and "skin effect" in the local eddy current circuit.

By decreasing the thickness of transformer plate, eddy current loss is diminished; but hysteresis loss is increased, since some iron is wasted and B is greater in the remainder. A thickness between 10 and 15 mils gives the least total loss, according to particular conditions; see *Elec. World*, Dec. 31, 1898.

† (§ 49b). Hysteresis loss, also, decreases with increase in temperature. The total core loss of a transformer when hot may be 6 or 8 per cent. less than when cold.

‡ (§ 50a). By so called "aging" due to heat, this coefficient increases in the course of time. Although not entirely eliminated, this effect has been reduced in the best steel now used.

(eddy currents). Hysteresis is the chief loss and has most weight in determining a and b. Hysteresis loss*, and hence the values of a and b, are affected by wave form. It will be understood that the hysteresis exponent 1.6 is not a constant but represents a fair average value for moderate ranges of flux densities; at high densities the hysteretic exponent may have a value as high as 2 or more.

§ 52. *One-Voltage and One-Frequency Method.*—If a and b are known, it is possible, having determined the core loss at one voltage for a particular frequency, to compute the core loss for any other voltage and frequency. It then becomes unnecessary to test a transformer at the *exact* rated voltage and frequency—which is indeed difficult to do.

Taking as average values $a = 1.666$ and $b = -.4474$, correction factors† for variation of core loss with frequency and voltage are given in the following tables.

CORRECTION TABLES.

VARIATION OF CORE LOSS WITH VOLTAGE.

Volts (per cent. normal)	90	95	96	97	98	99	100
Core loss	83.7	91.7	93.3	95	96.7	98.4	100
Volts (per cent. normal)	101	102	103	104	105	110	
Core loss	101.6	103.3	105	106.6	108.3	116.6	

VARIATION OF CORE LOSS WITH FREQUENCY.

Cycles	55	56	57	58	59	60	61	62	63	64	65
Loss	103.9	103.12	102.3	101.6	100.8	100	99.3	98.5	97.8	97	96.4

§ 53. **Separation of Hysteresis and Eddy Currents.**—To determine the eddy current loss in watts, the core loss is to be measured at two frequencies and at the same flux density. Let W' be total core loss at normal voltage E' and frequency n'. At a lower frequency n'' and a

* (§ 51a). When the wave of electromotive force is peaked, the maximum flux density and the core loss are less.

† (§ 52a). These are taken from a series of tests made in 1899 by W. F. Kelley and H. Spoehrer (see thesis, Cornell University Library). The transformers were small (1–15 K. W.) and were designed for 60 cycles and over. The writer has no data on most recent transformer iron. These tests also showed that each per cent. variation in voltage caused about .7 per cent. (.6945) variation in exciting current.

lower* voltage $E'' = \frac{n''}{n'} E'$, let the core loss be W''. We may compute the watts eddy current loss at the higher frequency (n') and normal voltage by the formula

$$\text{Watts eddy currents} = \frac{W' - \frac{n'}{n''} W''}{1 - \frac{n''}{n'}}.$$

Eddy current loss is substantially the same for all frequencies, but varies as the square of the voltage and so can be computed for any frequency and voltage. Hysteresis loss is found by subtracting eddy loss from total loss.

* (§ 53a). If the wave form of electromotive force for the two frequencies is different, $E'' = (n''f'' \div n'f')E'$, where the form factor f is the ratio of the effective to the average value. (For a sine wave, $f = 1.1$.) The eddy current loss in watts at the higher frequency n' and normal voltage is then

$$\left[\frac{n''}{n'} W' - W''\right] \div \left[\frac{n''}{n'} - \left(\frac{E''}{E'}\right)^2\right].$$

The above equations can be derived as follows: Eddy current loss, irrespective of frequency and wave form (§ 49), varies as E^2 and equals aE^2, where a is a constant. Similarly, for any wave form, hysteresis loss equals bnB^x, where b and x are constant; no assumption is made that $x = 1.6$. At the two frequencies the total losses are

(1) $$W' = a(E')^2 + bn'(B')^x;$$

(2) $$W'' = a(E'')^2 + bn''(B'')^x.$$

For B'', write B', this being the condition of the test; for E'', write $E'(E'' \div E')$. Multiply (2) by $n' \div n''$, subtract from (1) and solve for eddy current loss $a(E')^2$. When the wave form of electromotive force is the same at the two frequencies, $(E'' \div E') = (n'' \div n')$.

The separation of losses by measurements at two frequencies was first made by Steinmetz; the influence of form factor was introduced by Roessler. There are various methods for making the calculations, differing somewhat in detail. The formulæ here given are from a paper by the author before the Cornell Electrical Society, May 4, 1898. Note § 35, Exp. 5-A, and Appendix I., Exp. 2-B; also Bedell's *Transformer*, p. 312 *et seq.* (Some of these references, following Roessler, use form factor as the reciprocal of f, as defined above.) M. G. Lloyd has recently published a very complete investigation of the subject; see *Bull. Bureau of Standards*, February, 1909.

§ 54. **Insulation and Temperature Tests.**—These tests are of commercial importance but need no full discussion here. The *Standardization Rules* specify fully the conditions under which they are to be made; details of the tests are described in the usual handbooks.

§ 55. *Insulation.*—The insulation is tested between each winding and all other parts. The applied voltage is increased *gradually,* so as to avoid any excessive momentary strain. This is usually done by some means of primary control in a special testing transformer. Various companies make testing transformers for obtaining high potential for this test and furnish detailed instructions for their use. The voltage is preferably measured by means of a spark gap with a high protective resistance in series with it. The test consists in seeing that the apparatus withstands a specified over-voltage for a specified time without breaking down.

§ 56. *Heat Runs.*—These are made under full-load voltage and full-load current for a specified time, temperatures being found by thermometers and resistance measurements. The heat run could be made by actually loading the transformer, but is usually made by some kind of *opposition* or *pumping back* method, of which there are several. No load is then required and no power, except enough to supply the losses.

A common form of opposition run employs two similar transformers: the two secondaries (low potential side) are connected in parallel to source *A*, of normal frequency and normal voltage, which supplies the core loss; the two primaries are connected in series, opposed to each other, and are then connected to source *B*, which supplies the normal full-load current. (Source *B* requires a voltage equal to twice the impedance voltage of one transformer and can be of any frequency, *i. e.,* it may or may not be the same frequency as *A*.) All windings now have full-load current and normal voltage. Instruments in *A* will give, if desired, the core loss and exciting current; instruments in *B* will give copper loss and impedance voltage. (The two transformers need not be identical.)

Instead of connecting source *B* in the high potential side, a common modification is to connect the high potential windings of the two transformers directly in opposition and to insert source *B* in series with the low potential winding of one of the transformers. This has the advantage that all connections with supply lines and instruments are at low potential; see *Electric Journal*, p. 64, Vol. VI., and Fig. 322, Karapetoff's *Exp. Elect. Engineering*.

A modified form of opposition test can be applied to a single transformer; see Foster's Handbook.

§ 57. **Note on Efficiency.**—If the rated secondary voltage is $E_2 = 100$, the customary and most simple procedure is to take the core loss for this voltage from Fig. 5 (thus, $W_0 = 41.6$) and to compute the full load efficiency as in § 31. To be accurate, however, the secondary core voltage or flux voltage, E_s, should be taken as E_2 *plus* the secondary RI drop. Taking this drop as 1.28 ($=\frac{1}{2}r$ in § 35), we have $E_s = 101.28$ and the corresponding core loss, $W_0 = 42.5$; this gives the correct efficiency of 95.51 instead of 95.56. The difference between these values is so little that the method of § 31 is usually sufficiently correct.

Approached in another way, we might consider $E_s = 100$ and $W_0 = 41.6$. Then $E_2 = 100 - 1.28 = 98.72$. To get the rated output of 2 K. W., since E_2 is decreased, the current must be increased by the factor $1 \div 98.72$. The copper loss must then be increased by the factor $(1 \div 98.72)^2$, giving an efficiency of about 95.5.

EXPERIMENT 5-C. Circle Diagram for a Constant Potential Transformer.

§ 1. **Introductory.**—It has been seen, Exp. 4-B, that when the resistance is varied in a series circuit with constant reactance the vector representing the current follows the arc of a circle as a locus. In a similar manner, the primary current of a constant potential transformer follows the arc of a circle as a locus when the secondary resistance is varied. The same is true for an induction motor when its load is varied, and use is made of this fact in practical motor testing. The following experiment will, accordingly, serve to make clear certain principles of the induction motor as well as of the transformer; upon these principles is based the method of transformer testing developed in detail in Exp. 5-B.

In Part I. the general principles governing the action of a transformer will be discussed; in Part II. these principles will be applied in constructing a circle diagram. The practical results, so far as commercial testing is concerned, are all given in Exp. 5-B. The actual construction of a diagram *to scale* gives one a definite and concrete idea of what might otherwise be vague and abstract. Furthermore, the abstract diagrams given here (Figs. 1-11) and elsewhere are so grossly exaggerated that they give very wrong ideas of *real values.* Even Fig. 12, which is more nearly to scale, is much exaggerated.

§ 2. *Data.*—The same data are required as in Exp. 5-B. See § 25 of this experiment.

PART I. GENERAL DISCUSSION OF THE ACTION OF A TRANSFORMER.

§ 3. The action of a transformer will be most readily understood by considering its action first without a load—*i. e.,* on open circuit—and then with a load.

§ 4. **Transformer on Open Circuit.**—When a transformer is on open circuit, the secondary winding has no current flowing in it and it accordingly has no magnetizing effect on the core. A small current flows in the primary which magnetizes the core. Let us see what determines the magnitude and phase of this open-circuit primary current.

§ 5. *Assuming No Core Loss.*—The open-circuit diagram for a perfect transformer, in which there are no losses, is shown in Fig. 1. The primary electromotive force E_P causes a current I_0 to flow and this current sets up a flux ϕ. This flux, being alternating, causes a counter-electromotive force opposed to the primary impressed electromotive force. When the primary circuit is closed, the current I_0, and the flux ϕ which it sets up, assume such values that the counter-electromotive force is just equal* to the impressed electromotive force.

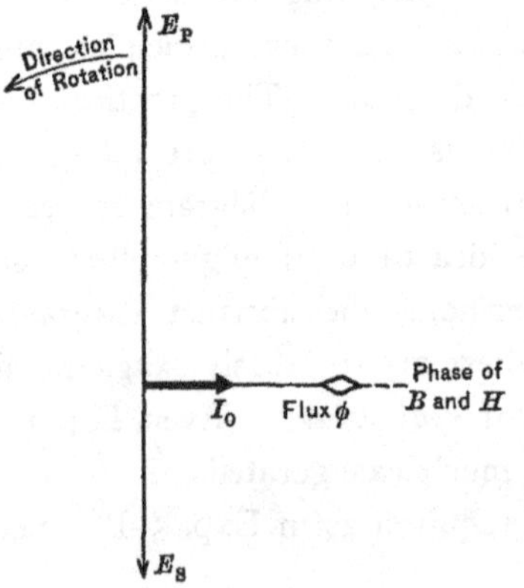

FIG. 1. Open-circuit diagram for a transformer with no core loss.

This primary counter-electromotive force has, at any instant, the value $e' = -S_1(d\phi \div dt)$, the equal and opposite impressed electromotive force being $e_P = S_1(d\phi \div dt)$. It will be seen that the electromotive force is zero when the flux is a maximum and that the flux ϕ lags 90° behind the impressed electromotive force E_P, as in Fig. 1.

§ 6. In the absence of core loss, the current I_0 is in phase with the flux ϕ, which it produces. When permeability is constant, magnetizing force H is proportional to I_0 and is in phase with

* The primary resistance on open circuit is very small and can be neglected.

and proportional to the flux density B. The B–H curve is a straight line, instead of the familiar hysteresis loop, and there is no hysteresis loss.

The current I_0, as shown in Fig. 1, is in quadrature with the electromotive force and is wattless.

§ 7. The flux ϕ links with the secondary circuit and induces in the secondary an electromotive force E_S, lagging 90° behind the flux. The instantaneous value of the secondary electromotive force is $e_S = -S_2(d\phi \div dt)$. It is seen that E_S is exactly opposite to E_P in phase and is equal to E_P, multiplied by $(S_2 \div S_1)$.

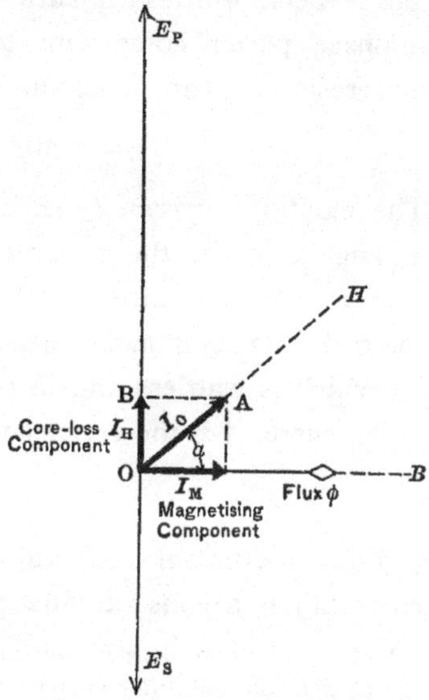

FIG. 2. Open-circuit diagram for a transformer with core loss, showing the two components of exciting current and the angle α of hysteretic advance.

§ 8. The flux ϕ throughout this discussion refers to the flux which links with both primary and secondary, and E_P and E_S are the induced or flux voltages,* proportional to ϕ. In an ideal transformer there is no other flux, but in an actual transformer there is, in addition to this main flux, a relatively small local or leakage flux, which links with the turns or part of the turns of one winding only and causes a reactance called leakage reactance. On account of the drop due to leakage reactance and the drop due to the resistance of the transformer windings, as discussed later, the terminal voltages, E_1 and E_2, are slightly different from the flux voltages E_P and E_S.

* (§ 8a). Strictly speaking E_P is not the flux voltage but is equal and opposite thereto.

§ 9. *With Core Loss.*—A transformer with an iron core differs from the ideal transformer just discussed because there is a loss in the iron due to hysteresis and eddy currents. The open-circuit diagram now becomes as shown in Fig. 2. The flux ϕ is still in quadrature with E_P and E_S, in accordance with Faraday's fundamental law of induced electromotive force, $e = -S(d\phi \div dt)$. The exciting current I_0, however, can no longer be a wattless quadrature current, for it must have an in-phase power component to supply the core losses due to hysteresis and eddy currents. This core loss component is

$$I_H = \text{watts core loss} \div E_P.$$

The exciting current I_0 is, accordingly, advanced in phase by an angle α, called the *hysteretic* angle of advance.*

It is seen, therefore, that I_0 consists of two components—the core loss component I_H and the true magnetizing component I_M which is wattless and in phase with the flux. The total exciting current† is the vector sum of these two components:

$$I_0 = \sqrt{I_H^2 + I_M^2}.$$

§ 10. A constant potential transformer (one in which E_P is constant) is a constant flux transformer. It therefore follows

* (§ 9a). As here defined, this angle includes the effect of eddy currents.

† (§ 9b). The exciting current of a transformer is distorted, *i. e.*, has a wave form different from that of the electromotive force, on account of harmonics introduced by hysteresis. (See Appendix II., Exp. 6–A.) These harmonics—currents of 3, 5, 7, etc., times the fundamental frequency—are necessarily wattless. They do not appear, therefore, in the power component I_H, but are included in the wattless component I_M. Strictly speaking, alternating currents in which harmonics are present can not be represented by vectors in one plane; for practical purposes, however, the plane vector diagram, as here given, is sufficiently accurate. (See § 47, Exp. 6–A; also "The Effect of Iron in Distorting Alternating Current Wave Form," by Bedell and Tuttle, A. I. E. E., Sept., 1906; and "Vector Representation of Non-Harmonic Alternating Currents," by B. Arakawa, *Physical Review*, 1909.) These harmonics have the same value at all loads; at full load they form such a small part of the total current that the distortion which they produce is very small.

that I_0, I_H and I_M are constant, and remain constant under all loads. This would not be quite true in a transformer in which the primary line voltage E_1 (and not E_P) is constant; the difference in the two cases is very small.

§ 11. **Transformer Under Load.**—The complete diagram for a constant potential transformer under non-inductive load is shown in Fig. 3. This will be seen to be exactly the same as Fig. 2, the open-circuit diagram, with certain additions. As in Fig. 2, we have the electromotive forces E_P and E_S opposite to each other in phase and in quadrature with the constant flux ϕ. On open circuit, the primary current I_0 flows as already discussed.

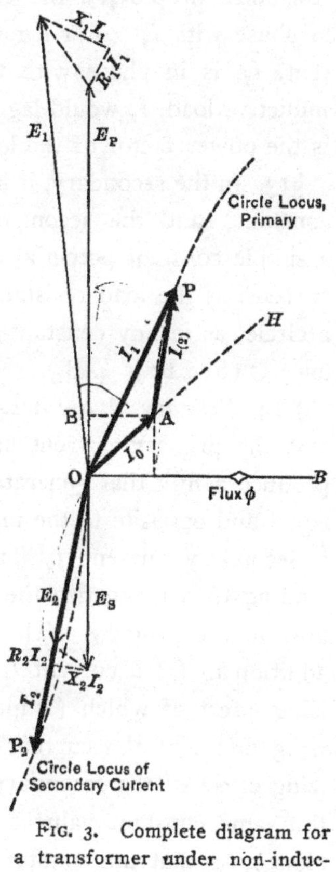

FIG. 3. Complete diagram for a transformer under non-inductive load. When a current I_2 flows in the secondary, a load current $I_{(2)}$ flows in the primary, opposite in phase and of equal ampere turns. $I_2 = I_{(2)} \times$ ratio of turns.

§ 12. *Secondary Quantities.*—When the secondary circuit is connected to a load, a secondary current I_2 flows, the value of which depends upon the load. With non-inductive load, this current would be in phase with E_S, if the transformer were perfect. On account of leakage reactance, X_2, the current I_2 lags a little behind E_S, as shown in Fig. 3. It is to be kept in mind that Fig. 3 and other diagrams here given are not at all to scale, being exaggerated in order to show more clearly the relations between the various quantities.

The secondary terminal voltage, E_2, is a little (perhaps one per cent.) less than E_S on account of reactance drop X_2I_2, and resistance drop R_2I_2, the former in quadrature and the latter in phase with I_2. For a non-inductive load, the secondary current, I_2, is in phase with the terminal voltage, E_2. (For an inductive load, I_2 would lag behind E_2 by an angle θ, where $\cos\theta$ is the power factor of the load.)

§13. In the secondary, it is seen that E_S is constant (flux being constant) and the secondary may, accordingly, be treated as a simple constant potential circuit. The locus of the secondary current, as the load resistance varies, is, accordingly, the arc of a circle, as in any constant-potential circuit with constant reactance. (See Exp. 4–B.)

§ 14. *Primary Quantities.*—It has been seen that on open circuit the primary current assumes a certain value I_0, so as to produce a flux that generates a counter-electromotive force just equal and opposite to the impressed electromotive force. When a secondary current I_2 flows, it disturbs this equilibrium by tending to demagnetize the core. This allows more current to flow in the primary. The primary current increases until (in addition to I_0) a current $I_{(2)}$ flows in the primary, the magnetizing effect of which (ampere turns) just balances the magnetizing effect of the current I_2 in the secondary. The magnetizing effect of the secondary being thus neutralized, the flux has the same constant value as before (as though produced by I_0 alone), so that the counter-electromotive force produced by the flux continues to be just equal and opposite to the impressed electromotive force.

In Fig. 3, the total primary current I_1, is seen to be composed of the constant I_0 (which is small) and the *load current* $I_{(2)}$, which is opposite to the current I_2 in the secondary and equal to I_2 multiplied by $(S_2 \div S_1)$. In a 1 : 1 transformer, the primary load current $I_{(2)}$ is equal to the secondary current I_2.

§ 15. Fig. 4 shows that, in a loaded transformer, the resultant ampere turns are constant; hence the flux is constant, so that the counter-electromotive force—irrespective of load—equals, the impressed electromotive force. As the load changes, the primary current assumes such a value that the resultant ampere turns remains constant and this condition of equilibrium is maintained.

§ 16. The primary electromotive force, thus balanced by the counter-electromotive force, is E_P. Referring to Fig. 3, it will be seen that the terminal impressed electromotive force, E_1, is a little greater (say one per cent. greater) than E_P, on account of the R_1I_1 and X_1I_1 drops, due to primary resistance and to leakage reactance.

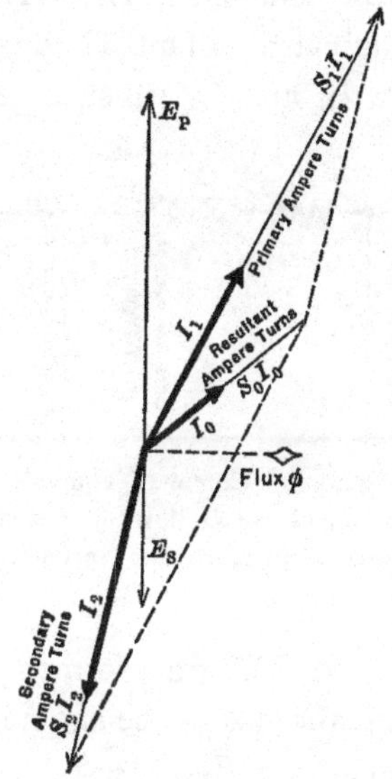

FIG. 4. Diagram of ampere turns; the resultant ampere turns are constant.

§ 17. The locus of the secondary current I_2 is the arc of a circle (§ 13). Hence the locus of the primary load current, $I_{(2)}$ in Fig. 3, is the arc of a circle. The total primary current, I_1, measured from O to P, follows this same locus.

Some simplified diagrams will now be discussed.

§ 18. **Representation of Transformer Circuits.**—From the foregoing discussion, it will be seen that the circuits of a transformer may be represented as in Fig. 5, in which the resistance and leakage reactance of the two windings are considered as external to the transformer. Furthermore, the exciting current, I_0, is considered as flowing in a shunt circuit, also external to the transformer. This shunt circuit consists of two branches:

a non-inductive branch for the in-phase component, I_H, and an inductive branch (without resistance) for the wattless quadrature component I_M. The currents which would flow in such equivalent shunt circuits correspond exactly to the currents I_0, I_H and I_M which actually flow in a transformer.

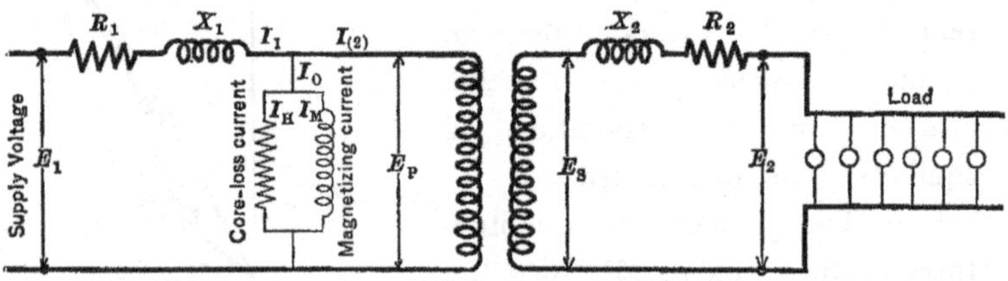

FIG. 5. Complete equivalent of a transformer. The exciting current I_0 is considered as flowing in a shunt circuit. The resistance and leakage reactance of primary and secondary are considered as external. Corresponds to Fig. 3.

§19. The transformer proper, in Fig. 5, is considered as ideal, all the losses being treated as external; $I_{(2)}=I_2(S_2 \div S_1)$; and $E_P=E_S(S_1 \div S_2)$. The voltage at the primary terminals, E_1, is more than E_P on account of the drop in X_1 and in R_1. Likewise, the voltage at the secondary terminals, E_2, is less than E_S on account of the drop in X_2 and in R_2.

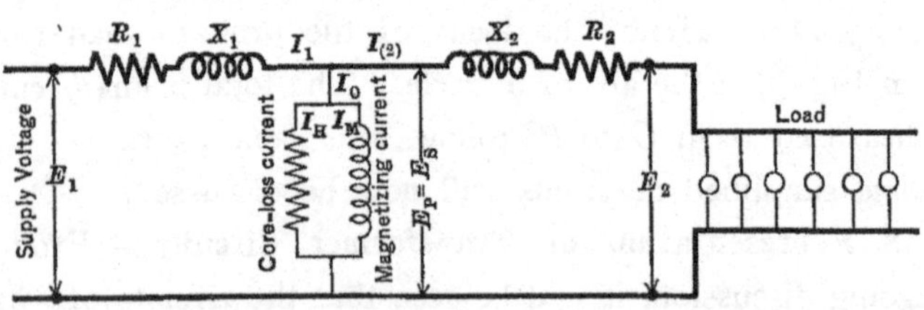

FIG. 6. Equivalent circuits as level (1 : 1) transformer. Corresponds to Fig. 7.

The total primary current I_1 is seen to be equal to the load current $I_{(2)}$, plus (vectorially) the small no-load current I_0.

§ 20. *Equivalent Circuits.*—The circuits of a transformer may be represented more simply by the equivalent circuits of Fig. 6, in which all quantities are expressed in terms* of the primary. This will be most readily understood by treating the transformer as a "level" (1:1) transformer; we have then, $E_P = E_S$; and $I_{(2)} = I_2$.

The diagram corresponding to Fig. 6 is shown in Fig. 7 and is seen to be the same as Fig. 3 with all secondary quantities expressed in terms of the primary and drawn in the first quadrant.

§ 21. *Simplified Circuits.*—The equivalent circuits so far considered (Figs. 5 and 6) and the corresponding diagrams (Figs. 3 and 7) are practically exact and may be used for the accurate solution of any transformer problem. It will be noted that the resistance and reactance for the two windings are treated separately, R_1X_1 in the primary and R_2X_2 in the secondary. By combining these into a single equivalent R and X, the transformer circuits can, with little error, be simplified in either of two ways:

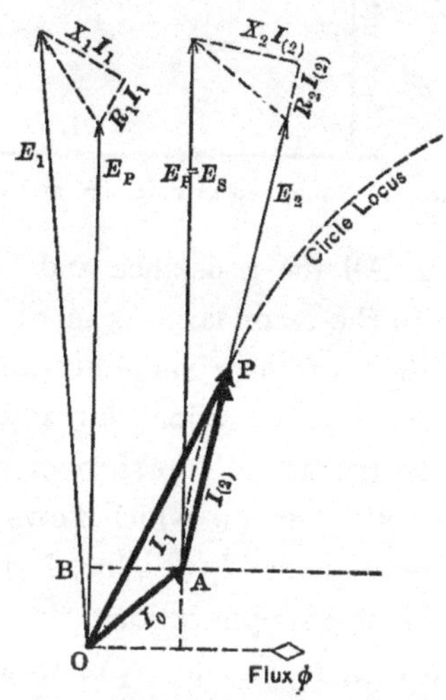

FIG. 7. Exact diagram as level transformer, corresponding to Fig. 6. The same as Fig. 3 with secondary quantities expressed in terms of the primary.

* (§ 20a). To express secondary quantities in terms of the primary: multiply current by $(S_2 \div S_1)$; multiply voltage by $(S_1 \div S_2)$; multiply X and R by $(S_1 \div S_2)^2$. See § 16a, Exp. 5-B. It will be understood that secondary quantities thus represented in the primary are not the real secondary quantities but the equivalent primary quantities which could produce the same results; thus, in a 10:1 transformer, 1 ohm in the primary is equivalent to 0.01 ohm in the secondary.

To express primary quantities in terms of the secondary, divide instead of multiply by these factors.

1. All the resistance and leakage reactance are considered to be in the primary, as in Figs. 8 and 10.

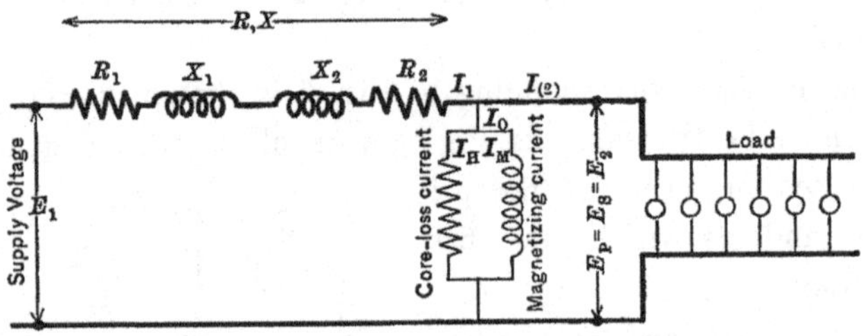

FIG. 8. Simplified circuits; R and X all in primary. Corresponds to Fig. 10.

2. All the resistance and leakage reactance are considered to be in the secondary, as in Figs. 9 and 11.

Each of these simplifications differ very little from the more exact representations already discussed.

In the actual transformer, as represented in Fig. 6, it is seen that the current which flows through X_2R_2 is $I_{(2)}$, while a different current I_1 (slightly larger, due to I_0) flows through X_1R_1.

In the simplifications, the *same* current is considered to flow through R_1X_1 and R_2X_2 which are now combined into a single

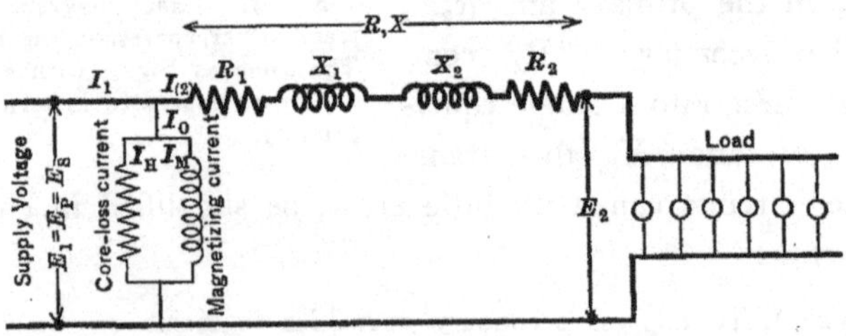

FIG. 9. Simplified circuits; R and X all in secondary. Corresponds to Fig. 11.

R and X, this current being either I_1 (as in Fig. 8) or $I_{(2)}$ (as in Fig. 9).

If I_0 were zero, Figs. 8 and 9 would not differ from Fig. 6, and all the representations would be identical. In fact, I_0 is so small that either simplification and its resultant diagram, Fig.

10 or 11, may, for most practical purposes, be considered as correct. This makes it possible to use the single equivalent values for R and X obtained by the short-circuit test of Exp. 5–B, and does not require separate values of R and X for the primary and secondary circuits.

§ 22. Again, the voltage which causes I_0 to flow is E_P, as is seen in Fig. 6. In the simplifications, this voltage is taken as E_2 (Fig. 8) which is, say, 1 per cent. less, or as E_1 (Fig. 9) which is, say, 1 per cent. more than the value of E_P in the actual cases of Fig. 6. This would make an insignificant change in the value of I_0 which is itself small. In the latter case I_0 depends only upon line voltage and is independent of load.

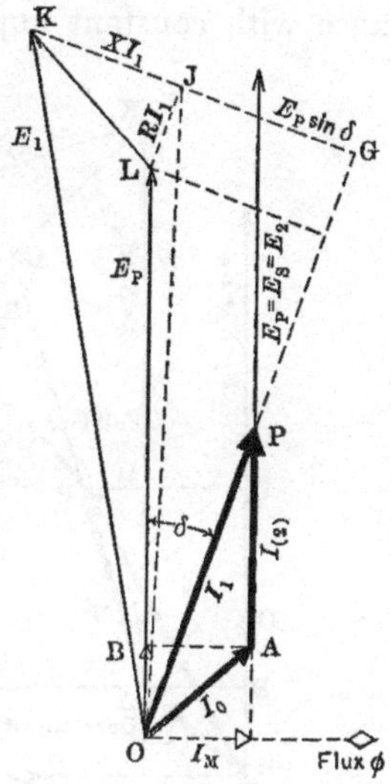

FIG. 10. Simplified diagram; R and X all in primary. Corresponds to Fig. 8.

§ 23. *Diagrams Compared.*—Let us compare the exact diagram, Fig. 7, with the simplifications, Figs. 10 and 11.

In Fig. 7, the primary and secondary RI drops are in phase with I_1 and $I_{(2)}$, respectively, the XI drops being in quadrature. The phase difference between I_1 and $I_{(2)}$ is small—much smaller in fact than shown in the figure. The primary and secondary drops may, accordingly, be combined with little error. This may be done by taking the combined resistance drop in phase with I_1 (Fig. 10), or in phase with $I_{(2)}$ (Fig. 11). The combined reactance drop is, in each case, at right angles to the combined resistance drop. In an actual case little error is introduced by these simplifications and either may be used, as is most convenient.

PART II. THE CIRCLE DIAGRAM AND ITS CONSTRUCTION.

§ 24. The circle diagram for a transformer shows the variation in the primary current for different values of load resistance with constant impressed voltage. In Fig. 11, the primary

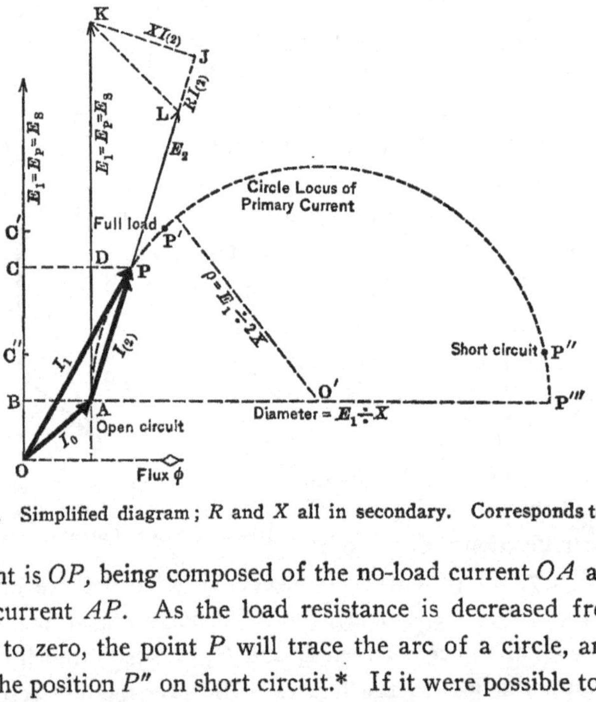

FIG. 11. Simplified diagram; *R* and *X* all in secondary. Corresponds to Fig. 9.

current is *OP*, being composed of the no-load current *OA* and the load current *AP*. As the load resistance is decreased from infinity to zero, the point *P* will trace the arc of a circle, and will take the position *P''* on short circuit.* If it were possible to eliminate the resistance of the transformer windings, the point *P*

* (§ 24a). If a transformer is constructed so as to have a large leakage reactance (or if a reactance is included in the circuit external to the transformer), the short-circuit current and the diameter, $E_1 \div X$, are reduced. The transformer may then be operated at or near short circuit, in which case the current will be nearly constant. This method is used for obtaining constant current from a constant potential line. (See § 4a, Exp. 5–A.) Large reactance or magnetic leakage in any apparatus tends towards constant current operation. See § 8, Exp. 3–A, §§ 27, 27a, Exp. 3–B, and § 14, Exp. 4–B.

would complete the semi-circle and assume the position P''', the current in this case $(E_1 \div X)$ being limited only by the leakage reactance, X. The short-circuit current of a transformer operated at full voltage would be, however, greatly in excess of the carrying capacity of the transformer windings, and, in actual operation, the point P does not go far beyond the full-load point P'. See also Fig. 12, which is more nearly to scale.

§ 25. **Data Necessary.**—The data necessary are the values of I_0, I_H and I_M, to locate the point A, and the leakage reactance X, to determine the diameter of the semicircle. These data are obtained from the open-circuit and short-circuit tests of Exp. 5-B.

All quantities are to be in terms of the primary (high-potential) side; thus, in Fig. 2. Exp. 5-B, the values of I_0, I_H and I_M, measured on the 100-volt coil, are divided by 20 to obtain the corresponding values for the 2,000-volt primary. This gives us:

$$I_0 = .03025; \quad I_H = .0208; \quad I_M = .0220.$$

The reactance X for the same transformer, is 35.2 ohms; see Fig. 7, Exp. 5-B.

§ 26. **Construction of Diagram from Experimental Data.**—From the data given above, lay off (Fig. 12):

$$OB = I_H; \quad BA = I_M; \quad OA = I_0.$$

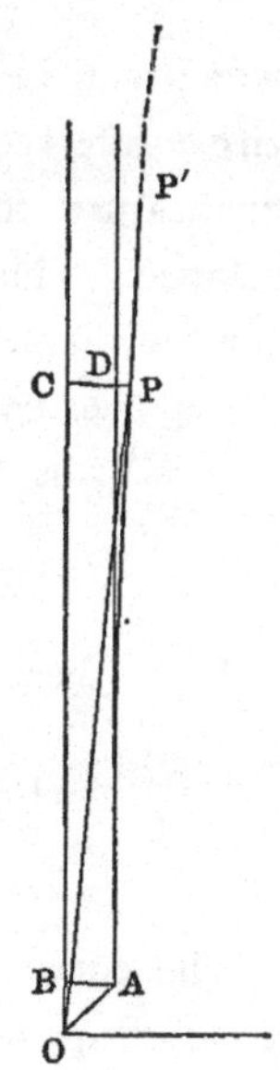

FIG. 12. Construction of circle diagram.

The diameter of the circle is $E_1 \div X = 2{,}000 \div 35.2 = 56.8$ amperes. The radius $\rho = E_1 \div 2X = 28.4$. These values are large compared with $I_0 = .03$ and full-load current $I_{(2)} = 1$ ampere. It is, accordingly, not practicable to construct the whole semicircle, as in Fig. 11, which is not at all to scale.

For a working range it can be readily constructed, as in Fig. 12, which is more nearly to scale, as follows:

Lay off $AD = I_{(2)}$ for $\frac{1}{100}$, $\frac{1}{10}$, $\frac{1}{4}$, $\frac{1}{2}$, $\frac{3}{4}$, 1 and $1\frac{1}{4}$ load. (It is to be noted that, in Fig. 12, the angle DAP is small; hence AD is taken as practically equal to AP or $I_{(2)}$.) Thus, for a 2,000-volt, 2 K.W., transformer, AD is laid off, successively, equal to .01, 0.1, .25, .50, .75, 1.0 and 1.25 amperes.

For each value of AD, the point P is located by laying off

$$DP = \rho - \sqrt{\rho^2 - \overline{AD}^2}$$

which can be derived from the figure and is the equation of a circle referred to A as an origin. The line DP represents the quadrature component of primary current due to leakage reactance. This is always small and would be zero when $X = 0$, for the diameter of the semicircle (see Fig. 11) is then infinite. The power component AD is, therefore, practically equal to AP.

It is to be noted that

$$CP = DP + BA; \quad \text{and } OC = OB + AD.$$

From these values, compute* for different loads

$$\text{Primary current} = OP = \sqrt{\overline{OC}^2 + \overline{CP}^2}.$$

$$\text{Power factor} = OC \div OP.$$

The curves in Fig. 4, Exp. 5–A were thus computed. Note §41a, Exp. 5–B.

---

* (§ 26a). It will be seen, also, that

$$\text{Watts input, } W_1 = OC \times E_1;$$
$$\text{Watts output, } W_2 = W_1 - \text{losses};$$
$$E_2 = W_2 \div AP.$$

This gives a possible method for determining the total voltage drop.

APPENDIX I.

NOTE ON REGULATION.

§ 27. **Definition of Terms.**—*Regulation* is defined by the Institute as follows:

In constant-potential transformers, the regulation is the ratio of the rise of secondary terminal voltage from rated non-inductive load to no load (at constant primary impressed terminal voltage) to the secondary terminal voltage at rated load. (Compare § 34, Exp. 5-B.)

If the secondary terminal voltage is E_0 at no load and E_2 at full load, the regulation is,

$$\text{Regulation} = (E_0 - E_2) \div E_2$$

The drop on which regulation depends is $E_0 - E_2$, which we may term the *regulation drop*. This drop, expressed as per cent. of E_2, gives the regulation.

§ 28. The *total voltage drop*, in terms of a 1:1 transformer, is $E_1 - E_2$ and is a little more than the regulation drop, because E_1 is a little more than E_0 on account of the drop due to exciting current in the primary winding.

The *per cent. voltage drop* is $(E_1 - E_2) \div E_2$, taking E_2 as 100 per cent.; or, $(E_1 - E_2) \div E_1$, taking E_1 as 100 per cent.

§ 29. Numerically, the difference between "regulation" and "per cent. voltage drop" is small. In earlier usage,* the term regulation was commonly employed to designate "per cent. voltage drop." This confusion is one cause for the differences between various methods which have been used (and still are used) for determining regulation. A difference arises, also, according to whether E_1 or E_2 is taken as 100 per cent.

* (§ 29a). See the following articles, in the *Elec. World*, on the predetermination of transformer regulation: Bedell, Chandler and Sherwood, August 14, 1897; A. R. Everest, June 4, 1898; F. Bedell, October 8, 1898. See also Foster's *Electrical Eng. Pocket Book*, p. 492, fifth edition, 1908.

14

§ 30. An illustration will make this clear. Let

$$E_1 = 100;\ E_0 = 99.9;\ E_2 = 97.$$

Regulation drop $= E_0 - E_2 = 2.9$ volts.
Per cent. regulation $= 2.9 \div 97 = 2.99$ per cent.
Total voltage drop $= E_1 - E_2 = 3$ volts.
Per cent. voltage drop $= 3 \div 100 = 3$ per cent.; or
$= 3 \div 97 = 3.1$ per cent.

§ 31. Regulation drop depends upon the difference between E_0 and E_2. This drop is due to load current, and does not include any drop due to exciting current, which affects E_0 and E_2 alike and, practically, does not affect their difference.

The total voltage drop depends upon the difference between E_1 and E_2. This drop is chiefly due to load current, but includes, in addition, a small drop due to exciting current which affects E_2 but not E_1 and so directly affects their difference.

§ 32. **Computations.**—To compute regulation drop, we have the problem: Given E_2, to compute E_0.

To compute total voltage drop, we have the problem: Given E_2, to compute E_1.

§ 33. *Regulation.*—For determing regulation, we compute

$$E_0 = \sqrt{(E_2 + p)^2 + q^2};\ \text{where}$$

$$\text{in-phase drop} = p = RI_{(2)},$$
$$\text{quadrature drop} = q = XI_{(2)}.$$

It is seen that exciting current does not enter. The various drops may be expressed either in volts or in per cent. The working details of the method are discussed in §§ 34–43, Exp. 5–B.

§ 34. *Total Voltage Drop.*—For determining total voltage drop, we compute

$$E_1 = \sqrt{(E_2 + p)^2 + q^2}.$$

The in-phase drop p, consists principally of $RI_{(2)}$, but includes the small additional terms $R_1 I_H$ and $X_1 I_M$, which are drops caused by the two components of the exciting current flowing through X_1, R_1. Without much error, X_1, R_1 may be taken as half of X, R. Hence

$$p = RI_{(2)} + R_1 I_H + X_1 I_M,$$
$$= RI_{(2)} + \tfrac{1}{2}RI_H + \tfrac{1}{2}XI_M.$$

In a like manner

$$q = XI_{(2)} + X_1 I_H - R_1 I_M,$$
$$= XI_{(2)} + \tfrac{1}{2}XI_H - \tfrac{1}{2}RI_M.$$

The last two terms are small and nearly cancel each other.

§ 35. Other methods of analysis may be employed for determining the total voltage drop, and the form in which the results are expressed will vary according to the manner in which the various terms are combined and the approximations which are introduced. In any case some small and troublesome terms are introduced, which affect the result very little and which do not enter in the determination of "regulation," as defined by the Institute. The results are affected less by these small terms than by variations in the value of R, depending upon whether, in its determination, load losses were included or not, and whether steady temperature conditions were maintained during the test.

§ 36. It might well be held that regulation should be so defined that magnetising current should enter into its determination, particularly since magnetising current has been much increased by the use of improved iron worked at higher densities; on the other hand, it is much simpler to define regulation independently of magnetizing current and to specify the value of the magnetizing or exciting current as a separate item.

CHAPTER VI.

POLYPHASE CURRENTS.

Experiment 6–A. A General Study of Polyphase Currents.*

PART I.

§ 1. **Introductory.**—In a polyphase system, several single-phase currents differing in phase from each other are combined into one system. The circuits for each phase may be *independent,* without electrical connection, or *interconnected.* The phase difference between the currents of the several phases is usually 90° or 120°, the corresponding systems being called *two-phase* or *three-phase.*

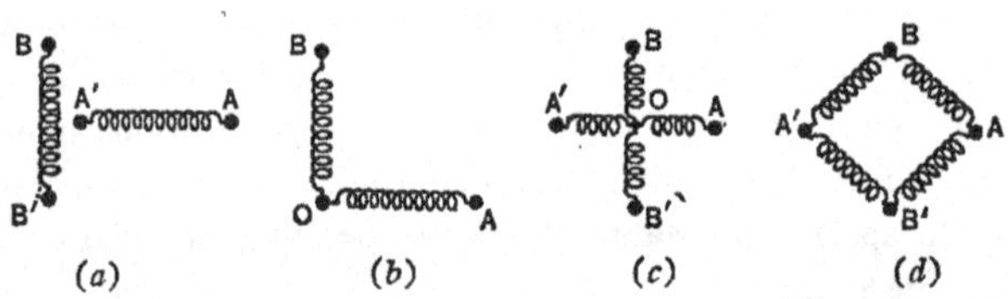

Fig. 1. Two-phase connections for generator or receiver circuits. *a,* 4-wire system with independent phases. *b,* 3-wire system. *c,* Quarter-phase, star-connected; or, 4-wire system with interconnected neutral. *d,* Quarter-phase, mesh-connected; or, ring-connected.

To form a polyphase system we must have several sources of single-phase electromotive force which differ in phase by proper amounts. For a *symmetrical* polyphase system these electromotive forces must be equal and differ from each other by equal phase angles, as in the 3-phase and quarter-phase systems soon

* (§ 1a). In making polyphase measurements, some form of voltmeter and ammeter switches will be found convenient, so that all readings can be made with one voltmeter and one ammeter. The same switches will serve to transfer one wattmeter from one circuit to another.

to be discussed. The sources of these electromotive forces are in principle several rigidly connected single-phase generators, but in practice they are generator coils on a single armature. The secondary coil of a transformer may be considered as a generator coil. The currents from these sources may be utilized separately as single-phase currents (as in lighting), or jointly as polyphase currents (as in an induction motor).

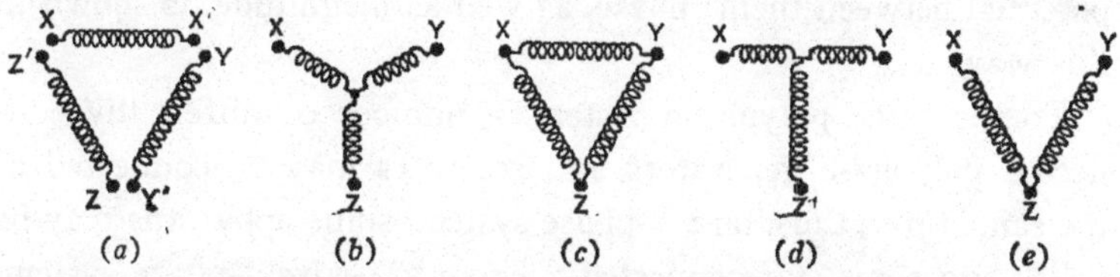

FIG. 2. Three-phase connections for generator or receiver circuits. *a*, Independent circuits; see § 3a. *b*, Star- or *Y*-connected. *c*, Mesh- or delta- (Δ) connected. *d*, *T*-connected. *e*, *V*-connected; or, open delta.

§ 2. The load on a polyphase system is *balanced* when each phase has an equal load with equal power factor. In a balanced polyphase system the flow of energy is *uniform*, which is a better and more general definition of such a system; (see Steinmetz, *Alternating Current Phenomena*). In a single-phase system or unbalanced polyphase system, the flow of energy is *pulsating*,—discussed further in § 1, Exp. 7-A. The torque is, accordingly, pulsating in all single-phase machinery; whereas it is uniform in polyphase motors and in polyphase generators on balanced load. Furthermore, a polyphase induction motor on account of its rotating field can be given a good starting torque, whereas a single-phase induction motor has none in itself and has only a small starting torque when auxiliary starting devices are used. Polyphase machinery has a greater output than single phase for a given size, or has a smaller size for a given output. These features, together with the copper economy of 3-phase as compared with single-phase transmission, all favor the use of polyphase systems; see Appendix III.

§ 3. *Methods of Connecting Phases.*—Generating or receiving coils or circuits may be combined in various ways, the common ones being shown* diagrammatically in Figs. 1 and 2. In Figs. 1 and 2, the relative positions of the various coils represent the relative phase positions of their several electromotive forces.† The black dots‡ may be taken as line wires in cross-section. On paper the distance between any two dots is the difference of potential between them; phase, as well as magnitude, is shown in this way.

To the same polyphase system, a number of differently connected polyphase generators and receivers may be connected at the same time; thus, on a 3-phase system, some apparatus may be delta- and some star-connected. From a 4-wire 2-phase system, induction motors may be run simultaneously when connected as (*a*), (*c*) or (*d*), Fig. 1. Connection (*b*) can be combined on the same system with (*a*), but not with (*c*) or (*d*). This is an objection to 3-wire 2-phase distribution, inasmuch as synchronous motors and converters as well as generators are frequently wound quarter-phase and so cannot be run from a 3-wire system. A further objection, that the line drop in the common wire makes the voltages unsymmetrical, is discussed later, § 14.

§ 4. *Object.*—In performing this experiment, the object is to gain a knowledge of the connections of polyphase circuits and polyphase apparatus, and to understand their electrical relations and various diagrammatic methods for representing them. Make a study of whatever polyphase supply circuits are available and by means of transformers obtain, so far as possible, all the systems indicated in Figs. 1 and 2.

* (§ 3a). The arrangement of Fig. 2 (*a*) is never used for independent 3-phase circuits; it is used only for connecting transformer secondaries to so-called 6-phase synchronous converters, § 27.

† (§ 3b). Although the diagram of connections can not in general be taken as the vector diagram of electromotive forces, this can be done in the simpler cases and makes the introduction to the subject more clear.

‡ (§ 3c). This representation by dots is called by Steinmetz the *topographic* method.

Note also the connections on various pieces of polyphase apparatus (as generators, motors, etc.) which may be available, and note for what kind of polyphase system the apparatus is intended.

PART II.

§ 5. **Two-phase Measurement.**—Take two transformers* with the same ratio of transformation (say 1 : 1). Connect the primary of one transformer to phase A of a 2-phase circuit,† and the primary of the other transformer to phase B. Measure the secondary voltages when the secondary circuits are independent, thus forming a 4-wire system with independent phases, Fig. 1 (*a*).

§ 6. *Addition of Electromotive Forces.*—Connect the two secondaries as a 3-wire system, Fig. 1 (*b*), and measure the voltage of each phase (E_A and E_B) and the voltage E between outside wires. Lay off these voltages as a triangle and note how nearly E_A and E_B are at right angles, so making a true 2-phase system. This triangle may be drawn as in Fig. 3, 4 or 5.

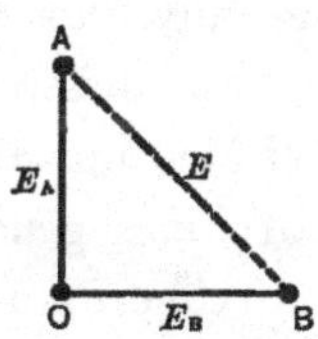

FIG. 3. Topographic method.

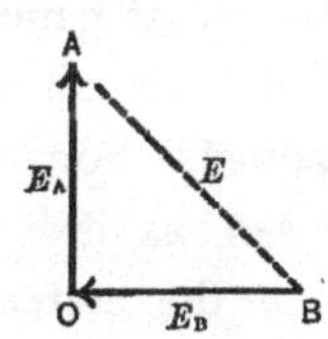

FIG. 4. Addition method.

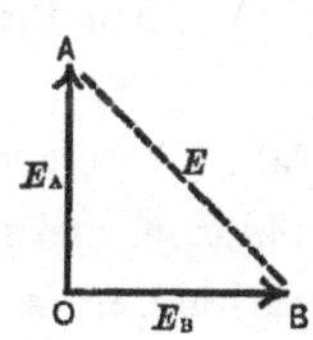

FIG. 5. Subtraction method.

§ 7. If we use the topographic method of Steinmetz and omit arrows, we can represent the electromotive forces of the 2-phase 3-wire system by Fig. 3 (see Appendix I.). This electromotive

* It is preferable that each secondary consists of two equal coils: thus, we might have primary 110 volts; secondaries 55 volts each, giving in series 110 volts with a middle or neutral point. Note the various possible voltage transformations for each transformer.

† It matters not whether the supply circuit is 3-wire or 4-wire, or how connected. If several kinds of supply circuits are available, use each one in turn. Compare Fig. 6.

force diagram is seen to be similar to the circuit diagram (*b*) of Fig. 1.

§ 8. If we take one outside line, say *B*, as our starting point (imagining if we wish that it is grounded, but this is unnecessary), we have the electromotive forces E_B and E_A represented by the vectors, *BO* and *OA*, in the direction shown by arrows in Fig. 4. The *sum* of these two vectors is *BA*.

§ 9. If, as is common, we take the neutral *O* as the starting point (say ground), the differences of potential between the side wires and ground are *OA* and *OB*, the direction of the vectors being from the starting point as in Fig. 5. The difference in potential between *A* and *B* is now the *difference* between *OB* and *OA*, Fig. 5; which is the same as the sum of *BO* (equal — *OB*) and *OA*, Fig. 4.

§ 10. In general, if we take electromotive forces *in sequence*—as *BO*, *OA*—they must be *added* (Fig. 4); if, however, we consider each electromotive force in a direction away from a common joining point—as *OB*, *OA*—they must be *subtracted* (Fig. 5). For the simple case at hand, involving only two electromotive forces connected to a common point, the difference method may be readily applied. For more complicated networks the addition method is used, as it is capable of more general application; it is based on the statement of Kirchhoff's Law (§ 32, Appendix I.) that the differences of potential around any mesh add up to zero. For this addition method, all arrows are taken consecutively, from feather to tip.

§ 11. If each transformer has a secondary winding, consisting of two equal coils, connect the secondary coils of the two transformers so as to form a *star-connected* and a *mesh-connected* quarter-phase system, as in (*c*) and (*d*) of Fig. 1. Measure all voltages and draw diagrams of voltages for the star and for the mesh connection.

In the mesh connection, the two secondaries of one transformer are connected as the opposite sides of a square, due attention being given to polarity; the two secondaries of the other transformer form the remaining two sides. Before closing the square, connect a voltmeter between the two points about to be connected and proceed to connect them only in case the voltmeter reads zero. This precaution should be taken in making any mesh connection.

§ 12. A convenient laboratory supply board is obtained from 2-phase secondaries, the secondary circuit on each phase consisting of four equal coils in series so as to form a 5-wire system on each phase. With the neutrals of the two phases interconnected, this gives supply voltages, as Fig. 6. If the total voltage of each phase is 220 volts, this gives 2-phase voltages as follows: 4-wire 110 and 220 volts; 3-wire 55, 110, 77.8 and 155.6 volts; also additional single-phase voltages, 123 and 165 volts. The voltage between any two points can be scaled off from the drawing in Fig. 6, as shown in the discussion of Figs. 3, 4 and 5. When the transformer secondaries cannot be so subdivided, the result can be obtained by connecting across each phase of a 4-wire system an autotransformer made of four equal coils. Verify these voltages by calculation or by measurement.

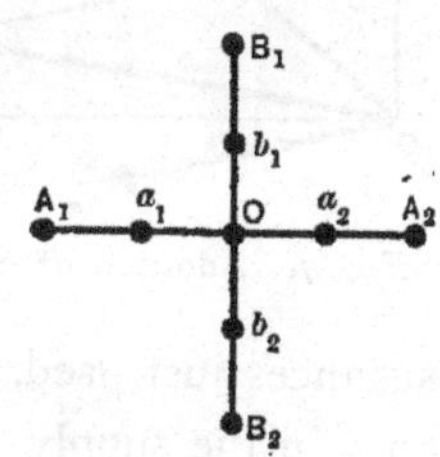

FIG. 6. Two-phase laboratory supply voltages.

The preceding study has brought out the fact that in polyphase circuits, the single-phase voltages of interconnected generator or receiver coils are combined geometrically to give resultant voltages. Although this was shown particularly for 2-phase circuits, it will be understood to be general and to apply as well to a 3-phase circuit or to any circuit whatsoever.

§ 13. *Addition of Currents.*—Currents, also, when of different phases, are added* vectorially to obtain the resultant current. To show this proceed as follows:

* Branch currents, flowing to or from a common point, always combine by addition—not by subtraction—to give the total current. See Appendix I.

From a 3-wire 2-phase supply, connect two resistances as load, one on each phase. Measure the currents, I_A and I_B, in each resistance and the total current I in the common conductor. If the two currents I_A and I_B differ in phase by 90°, we will have $I = \sqrt{I_A^2 + I_B^2}$. This will be true for an inductive as well as for a non-inductive load, provided the load on each phase has the same power factor,—*i. e.*, $\theta_A = \theta_B$.

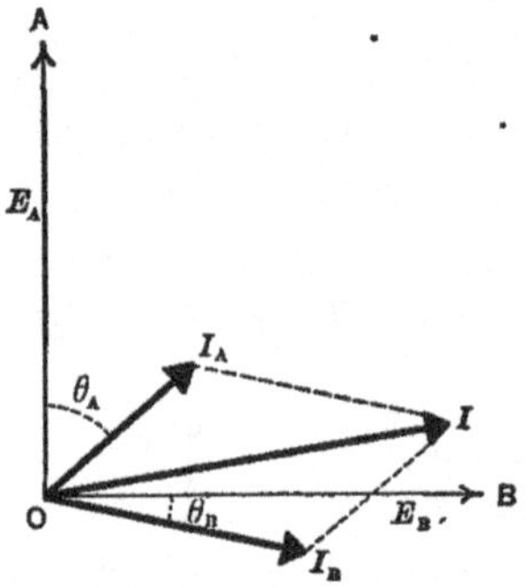

FIG. 7. Addition of currents.

If E_A and E_B are not at right angles, or θ_A and θ_B are not equal, the currents I_A and I_B will no longer be at right angles; the branch currents will still, however, add as vectors to give the total current, as in Fig. 7.

§ 14. *Line drop.*—To illustrate line drop, with the same circuits and resistances just used, insert a small additional non-inductive resistance in the supply wires to represent resistance in a long supply line.

Construct a triangle OAB for the supply voltage and $O'A'B'$ for the delivered voltage for the following three cases:

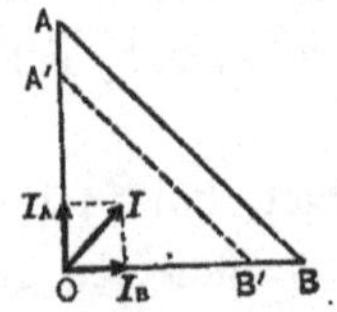

FIG. 8. Resistance in lines A and B.

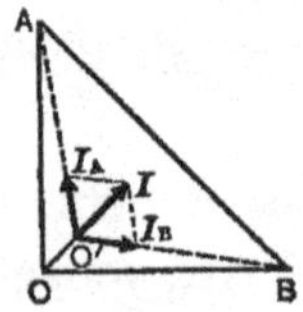

FIG. 9. Resistance in common conductor.

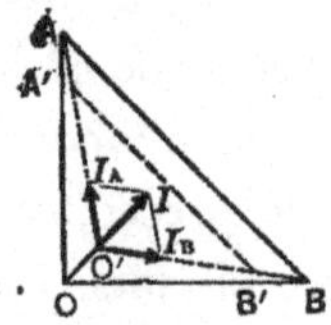

FIG. 10. Resistance in all three lines.

With resistances in lines A and B only, Fig. 8;

With a resistance in the common conductor O only, Fig. 9;

With resistances in all three lines, Fig. 10; in this third case measurements of voltages $O'A$ and $O'B$ are also to be taken.

For the first case (Fig. 8), the supply voltages, OA and OB, shrink to the delivered voltages, OA' and OB'; the drop due to resistance in lines A and B is in phase with the currents I_A and I_B. There is the same phase difference (90°) between the delivered voltages as between the supply voltages.

For the second case (Fig. 9), the line drop in the common conductor is in phase with I, and it is seen that, on account of this drop, the phase angle between the delivered voltages is greater than between the supply voltages. This, also, is true in Fig. 10. This lack of symmetry in delivered voltages is one disadvantage of the 3-wire system; see § 3.

§ 15. These diagrams illustrate the topographic or mesh method for representing electromotive forces. The direction assigned to any line depends upon the sense in which it is taken. Resistance drop *consumed by* resistance is in phase with current; resistance drop *produced by* a resistance is opposite to the current, as discussed in Exp. 4-A. It is taken in this latter sense in applying the mesh principle,—Law (1) of Appendix I.—that the electromotive forces around any mesh have a vector sum of zero and can be represented as a closed polygon. Thus, in Fig. 10, proceeding around the mesh $OAA'O'$, we have the following electromotive forces: OA produced by the generator; AA' produced by resistance in line A and opposite to I_A; $A'O'$ the counter electromotive force produced by the load (the electromotive force delivered to the load being $O'A'$); $O'O$ produced by resistance in the common line O and opposite to I. The line drop for a single-phase circuit can be similarly represented.

With inductance in the lines, besides the resistance drop just discussed, there is a reactance drop at right angles to the current; this reactance drop is 90° ahead of the current when considered as consumed by reactance, and 90° behind the current when considered as produced by reactance. (See § 18*a* and Fig. 2, Exp. 4-A, and Figs. 3, 4 and 5, Exp. 3-B.)

§ 16. The line drop diagram, Fig. 10, is true for any 3-wire system, and may be applied to a 3-phase system by making the triangles more or less equilateral. A 4-wire system or any other system can be treated in a similar manner.

Furthermore, the method just discussed for treating the effect of resistance drop and reactance drop in line conductors is not limited to non-inductive loads, but is applicable as well to other loads, either with leading or lagging currents. (See § 56, Exp. 3–B.)

§ 17. *Conclusion.*—In the main it has been seen that 2-phase circuits are essentially the same as two single-phase circuits and can be so treated. Three-phase circuits are likewise essentially three single-phase circuits and the conception of polyphase circuits is thus made simple. In any polyphase circuit the fundamental principles for the vector addition of currents and electromotive forces apply as in single-phase circuits. For 3-phase circuits, however, there are modified forms of treatment that are found practically convenient; these will now be considered.

PART III.

§ 18. **Three-phase Measurement.**—The most important 3-phase connections (Fig. 2) are the star and delta connections, the electrical relations of which will first be studied. Other 3-phase connections will then be studied with reference to various arrangements of transformers on 3-phase circuits.

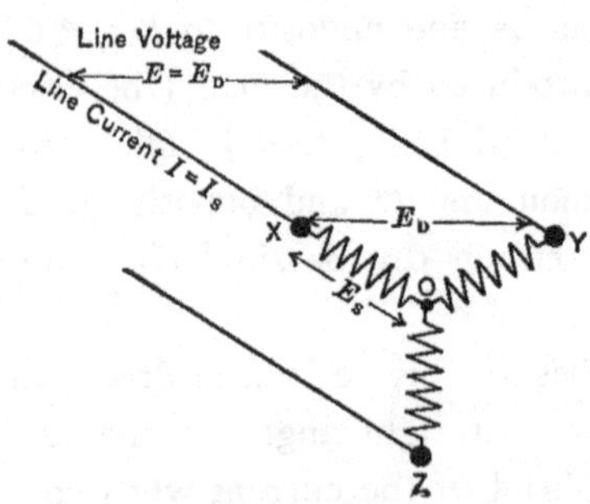

FIG. 11. Star- or *Y*-connection of load resistances.

§ 19. *Star-connection.*—On a 3-phase line, connect three approximately equal resistances* in star-connection; see Fig. 11. Measure the *line voltages XY, YZ* and *ZX*; these are also called *delta voltages*

* Some measurements should also be made with unequal resistances.

and for clearness may be designated by the subscript D—thus, E_D. When nothing further is specified than *the* voltage E of a 3-phase line or machine, it is this delta or line voltage that is meant.

Measure* the *star voltage* E_S (called also voltage per phase or phase voltage, § 30) from each line to the junction O, Fig. 11. Also measure the *star current* I_S for each phase. The *line current* is always the star current, as is evident for this case.

Compare the measured values of E_D and E_S with the expression (which should be proved)

$$E_D = \sqrt{3}\, E_S.$$

§ 20. Compute the power for each resistance. This is obviously, as in a single-phase circuit, equal to the product of volts × amperes (for a non-inductive load), *i. e.*, the product of star voltage and star current ($E_S I_S$) for each phase. For an inductive load in which the current lags by an angle θ, as in Fig. 12, the power for each star circuit is $E_S\, I_S \cos\theta$. When E_S, I_S and θ are the same for each phase, we can multiply the power for each phase by 3 to obtain the total power; thus,

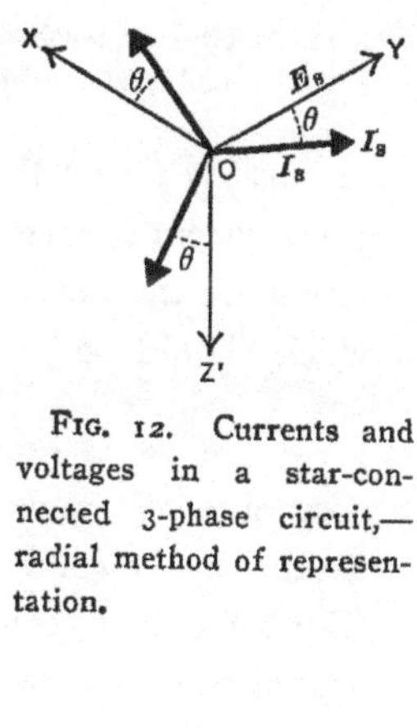

FIG. 12. Currents and voltages in a star-connected 3-phase circuit,—radial method of representation.

$$\text{Total power} = 3E_S I_S \cos\theta.$$

But

$$E_S = E_D \div \sqrt{3};$$

hence

$$\text{Total power} = \sqrt{3}\, E_D I_S \cos\theta.$$

* (§ 19a). If the neutral point of the supply is available, measure the voltage between it and O, and test with a telephone as described in Appendix II., § 44. This can be done either in connection with the present test or later in connection with Appendix II.

Since line voltage is E_D and line current is I_S, we may drop the subscripts and write

$$\text{Total power} = \sqrt{3}\, EI \cos\theta = \sqrt{3}\, EI \times \text{power factor},$$

where E is line voltage and I is line current. This is the customary formula for power in any balanced 3-phase system, no matter how connected. In the next paragraph it will be derived for a delta-connection.

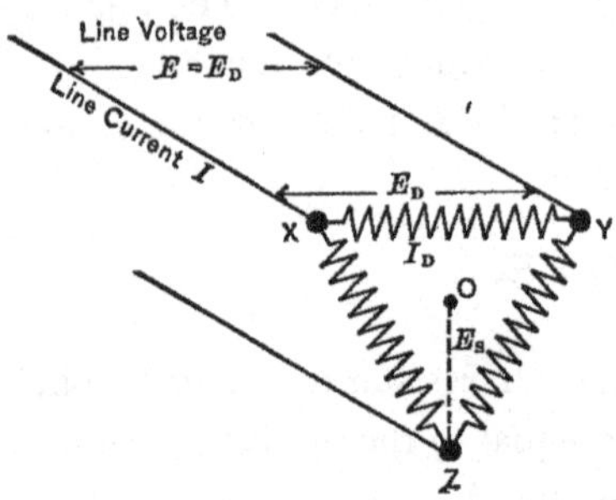

FIG. 13. Delta- or mesh-connection of load resistances.

§ 21 *Delta-connection.*—Connect the same three equal* resistances in delta to a 3-phase supply, as in Fig. 13. Measure the current and voltage for each resistance,—namely the delta current I_D and the delta (line) voltage E_D. Also measure the line current I and the star voltage E_S, if the neutral O of the supply system is accessible. It is seen, as above, $E_S = E_D \div \sqrt{3}$.

Compare the measured values of I and I_D with the expression (which should be proved)

$$I = \sqrt{3}\, I_D.$$

Compute the power for each resistance $E_D I_D$, and compare with the power found for the same resistances in star-connection.

For an inductive load, we should multiply by cos θ to obtain the true power in each resistance. If E_D, I_D and θ are the same for each phase, we find total power by multiplying by 3; hence

$$\text{Total power} = 3E_D I_D \cos\theta.$$

But

$$I_D = I \div \sqrt{3};$$

hence

* Some measurements should also be made with unequal resistances.

$$\text{Total power} = \sqrt{3}\, E_D I \cos\theta,$$
$$= \sqrt{3}\, EI \cos\theta,$$
$$= \sqrt{3}\, EI \times \text{power factor},$$

where E and I are line voltage and line current. This is the customary power formula for any balanced 3-phase system, as has already been found for the star-connection.

§ 22. The currents and voltages for the delta-connection can be laid off by the radial method (see Appendix I.) from a common center, giving a diagram similar to Fig. 12.

Another method is shown in Fig. 14, in which the voltages are laid off as a triangle (polygon method) and the currents radially from the corners. The currents in Fig. 14 are drawn as lagging. These currents are I_{XY} (from X to Y), I_{YZ} (from Y to Z), and I_{ZX} (from Z to X). With sign reversed, the latter becomes I_{XZ}, measured from X to Z. The sum* of I_{XY} and I_{XZ} gives I. If we wish to select signs so that the sum of these three vectors is zero, we must reverse the sign of I so as to give the line current I'; we now have I', I_{XY} and I_{XZ} all measured from X, so that Law (3) of Appendix I. is satisfied.

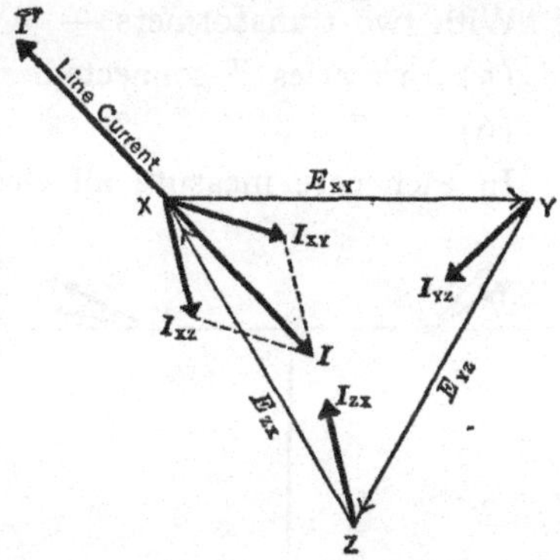

FIG. 14. Currents and voltages in a delta-connected 3-phase circuit,—polygon or mesh method of representation.

§ 23. **Transformer-connections on 3-Phase Circuits.**—Transformer secondaries and primaries—like any generating or receiving circuits—can be connected to a 3-phase circuit by Δ-, Y-, T- or V-connections, shown in Fig. 2.

* (§ 22a). The current I is the *sum* of I_{XZ} and I_{XY} (both measured from X), or the difference between I_{ZX} and I_{XY} (measured one towards and the other away from X). See Laws (3) and (4), Appendix I.

The most convenient and instructive method for studying the electrical relations of these connections is to use three transformers with the same ratio of transformation (say 1 : 1), the primaries and secondaries of which can be connected in any desired manner.

With three such transformers and with a 3-phase supply given, make connections in the following six ways:

With three transformers:—

(1)	Primaries star-connected.	Secondaries	star-connected.
(2)	" " "	"	delta "
(3)	" delta "	"	star "
(4)	" " "	"	delta "

With two transformers:—

(5)	Primaries *T*-connected.	Secondaries	*T*-connected.
(6)	" *V* "	"	*V* "

In each case measure all electromotive forces and construct electromotive force diagrams, comparing computed and measured results.

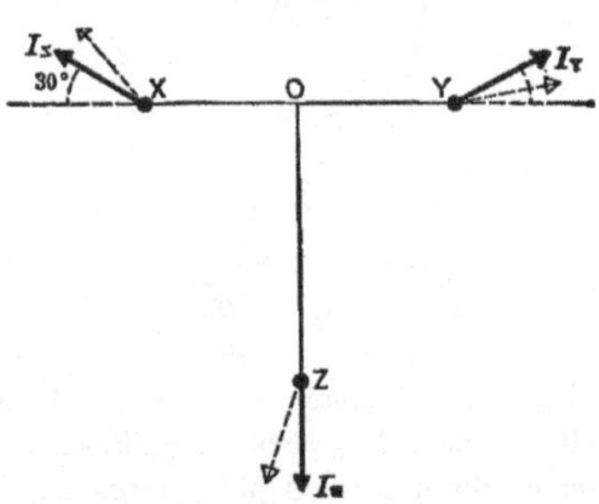

FIG. 15. Relation between currents and voltages in a *T*-connection.

The star- and delta-connections have already been discussed; the special relations of the *T*- and *V*-connections, will now be considered.

§ 24. *T-connection.*—For the *T*-connection, measure the voltage *OZ*, Fig. 15, and note that it is 86.6 when *XY* is 100. For a balanced load, the three currents, I_X, I_Y, I_Z, are equal. For a non-inductive* load, Fig. 15, the current in transformer *XY* is out of phase with the elctromotive force by 30° and the power factor (cos 30°) is 0.866;

* (§ 24a). For an inductive load, the currents take the positions shown by dotted lines in Fig. 15; I_X is now out of phase more than 30°, and I_Y less than 30°.

in transformer OZ the current is in phase with the voltage, giving unity power factor. For a current of 100 amperes, on non-inductive load, we have

	E	I	Power Factor.	Watts.	Volt-amperes.
Transformer XY	100	100	0.866	8,666	10,000
Transformer OZ	86.6	100	1.00	8,666	8,666
				17,333	18,666

This shows that the power output of each transformer is the same; for non-inductive load the two transformers require about 8 per cent. more transformer capacity (volt-amperes) than watts power transmitted. For delta- and star-connection, on non-inductive load, no excess of transformer capacity is needed.

The T-connection is discussed further under Polyphase Transformation, Exp. 7–A, where it is shown (§ 8) that, for good regulation, the windings OX and OY on one transformer must be interspaced so as to reduce the magnetic leakage between them.

The neutral point in a T-connection can be obtained by a tap at N in the coil OZ (see Fig. 9, Exp. 7–A), dividing the coil into $\frac{1}{3}$ and $\frac{2}{3}$.

§ 25. *V-connection or Open Delta.*—Draw a diagram similar to Fig. 15, for the V-connection, and from the power factor of each transformer show that for non-inductive* load this connection requires $15\frac{1}{2}$ per cent. more transformer capacity than power transmitted. Obviously a V-connection can be replaced to advantage by a T-connection; even using the same two transformers, there will be an advantage in the change, for there will be less voltage on one of the transformers and hence less core loss.

It is seen that, as a general principle, apparatus in which currents and voltages are out of phase require greater volt-ampere

* (§ 25a). Note that an inductive load will cause the power factor for one transformer to become more and for the other less than cos 30°, which will make the regulation better on one and worse on the other.

15

capacity for the same power than apparatus in which currents and voltages are in phase.

§ 26. *A Comparison.*—In comparing the relative advantages of transformer-connections, it is to be borne in mind that three transformers (even though of somewhat smaller aggregate capacity) will usually cost more than two. The *Y*-connection gives the least voltage per transformer and the least insulation strain, particularly if the neutral is grounded; for this reason it is to be preferred on high potential lines, say, 20,000 volts or over. On the other hand, the delta-connection has the advantage that, if one transformer breaks down, the remaining two will operate *V*-connected; at moderate voltages (say, under 20,000 volts) the delta-connection is accordingly to be preferred.

In the delta-connection, if one transformer breaks down, each remaining transformer will have $\frac{1}{2}$ instead of $\frac{1}{3}$ of the whole power and will have to carry the line current instead of the delta current. By what per centages are current and power in each transformer thus increased? This increase would cause abnormal heating. For the same heating (same current) show that the two transformers *V*-connected will carry $57\frac{2}{3}$ per cent. as much load as the three delta-connected transformers.

With transformers delta-connected, the voltage of the system can be increased by using the same transformers *Y*-connected. In a new system, the delta-connection is sometimes installed with a view to changing later to a *Y*-connection and a higher voltage.

A single 3-phase transformer requires less material than three single-phase transformers of the same aggregate capacity, and is more efficient. (See Handbooks.) The three single-phase transformers may be cheaper or more readily obtained because more nearly standard, and in case of breakdown one third and not all the equipment needs be replaced; in other respects the single 3-phase transformer is preferable and is coming more and more into use.

§ 27. *Six-Phase Circuits.*—A 6-phase circuit is a 6-wire circuit, the potential diagram of which forms a hexagon. Its only use is in connecting transformer secondaries to 6-phase synchronous converters.* The usual and best method for obtaining a 6-phase circuit is by means of the *diametrical*-connection, as follows. Three transformers have primaries connected to a 3-phase circuit. The six wires of the 6-phase circuit may be represented by the apices of a hexagon; the three transformer secondaries, Fig. 2 (*a*), are connected so as to form diagonals or diameters of the hexagon. The three neutral or middle points of the secondaries may, or may not, be interconnected. Connect transformers in this manner, with the neutrals interconnected, and test with a voltmeter; for present purposes this one test will be sufficient.

If each transformer has two separate secondaries of equal voltage, these six coils can be used as a 6-phase supply by a ring- or *mesh-connection* (each coil forming diagrammatically one side of a hexagon); or, a 6-phase supply can be obtained, by a *double T or double delta,* one *T* or delta being reversed with respect to the other. A double *Y*-connection is the same as the diametral-connection. One advantage of the diametral-connection is that it gives a neutral which may be used as a "derived neutral" for a 3-wire system on the direct current service from the converter; this is particularly useful in lighting systems.

PART IV.

§ 28. **Equivalent Single-phase Quantities.**—Polyphase quantities are sometimes reduced to *equivalent single-phase* values for

* (§27a). A 3-phase converter may be increased in rating 40 or 50 per cent. with no increased losses and with a corresponding higher efficiency when changed to 6-phase by the addition of three more collector rings and (if necessary) an extension of the commutator. A most valuable paper on this subject is one by Woodbridge (A. I. E. E., February 14, 1908), who states that of 1,000,000 K. W. of railway converters, one third are 6-phase; above 500 K. W. one company makes all converters 6-phase. See also Chap. XI., *Alternating Current Motors,* by A. S. McAllister, where 6-phase transformer connections are given in detail.

simplicity in working up and comparing data relating to polyphase machinery.

The *equivalent single-phase current* I' (sometimes called *total current*) in any balanced polyphase system is the current which, multiplied by the line voltage and power factor, gives the true (total) power; hence

$$\text{Total power} = EI' \times \text{power factor}.$$

For a 2-phase circuit, the equivalent single-phase current I' is evidently twice the line current.

For a 3-phase circuit, the equivalent single-phase current is $\sqrt{3}$ times the line current. (In a delta-connection, it is seen that this is three times the delta current,—hence the significance of *total* current.)

§ 29. *Equivalent single-phase resistance* R' is the resistance which, multiplied by the square of the equivalent single-phase current, gives the total copper loss ($=R'I'^2$). It will be found* that for star- or mesh-connection, or any symmetrical combination of star and mesh,—2-phase as well as 3-phase,—R' is one half the resistance measured between lines of one phase.

For a 2-phase circuit, this becomes apparent upon inspection.

For a 3-phase circuit, with the three equal resistances r under test connected star and connected delta, determine R' and I'; in each case compare R' with r and with the resistance measured between any two line-wires.

Equivalent single-phase reactance and impedance are likewise one half the measured values between lines of one phase.

§ 30. *Current and Voltage per Phase.*—Current per phase and voltage per phase (or phase voltage) are more commonly used than equivalent single-phase quantities; the meaning is not so definite, but can generally be told from the context. The terms

* See *Standard Electrical Handbook;* or *Alternating Current Motors,* by A. S. McAllister, in which equivalent single-phase quantities are extensively used.

are usually so used that the total power in a 2-phase circuit is twice the product of current per phase, voltage per phase and power factor; the total power in a 3-phase circuit is three times the product of current and voltage per phase, and power factor.

In a 2-phase system, there is little chance for ambiguity.

In a 3-phase system, the current and voltage per phase (as defined above) may be either the star (line) current and star voltage, or the delta current and delta (line) voltage. In either case, the total power is three times the *power per phase.* Using line current, we must use star voltage; using line voltage, we must use delta current. It will be remembered that, if line current and line voltage are used, the total power is $\sqrt{3}$ times their product multiplied by power factor.

APPENDIX I.

VECTOR ADDITION OF ALTERNATING CURRENTS AND ELECTROMOTIVE FORCES IN A NETWORK OF CONDUCTORS.

§ 31. **Laws of Vector Addition and Subtraction.**—Any hill may be considered to be up or down according to the direction in which one is walking; the difference in level may be considered positive or negative. In the same way difference of potential may be considered as positive or negative according to the sense in which it is taken—that is, according to the direction one takes in proceeding around a circuit or from point to point in a circuit.

Consider a network of highways in a hilly country. If from any starting point one proceeds by any route or circuit back to the starting point, he will find himself at the original level—the plus hills and the minus hills adding up to zero. On different trips he may traverse the same hill in opposite directions, giving it one time a plus and the other time a minus sign. This would be true *at any instant,* even if the surface were rising and falling, as in an imaginary earthquake or on the surface of the ocean.

Consider now a network of conductors. If from any starting point one proceeds by any route or circuit back to the starting point, he will reach the original potential; the algebraic sum of the potential differences at any instant, taken in the proper sense, adding up to zero.

For an alternating current circuit in which currents and potential differences vary harmonically and can be represented by vectors, algebraic addition is used for instantaneous values and vector addition for maximum or for effective values; hence, for maximum or effective values we have the modified statement of Kirchhoff's Law:

§ 32. *Law (1). Vector Addition of Electromotive Forces: General Law.*—In proceeding completely *around* any mesh or number of meshes in an alternating current system of conductors, the *vector sum* of all the differences in potential is zero; such vectors form a closed polygon.

For this vector addition, electromotive forces are represented by arrows, the tip of one to the feather of the next, which must be in sequence according to the direction in which we proceed around the circuit. A coil xy may have an electromotive force represented by a vector XY, as measured from x to y. Taken in the opposite sense (by traversing the circuit in the opposite direction) the electromotive force would be YX, the same vector with arrow reversed.

To illustrate* further this addition, from a point O on the side of a hill, let two paths ascend: one to the point A (elevation 100); the other to B (elevation 90). If a man starts at A, descends to O, ascends to B and back to A, the ascents and descents add to zero (-100; $+90$; $+10$).

To illustrate the special case of subtraction, if the sense or sign of one quantity be reversed: let two men start from O, one ascending to A ($+100$) and the other to B ($+90$). The difference in their level is now the difference between $+100$ and $+90$, which illustrates the following law:

§ 33. *Law (2). Vector Subtraction of Electromotive Forces: Special Law.*—In an alternating current system, if two electromotive forces are separately measured *away from* a common point (as OA and OB) the difference in potential between their outer ends (A and B) will be the *vector difference* of the two electromotive forces (OA and OB).

* For unvarying potentials or instantaneous values of varying potentials this is a correct analogy; for the vector addition of varying quantities it is merely an illustration.

The discussion of Figs. 3, 4 and 5 illustrates the application of Laws (1) and (2).

The modified form of Kirchhoff's Law for current becomes:

§ 34. *Law (3). Vector Addition of Currents: General Law.*—At any point in an alternating current system the vector sum of the currents measured all towards or all away from that point is zero; such vectors form a closed polygon.

§ 35. *Law (4). Vector Subtraction of Currents: Special Law.*—At any point in an alternating current system where three currents come together, if one current is measured towards and the second away from that point, the third current will be the *vector difference* of the two.

The discussion of Fig. 14 illustrates the application of Laws (3) and (4).

§ 36. **Notation.**—There is no universally adopted notation for polyphase circuits. The most complete and least ambiguous method is to letter every junction or point on the diagram of connections and to use two letters (as subscript if desired) in the proper sequence to designate the vector current or electromotive force between two points. Thus, from X to Y we may have electromotive forces XY or E_{XY}; in the reverse sense, YX or E_{YX}; similarly, we may speak of the currents XY or I_{XY} and YX or I_{YX}. This makes definite the direction or sign of the vector quantity in every case. In some cases, particularly the simpler ones, the complete definiteness is not needed (being unessential or obvious) and a single subscript is then simpler, as $\dot{E}_D$, E_S, I_A, I_B. In general the double-subscript notation is to be recommended on account of its exactness, as illustrated in the discussion of Fig. 14.

§ 37. In applying Law (1) it is necessary, in order to obtain a vector sum of zero in proceeding from a generator around a circuit and back to the generator, to take the *generated* electromotive forces or counter electromotive forces in each part of the circuit: thus, the electromotive force *produced by* self-induction 90° behind the current (not that to overcome self-induction 90° ahead of the current); and the electromotive force *produced by* resistance, in direction exactly opposite to the current (not the electromotive force to overcome resistance which is in phase with current). This becomes obvious upon inspection of the triangle for the electromotive forces in a simple circuit, the hypotenuse of which is E, one side RI and the other side XI; the principle is applied in the discussion of Fig. 10.

§ 38. **Polygon or Mesh Method of Representation.**—As applied to electromotive forces, there is in this method of representation a certain similarity between the diagram of connections and the diagram for electromotive forces. It seems a natural method to apply in many cases, as in Figs. 8, 9, 10. There is no essential difference between it and the topographic method. Law (1), above, applies directly and the electromotive forces around any mesh have a vector sum of zero, introducing arrows with feather to tip in sequence. (Compare analogy of network of highways, § 31.)

As applied to currents, the three currents drawn radially in Fig. 12 may be drawn as a closed polygon. So also in Fig. 7. Compare likewise Fig. 14.

§ 39. **Radial Method of Representation.**—In this method all vectors for currents and electromotive forces are drawn radially from a common center. This method is advocated by some for all cases (Porter, *Electric Journal,* September, 1907), together with the double subscript notation, in order that in involved problems ambiguity can be minimized. For a star-connection the application is obvious. For a delta-connection, we have the same radial diagram as for the star-connection. See Fig. 12.

A modified radial method, with vectors from several centers, is illustrated in Figs. 14 and 15, and for particular cases, as in those illustrated, possesses some advantages.

§ 40. **Preferred Method.**—It is not proposed to advocate here a particular convention but rather to assist in making underlying principles clear. One may choose or develop one method and apply it in all cases; or he may select the method which is simplest or clearest for each particular case. The important point is to see clearly the significance of whatever method is used.

APPENDIX II.

TRIPLE HARMONIC IN DELTA AND STAR CONNECTION.

§ 41. In a circuit supplying current to a transformer, induction motor or similar apparatus with iron, hysteresis in the iron introduces* in the exciting current odd harmonics of 3, 5, 7, 9, etc., times the fundamental frequency.

In a 3-phase system, if three transformers have their primaries either star- or delta-connected, the currents in the three transformers will have a phase difference of one third of the fundamental period. The third harmonic due to hysteresis will accordingly have the same phase in each of the three transformers. This will be seen by sketching curves for the fundamental and third harmonic, and shifting the curves to left or right one third of the fundamental period, which is one full period for the third harmonic. In a 3-phase system all harmonics divisible by 3, as the 9th, 15th, etc., will likewise have the same phase in each transformer.

For a 5- or 7-phase system, the harmonics thus appearing would be 5, 15, 25 and 7, 21, 35, etc., respectively.

In an even phase (single- or 2-phase) system, even harmonics only could appear; but no even harmonics are produced by hysteresis.

These facts can be shown by curves taken by the method of instantaneous contact or the oscillograph. A set of such curves has been published and discussed by E. J. Berg (*Electrical Energy*, p. 154).

§ 42. **Third† Harmonic in Delta-connection.**—If the transformer primaries are delta-connected, the harmonics due to hysteresis for the three transformers are in phase and form a current which circulates around the delta but does not appear in the line. The delta current I_D may accordingly be 5 or 10 per cent. more than the line current I, divided by $\sqrt{3}$. If H is the current (third and higher harmonics) caused by hysteresis, we have‡ $I_D = \sqrt{(I \div \sqrt{3})^2 + H^2}$.

* Compare "The Effect of Iron in Distorting Alternating Current Wave Form," A. I. E. E., September, 1906, and its discussion by Steinmetz.

† The third harmonic is mentioned, being most important; it will be understood that the ninth, fifteenth, etc., are included when only the third is mentioned.

‡ (§ 42a). If A and B are currents or voltages of any two frequencies, the total effective value is $\sqrt{A^2 + B^2}$. This is easily shown experimentally

Suppose, for example, I is 173 and I_D is 105 instead of 100 (an increase of 5 per cent.); then $H=\sqrt{105^2-100^2}=32$.

In the laboratory measure I and I_D and calculate H. Compute H as per cent. of $I \div \sqrt{3}$; also compute the per cent. increase in I_D over $I \div \sqrt{3}$. Although noticeable at no load, the percentage difference practically disappears under load, for H remains constant and hence is relatively smaller when I and I_D become large.

§ 43. **Third Harmonic in Star-connection.**—If the transformer primaries are Y-connected, the third harmonic caused by hysteresis will be in the same phase in the three transformers and will tend to flow to or from the neutral simultaneously in the three. The star voltage E_S will thus be more than the line voltage E divided by $\sqrt{3}$, thus $E_S=\sqrt{(E \div \sqrt{3})^2+E_H^2}$, where E_H is voltage due to hysteresis. If the neutral is insulated no current due to these harmonics can flow. If there is a return circuit from the neutral, through ground or a fourth wire, a current of triple frequency will flow; but no current of fundamental frequency will flow in the neutral if the line voltages are symmetrical.

§ 44. The third harmonic in the neutral can be prettily shown in the laboratory by means of a telephone, which should be protected by a resistance in series, or in shunt, or both, or by connecting through a transformer. Connect to a 3-phase supply three transformers or other coils* with iron; the more nearly similar these are the better. Let O' be the neutral of the three coils and let O be the neutral of the supply system. (If the supply system has no neutral, one may be obtained by three Y-connected resistances.) Connect the telephone between O and O'. If the coils are well balanced, the fundamental will be perhaps scarcely discernible; the third harmonic will sound very clearly an octave and a fifth (*do* to *sol*) above the fundamental; the ninth, if discernible, is the same interval above the third. On a 64-cycle circuit, the fundamental is C with harmonics g, d'', b'', etc.

If there is any question as to what is the fundamental, it can usually be told by listening to various apparatus in the laboratory; or by connecting the telephone, with a series resistance, to the supply circuit.

---

by measuring the total and separate voltages when two sources are put in series. Do not short circuit one source on the other.

* Shunt choking-coils used for series lighting are suitable for this.

If, instead of coils with iron, three resistances are used, the harmonics cannot be heard; the fundamental will no doubt be heard due to lack of perfect symmetry, and will become louder if the circuits are thrown more out of balance.

§ 45. If an electrostatic voltmeter is available, connect it between O and O' (in place of the telephone) and measure the hysteresis voltage, E_H. Measure also from one line wire X, the voltages OX and $O'X$, the latter being the larger on account of hysteresis harmonics.

Compute $O'X$ from the formula $O'X = \sqrt{(OX)^2 + E_H^2}$ and compare with the measured value. It is to be noted that $OX = E \div \sqrt{3}$ and $O'X = E_S$. By what per cent. is E_S greater than $E \div \sqrt{3}$? What per cent. is E_H of $E \div \sqrt{3}$?

§ 46. With a voltmeter* measure the line voltage XYZ and construct a triangle as in Fig. 16. Measure also the three star voltages $O'X$, $O'Y$, $O'Z$ and lay them off as shown, each one twice. A supply neutral is not necessary for this test. Cut out the diagram on the heavy lines and fold on the light lines, bringing the three points O' together so as to form a pyramid. The height of the pyramid represents the voltage E_H due to hysteresis harmonics.

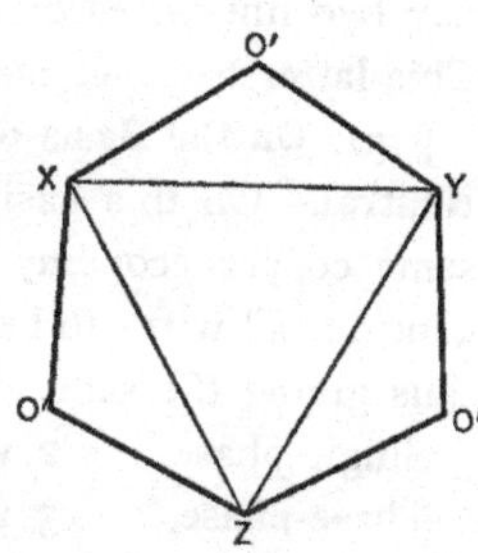

FIG. 16.

§ 47. The foregoing illustrates the fact that vectors in a plane can exactly represent only currents and electromotive forces which are simple sine functions; the error due to harmonics is commonly neglected.

§ 48. **Generator Coils.**—If there is a third harmonic in the generated electromotive force, with the generator coils delta-connected it cannot appear in the line but will appear as a circulating current in the delta. This may cause appreciable heating if the harmonic is large.

§ 49. The third harmonic can appear on the line only in case the generator coils are Y-connected and have the neutral connected to ground or a 4th wire. If the line is not grounded also at the receiv-

* Use an electrostatic voltmeter; although this is not important with large transformers, it becomes necessary in case the coils or transformers are small, as the current taken by an ordinary voltmeter may cause considerable error.

ing end or a 4th wire return used, the potential of the line as a whole will be raised by this electromotive force of triple frequency.

APPENDIX III.

COPPER ECONOMY OF VARIOUS SYSTEMS.

§ 50. In figuring copper economy, it is to be assumed that all systems compared are to have the same line loss and per cent. resistance drop.

As a general principle, in any given system, the amount of copper necessary varies inversely as the square of the voltage; thus, if the voltage is doubled, the current will be halved and the copper reduced to one fourth, increasing R four-fold. This gives the same RI^2 loss in the line and the same per cent. RI drop.

Any comparison of systems should, therefore, be made on the basis of equal voltage; this may mean either the greatest voltage between any two line wires or the voltage between any wire and the neutral. This latter becomes more significant when the neutral is grounded.

§ 51. **On the Basis of the Same Voltage E_S from the Line Wire to Neutral.**—On this basis all symmetrical alternating systems give the same copper economy, as will be seen from the following. Let us consider all wires to be of a given size and to carry a given current I, thus giving the same drop and loss per wire. We then have

Single-phase, 2 wires: amount of copper 2; power $= 2\ E_S I$.
Three-phase, 3 wires: amount of copper 3; power $= 3\ E_S I$.
Quarter-phase, 4 wires: amount of copper 4; power $= 4\ E_S I$.
n-phase, n wires: amount of copper n; power $= n\ E_S I$.

The amount of power is seen to be proportional to the amount of copper, giving therefore equal copper economy for all systems on the basis of equal voltage between the line and the neutral or ground.

§ 52. **On the Basis of the Same Voltage Between Line Wires.**—Between line wires the voltage is $2E_S$ for the single-phase (or quarter-phase) system and $\sqrt{3}E_S$ for the 3-phase system. To make the voltage between line wires equal in these systems, the voltage in the 3-phase system can be increased in the ratio $\sqrt{3}:2$. The amount of copper can accordingly be reduced (see § 50) inversely as the square of this ratio, namely, 4:3. Hence, for the same line voltage, a 3-phase system requires 75 per cent. as much copper as a single-phase or quarter-phase system.

§ 53. **Direct Currrent System.**—A direct current system has the same copper economy as a single-phase system, when the direct current voltage is made equal to the effective (sq. rt. of mean sq.) value of the alternating voltage.

If, however, the direct current voltage is increased so as to equal the maximum value of the alternating current voltage, the direct current voltage is increased in the ratio of $1:\sqrt{2}$ and the copper is decreased as the inverse square of this ratio. The direct current system then requires only one half the copper of a single-phase or two thirds the copper of a 3-phase system, on the basis of equal voltage between wires.

A direct current system would, therefore, be more economical of copper than any other system, *at the same voltage.*

§ 54. **Choice of Systems.**—On account of copper economy and the simplicity due to the use of only two wires, direct current would be superior to any alternating current system, if it were not for lack of simple and suitable means for transforming direct current so as to obtain the advantage of high potential transmission with low potential generation and utilization. In the case of alternating currents, these means are provided for by the *transformer* which makes alternating current systems so flexible that they are practically always* used for long distance transmission, instead of direct current.

In comparing alternating current transmission systems, the choice is to be made between single-phase—with its simpler line construction, fewer insulators, etc.—and 3-phase, requiring only 75 per cent. as much copper. If these were all the factors, single-phase transmission systems would be more common than they are, the simplicity offsetting the poorer copper economy. An important and perhaps a determining factor, however, is the superiority of polyphase as compared with single-phase machinery (§ 2); for this reason a polyphase system is commonly preferred, quite aside from considerations of copper economy. Of polyphase systems, the 3-phase system is most economical and is therefore the system in general use.

* (§ 54a). In a few cases high potential direct current has been used for power transmission, notably in the Thury system. This is essentially a constant current system. The high potential is obtained by generators in series; the motors are likewise in series. See *Lond. Electrician,* March 19, 1897; New York *Elect. Rev.,* January, 1901.

EXPERIMENT 6–B. Measurement of Power and Power Factor in Polyphase Circuits.

PART I. GENERAL DISCUSSION.

§ 1. **Preliminary.**—For measuring power in *any* 3-wire system, the best method is the two-wattmeter method § 23; for the particular case of a balanced 3-phase load, some one-wattmeter method, §§ 32–9, *may* be used.

For measuring power in systems with more than three wires, the $n - 1$ wattmeter method of § 16 is correct for all cases; for the particular case of a balanced 2-phase load, on a 4-wire system, the method of § 10, employing two wattmeters, may be used.

An unknown load should not be assumed to be balanced.

It will be understood that, in cases where several wattmeters are described as being required, a single instrument may be used and shifted by suitable switches from circuit to circuit, readings being taken successively in the different positions.

§ 2. **Separate Phase Loads.**—In any single-phase system power is measured by means of a wattmeter, the current coil being connected in series and the potential coil in parallel with the circuit, as discussed in Appendix III., Exp. 5–A. An extension of this method can be applied to a polyphase system, if the phases are separately accessible so that the load of each phase can be separately measured. A wattmeter is then used for each phase load, with current coil in series and potential coil in parallel with the particular load being measured, the total power being the arithmetical sum of the several wattmeter readings.

For example, to measure the power in three star-connected resistances on a 3-phase circuit by this method, three wattmeters would be required, each current coil carrying the star (or line) current and each potential coil being subjected to the star voltage. With three resistances delta connected, three wattmeters would also be required, each current coil carrying the delta current and each potential coil being subjected to the delta (or line) voltage.

§ 3. This method of measuring the separate phase loads is simple in principle and is commonly used on a 2-phase circuit (§ 6), but it is not capable of general application inasmuch as phase loads are not always separable. On a 3-phase circuit—in testing, for example, a 3-phase induction motor—it may be impossible to measure delta current or star voltage, so that some method not requiring either of these measurements becomes necessary; furthermore, the method is open to objection on account of the number of measurements required,—unless the assumption is made that all phases are alike, so that measurements are necessary on one phase only.

§ 4. **Polyphase Power Factor.**—A polyphase system is a combination of single-phase elements. If E, I and W are, respectively, the voltage, current and power for any separate element, the power factor for that element is $W \div EI$, by definition. When the separate elements or phases of a polyphase system have the same power factor, this is the power factor for the whole system.

§ 5. When, however, the separate elements have different power factors, there is no one power factor that has a definite value or physical significance for the whole system. It is convenient, however, to obtain a kind of average power factor for the system, the value of which will depend upon the method used in its determination.* An average power factor may be satisfactorily determined when the separate phases are nearly alike, but has little meaning when they are widely different.

§ 6. **Two-phase Load.**—Two-phase power is usually measured by two wattmeters, one on each phase, as just described.

§ 7. When the phases are independent, as in *a*, Fig. 1, Exp. 6-A, the measurements differ in no respect from measurements made on single-phase circuits.

* (§ 5a). See A. S. McAlliser, *Alternating Current Motors,* p. 12; A. Burt, *Three-phase Power Factor,* A. I. E. E., p. 613, Vol. XXVII., 1908.

§ 8. On a 3-wire, 2-phase circuit, as in *b*, Fig. 1, Exp. 6–A, the same method may also be used, the two wattmeter current coils being located in the two "outer" conductors, *A* and *B*, respectively. With the wattmeters thus located, the sum of their two readings will give the true power (§ 23) for any load whatsoever, even when part of the load is between *A* and *B*. (These connections are seen in Fig. 1, in which *X* and *Y* are the outer conductors and *Z* is the common conductor or return.)

§ 9. When the load in a 3-wire 2-phase system is balanced and there is no load between the two outer conductors *A* and *B*, one wattmeter may be conveniently used by connecting the current coil in the common conductor; one end of the potential coil is connected to the common conductor and the other end connected first to one and then to the other outer conductor. A reading* is taken in each position and the algebraic† sum gives the total power. (The connections are seen in Fig. 7, in which *Z* is the common conductor.) A 3-wire 2-phase circuit is likely not to be balanced (§ 14, Exp. 6–A) and the method should be used with caution.

§ 10. On a 4-wire, quarter-phase, 2-phase system, as in *c* and *d*, Fig. 1, Exp. 6–A, two wattmeters, one on each phase, will give the correct power only when the load is balanced. The method may be used for testing a single machine, but not for measuring the power of a circuit when the character of its load is unknown.

* (§ 9a). For a balanced load, power can be determined from a single reading of the wattmeter by connecting the current coil in the common conductor and connecting the potential circuit from the common conductor to the middle point of two approximately equal non-inductive resistances, R_1 R_2, connected across the two outer conductors as in Fig. 5. A single reading of the wattmeter gives one half the total power, if the wattmeter is calibrated as a single-phase instrument with R_1 and R_2 connected in parallel with each other and in series with the potential circuit (§ 36a). See also § 33a.

† (§ 9b). For low power factors, when θ exceeds 45°, the reading of the wattmeter in one position is negative. The similar case for a 3-phase circuit is fully discussed later.

That the method is not generally correct will be seen by assuming the current coils of the two wattmeters to be connected in two of the lines, as A and B; neither wattmeter would then record a single-phase load drawing current from the other two lines, $A'B'$.

On a 4-wire system, with unbalanced load, at least three wattmeters must be used, § 16.

§ 11. *Power Factor in a Two-phase Circuit.*—If E, I and W are measured on one phase of a 2-phase circuit, $W \div EI$ is the power factor for that phase, § 4. This may be called the ***cosine method*** for determining power factor, since $W \div EI = \cos\theta$ when currents and electromotive forces are represented by sine waves.

§ 12. The following *tangent method* for determining power factor from two readings of the wattmeter will be found simple and often convenient.

The current coil of the wattmeter is connected in one line of phase A; the potential coil is connected across phase A, whose voltage is E_A. The wattmeter now reads the power volt-amperes or true watts

$$W_1 = E_A I_A \cos\theta. \tag{1}$$

Transfer the potential coil to phase B, whose voltage is E_B. The wattmeter now reads the wattless or quadrature volt-amperes (sometimes called wattless, or quadrature, watts),

$$W_2 = E_B I_A \sin\theta. \tag{2}$$

Dividing the second reading by the first,

$$\frac{W_2}{W_1} = \frac{E_B}{E_A} \tan\theta. \tag{3}$$

Tan θ, and hence power factor ($\cos\theta$), is determined by the ratio of the two readings. Usually $E_B = E_A$, so that $\tan\theta = W_2 \div W_1$. The power factor thus determined is the power factor of phase A; θ is the phase difference between I_A and E_A. The method assumes that E_A and E_B differ 90° in phase and that electromotive forces and currents follow a sine law.

The advantage of the tangent method is its simplicity and independence of the calibration of instruments. The method can be used for determining the power factor of a single-phase load, drawn from a 2-phase supply, and a somewhat similar method can be used for determining the power factor of a 3-phase load, §§ 28, 38, 41.

§ 13. The value of θ and power factor can be found by the *sine method* directly from (2); thus, $\sin\theta = W_2 \div E_B I_A$. For a single-phase or 2-phase load there is little advantage in this method, which is useful, however, on 3-phase circuits, § 43.

§ 14. The "cosine" method gives correct power factor by definition and is general, being independent of wave form. The "tangent" and "sine" methods are based on the assumption that voltages and currents follow a sine* law. The "cosine" and "sine" methods require carefully calibrated instruments.

§ 15. The three methods are seen to be based on the relation,

$$\cos\theta = \frac{\text{power volt-amperes}}{\text{total volt-amperes}}.$$

$$\sin\theta = \frac{\text{wattless volt-amperes}}{\text{total volt-amperes}}.$$

$$\tan\theta = \frac{\text{wattless volt-amperes}}{\text{power volt-amperes}}.$$

§ 16. **General Method for Measuring Power; n — 1 Wattmeters.**—This method consists in selecting any one conductor of a system and considering it as a common return for all the others. One wattmeter is then used for each conductor, except

* (§ 14a). With non-sine waves, the value of power factor by the tangent method would, theoretically, be a little larger than the true value by the cosine method; the value by the sine method would be a little larger than the value by the tangent method.

this common return. No wattmeter is required for a return circuit; thus, for a 2-wire system, one wattmeter only is needed, no wattmeter being needed in the return conductor; in a 3-wire system, two wattmeters are used, none being needed in the return conductor, etc. If n is the number of line conductors, $n - 1$ wattmeters are, accordingly, required. For a 3-wire system, the connections are shown in Fig. 1.

To measure power in any system, connect a wattmeter in every line circuit except one (considered as the return conductor), each wattmeter having its current coil in series with one of the lines and its potential coil connected from this line to the return conductor. One less wattmeter is required than the number of line wires; the total power is the algebraic sum of the individual wattmeter readings.

§ 17. To read positive power each wattmeter is to be connected in the positive sense,—that is, connected in the same way as for measuring power in a 2-wire system, direct or alternating. If, when connected in this manner, the needle of any wattmeter deflects the wrong way, the connections of its potential or current coil are to be reversed and its reading is to be considered negative. Compare § 25.

§ 18. This method of measuring power is absolutely general; the current may be direct or alternating and may vary by any law whatsoever; the system may be single-phase or polyphase, balanced or unbalanced, symmetrical or unsymmetrical.

As a particular case, the two-wattmeter method for a 3-wire system is of special importance with reference to 3-phase circuits and will be considered later (§ 23) in detail.

§ 19. The foregoing method has been explained by considering one conductor as a common return for all the others, and for most purposes this explanation is sufficient. The method with $n - 1$ wattmeters can be rigorously established (§ 22) by first developing the method with n wattmeters, § 20.

§ 20. **General Method, n Wattmeters.**—In any star-connected system, if a wattmeter is connected in each line—the current coil connected in series with the line and the two ends of the potential coil connected, respectively, to the line wire and to the junction or neutral point of the system—the total power of the system will be the sum of the separate wattmeter readings, as discussed in § 2.

§ 21. This arrangement of wattmeters, however, is not limited to star-connected circuits; nor is it necessary to have the neutral point accessible. *The true power of any system whatsoever may be measured by connecting one wattmeter in each line, with current coil in series with the line and potential coil with one end connected to the line and the other end to any point P of the system, which may or may not be the neutral.* To this potential point *P* is connected the potential coil of every wattmeter. The algebraic sum of the wattmeter readings gives the true power. A general proof of this is given in § 53; it can be verified by experiment, §§ 45–49.

§ 22. The fact that *any* point of the system may be taken as the potential point *P* leads to the practical simplification by which one wattmeter is omitted. In a system of line wires, *a, b, c* ··· *n*, let the line wire *n* be taken as the potential point. Wattmeters *A, B, C*, etc., will have current coils connected in series with *a, b, c*, etc., and potential coils connected from *a* to *n*, from *b* to *n*, etc. Wattmeter *N* would, accordingly, have its potential coil connected from *n* to *n*; as both ends of the pressure coil would thus be connected to the same point, this wattmeter would always read zero and, accordingly, can be omitted. The method of $n - 1$ wattmeters, § 16, is thus established.

§ 23. **Two Wattmeter Method for any 3-wire system.**—This is the method generally used for measuring 3-phase power. Being a particular application of the $n - 1$ wattmeter method, § 16, the two-wattmeter method can be applied to any 3-wire system* and is independent of any assumptions as to wave form or the nature of the load.

§ 24. The arrangement of instruments is shown in Fig. 1. The wattmeters† are inserted in any two lines, as X and Y, the third wire Z being considered as a common return.

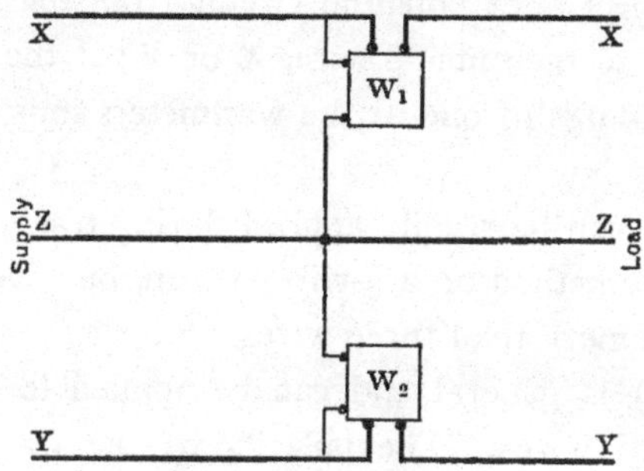

FIG. 1. Two-wattmeter method for measuring power in any 3-phase or other 3-wire circuit.

The total power is the algebraic sum of the readings of the two wattmeters. For high power factors (more than 0.5) this will be the arithmetical sum, both wattmeter readings being positive. For low power factors (less than 0.5), the reading of one wattmeter is to be considered negative, the total power in this case being the arithmetical difference of the two readings, as shown later in § 31.

§ 25. There are several ways for telling whether one reading is negative or not, the principal ones being as follows:

(a) From the sense of the connections, § 17.

* (§ 23a). If each end of a 3-phase line has its neutral well grounded, it becomes virtually a 4-wire system; the ground circuit can not be neglected unless the load is practically balanced.

† (§ 24a). *Polyphase Wattmeter.*—Instead of two single-phase wattmeters, a single instrument combining the two is commonly used. This consists of two wattmeters, one above the other, with the moving elements mounted upon a common shaft. The reading of such an instrument gives the total power. The electrical connections are the same as for two separate instruments.

(*b*) For the given load substitute a load that is non-inductive or is known to have high power factor; if, with certain connections, both wattmeters deflect properly, their readings for these connections are positive. When one connection needs to be reversed to obtain proper deflection, one reading is negative.

(*c*) Disconnect one* potential circuit from the middle wire Z and connect it to the outside wire, X or Y; if the wattmeter reverses, the readings of one of the wattmeters must be considered negative.

Method (*c*) can be readily applied during test, when using the two-wattmeter method on a 3-wire system, but does not apply to a system with more than three wires.

Method (*a*) is general and can be applied to a system with any number of wires. The polarity of the wattmeter circuits may be marked once for all, instruments of one make being similar. The instruments can be properly connected in the positive sense† in advance and confusion during the test avoided.

§ 26. **Two-wattmeter Method with Balanced Three-Phase Load.**—As has been already stated, the two-wattmeter method is general for any kind of 3-wire circuit. Detailed proof for each particular case is, accordingly, unnecessary. A discussion of its application to measuring a balanced 3-phase load will, however, prove instructive as an illustration and will serve to make clear the negative reading of one wattmeter at low power factors. Furthermore, it will show a method for obtaining 3-phase power factor.

§ 27. Fig. 2 is the diagram for a balanced 3-phase load, it being assumed that currents and voltages follow a sine law. For unity

* (§ 25a). On a 3-phase circuit it is sufficient to do this with one potential circuit only; but in general it should be done, successively, with each potential circuit, a reversal of either instrument indicating that one reading is negative.

† (§ 25b). This also indicates the direction of the flow of power; see "Polyphase Power Measurements," by C. A. Adams, *Elect. World*, p. 143, January 19, 1907.

power factor ($\theta=0$), the three line currents are shown by the heavy arrows I_X, I_Y, I_Z. The dotted arrows show these currents for lower power factors, $\theta=30°$, $\theta=60°$ and $\theta=90°$.

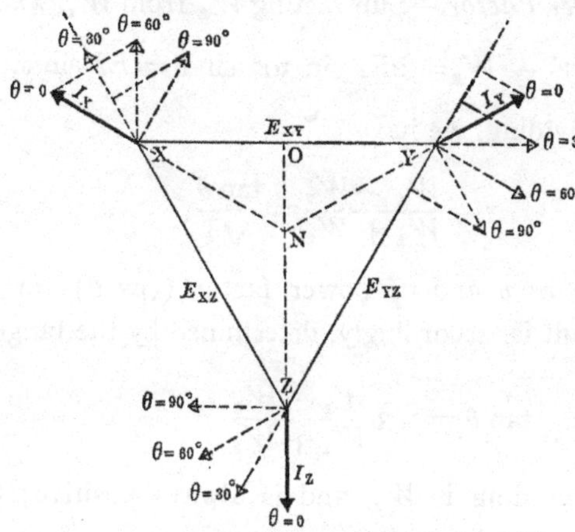

FIG. 2. Currents and voltages in a balanced 3-phase system.

If two wattmeters are connected as in Fig. 1, wattmeter (1) has a current I_X in its current coil and a voltage E_{XZ} across its potential coil, the phase difference between this current and voltage being $\theta-30°$. The component of I_X in phase with E_{XZ} is, accordingly, $I_X \cos(\theta-30°)$; hence—writing E for E_{XZ} and I for I_X—wattmeter (1) reads

$$W_1=EI\cos(\theta-30°)=EI(\cos 30°\cos\theta+\sin 30°\sin\theta).$$

In a like manner, wattmeter (2) has a voltage E_{YZ} and a current I_Y, having a component $I_Y \cos(\theta+30°)$ in phase with E_{YZ}. Hence—writing E for E_{YZ} and I for I_Y—wattmeter (2) reads

$$W_2=EI\cos(\theta+30°)=EI(\cos 30°\cos\theta-\sin 30°\sin\theta).$$

Adding W_2 to W_1, we have

$$W_1+W_2=2EI\cos 30°\cos\theta=\sqrt{3}EI\cos\theta,$$

which is seen to be the expression for the total power in a 3-phase circuit (§§ 20, 21, Exp. 6–A). The two-wattmeter method for a balanced 3-phase load is thus established.

§ 28. *Power Factor.*—Subtracting W_2 from W_1, we have

$$W_1 - W_2 = 2EI \sin 30° \sin \theta = EI \sin \theta.$$

Hence, by dividing, we have

$$\frac{W_1 - W_2}{W_1 + W_2} = \frac{\tan \theta}{\sqrt{3}}.$$

The value of θ and of power factor ($\cos \theta$) for a balanced 3-phase circuit is, accordingly, determined by the tangent formula

$$\tan \theta = \sqrt{3}\,\frac{W_1 - W_2}{W_1 + W_2}.$$

The larger reading is W_1, and is always positive; the smaller reading, W_2, may be positive or negative.

§ 29. To save labor in computation, it is convenient to plot a curve, Fig. 3, with power factor ($\cos \theta$) as ordinates and the ratio of wattmeter readings, $W_2 \div W_1$, as abscissæ. Points on this curve are determined by the relation

$$\frac{W_2}{W_1} = \frac{\cos(\theta + 30°)}{\cos(\theta - 30°)}.$$

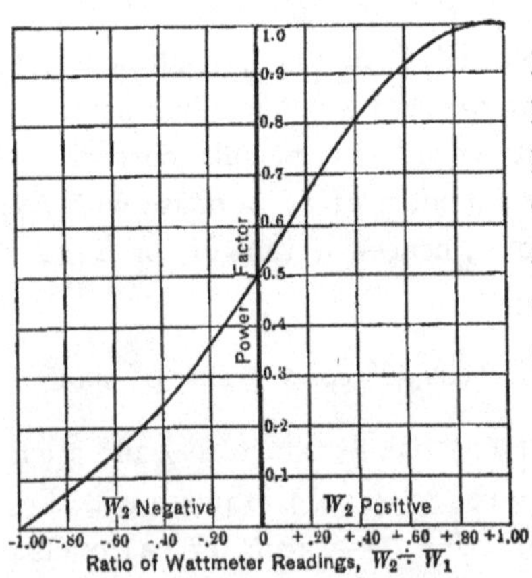

FIG. 3. Power factor of balanced 3-phase circuit for different ratios of wattmeter readings in two-wattmeter method.

By means of this curve, the power factor for a balanced load is readily determined from the ratio of the two wattmeter readings. For plotting the curve in Fig. 3, the following points were determined:

$W_2 \div W_1$	—1.	—.80	—.60	—.40	—.20	0	+.20	+.40	+.60	+.80	+1.
cos θ	0	.064	.143	.240	.359	.5	.655	.803	.918	.982	1.

Intermediate values can be found by interpolation. It is seen that the curve is not symmetrical.

§ 30. Errors of calibration are avoided if one wattmeter is used, successively, in the two positions to determine W_1 and W_2. Since the assumption is made that the current in the wattmeter is the same for the two readings ($I_X = I_Y = I$), greater accuracy is obtained if the current in the two cases is actually the *same* current. This is accomplished by using the one wattmeter method of Fig. 7, which is more accurate for determining power factor than is the two-wattmeter method. In either case, corrections may be made (§ 42) for slight variations in voltage.

§ 31. *Negative Reading of Wattmeter.*—Referring to Fig. 2, it is seen that for all values of θ from 0 to 90°, the projections of I_Y upon E_{ZX} have the same sign; the wattmeter reading W_1 is, therefore, in all cases positive.

The projection of I_Y upon E_{YZ} decreases as θ increases, becomes zero when $\theta = 60°$, and then changes sign. The wattmeter reading W_2, accordingly, changes sign, being positive when θ is less than 60° (power factor more than 0.5) and negative when θ is more than 60° (power factor less than 0.5). In all cases W_1 is the larger, and W_2 is the smaller, reading.

On non-inductive load, $\theta = 0$ and $W_1 = W_2$; each wattmeter reads half the total power. When $\theta = 90°$, $W_1 = -W_2$ and the total power is zero.

§ 32. **Three-Phase Power with One Wattmeter.**—With only one wattmeter, 3-phase power can be measured by the two-wattmeter method (§ 23) by using suitable switches for throwing the wattmeter from one position to the other. This procedure gives the true power for unbalanced as well as balanced loads and is generally the best one to follow.

The transfer of the current coil of the wattmeter from one line to another is not always convenient or possible and, *when the load is balanced,* the power in a 3-phase system can be measured with only one wattmeter without such transfer by one of the following methods.

§ 33. *With Neutral Available.*—When the neutral is available, the current coil of the wattmeter can be connected in any one line circuit and the potential coil connected from that line to the neutral. For a balanced load, the total power will be three times the reading* of the wattmeter. The power factor is equal to $W \div EI$, where I is the line current and E is the star voltage. When the load is not balanced the total power will be the sum of three readings, one on each phase.

§ 34. *With Artificial Neutral.*—When the neutral is not available, an artificial neutral can be created, as by means of three equal star-connected non-inductive resistances, R_1, R_2, R_3 in Fig. 4. The method of § 33 can then be applied.

It is necessary that these resistances be relatively low, as compared with the resistance of the potential circuit Rw of the wattmeter. The current in them will then be relatively large, so that the potential of the neutral will not be disturbed by the connection of the potential circuit of the wattmeter. The power taken in the resistances may be included or not in the measured power as desired; correction for this power can be made when necessary.

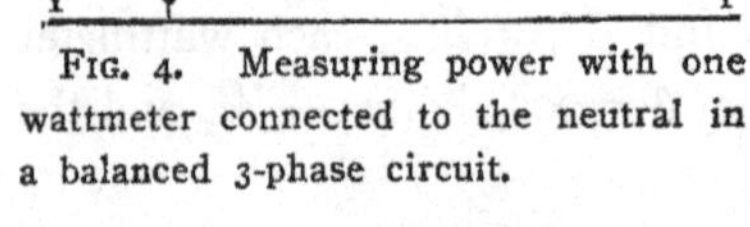

FIG. 4. Measuring power with one wattmeter connected to the neutral in a balanced 3-phase circuit.

§ 35. Strictly speaking R_1 and R_2 should each be equal to the joint resistance of R_3 and R_W in parallel. In this case there is no need of making the resistances low; this leads to the method of § 36 in which R_3 is omitted entirely, that is, $R_3 = \infty$.

* (§ 33a). *Calibration for Total Power.*—In this method, or in any method depending upon a single reading, the wattmeter can be calibrated to read total power.

§ 36. *With Y-Multiplier.*—With a balanced load and with one wattmeter, the current coil of the wattmeter is connected in one line. One end of the potential circuit is connected to the same line, the other end being connected to the junction of two resistances R_1 and R_2, which are connected to the other two lines, as shown in Fig. 5. The resistances R_1 and R_2 are non-inductive and are each equal* to R_W, the resistance of the potential circuit of the wattmeter.

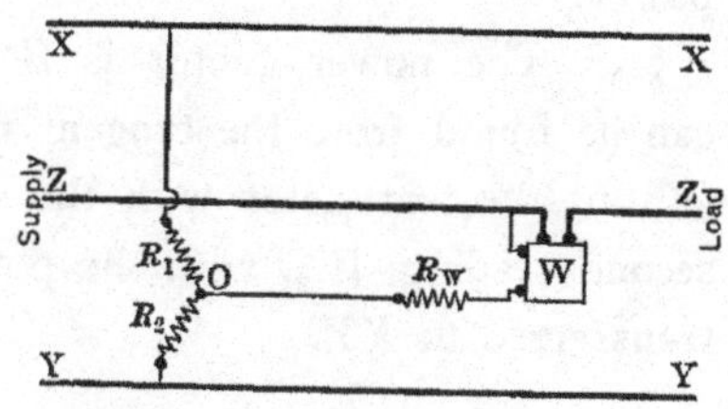

FIG. 5. Measuring power with one wattmeter and a Y-multiplier in a balanced 3-phase system.

True power is three times the reading of the wattmeter, calibrated as a single-phase instrument. The resistances are sometimes put up in a special volt-box or Y-multiplier for 3-phase circuits; the instrument may then be calibrated so as to read total power, § 33a.

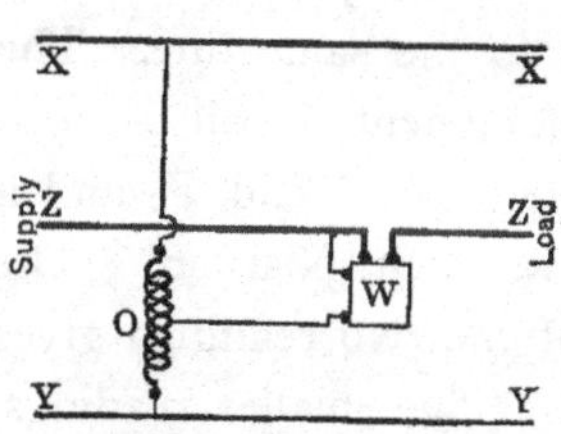

FIG. 6. Measuring power with one wattmeter, *T*-connected, in a balanced 3-phase circuit.

§ 37. *By Means of T-connection.*—In a 3-phase system, with three lines X, Y, Z, connect the current coil of the wattmeter in any one line, as Z, Fig. 6. Connect the potential coil from Z to a point O, the middle point of a transformer coil across XY. See Fig. 15, Exp. 6-A. In any balanced 3-phase system, however the load is connected, the wattmeter will now give one half the total power. (This may be seen as follows: If the wattmeter

* (§ 36a). Provided R_1 and R_2 are approximately equal to each other, this same method may be used without having R_1 and R_2 equal to R_W. The instrument is calibrated as a single-phase wattmeter with R_1 and R_2 in parallel with each other and in series with R_W; a single reading then gives one half the total power. Compare §§ 9a, 33a.

were connected with its potential coil on the star voltage, the wattmeter would read one third the total power; with its potential increased 50 per cent.—see Fig. 2,—it will read one half the total power.)

§ 38. The power factor is $W \div E_{OZ}I_Z$. The power factor can be found from the tangent formula, by taking one reading, W_1, of the wattmeter with the connections as described and a second reading, W_2, with the potential circuit of the wattmeter transferred to XY.

$$\frac{W_2}{W_1} = \frac{E_{XY}I_Z \sin\theta}{E_{OZ}I_Z \cos\theta}\text{; hence } \tan\theta = 0.866\frac{W_2}{W_1}.$$

§ 39. *Two-reading Method.*—This is one of the simplest and most satisfactory methods for measuring power and power factor with one wattmeter in a balanced 3-phase circuit. The current coil is connected in one line, as Z, Fig. 7, one end of the potential circuit being connected to the same line. The other end of the potential coil is connected, successively, to X and Y, and a reading taken in each position. The algebraic sum of the two readings gives the total power. (The smaller readings, W_2, is considered negative whenever it is necessary to reverse the potential or current coil of the wattmeter to obtain a proper deflection.)

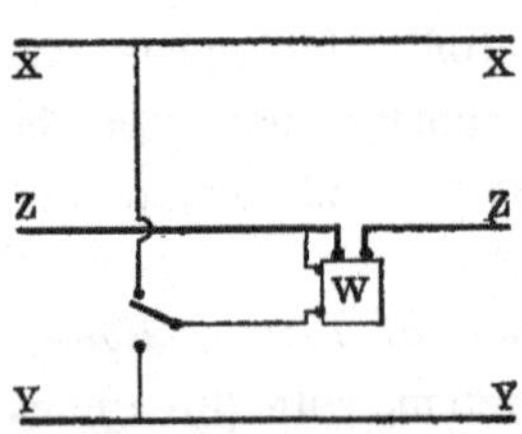

FIG. 7. Measuring power by two readings of one wattmeter in a balanced 3-phase circuit.

§ 40. The proof of the method will be seen by referring to Fig. 2, which assumes that voltages and currents follow a sine law. The two readings of the wattmeter are

$$W_1 = EI\cos(\theta - 30°); \quad W_2 = EI\cos(\theta + 30°).$$

Hence, the sum of the two readings gives the total power, § 27.

§ 41. The power factor ($\cos\theta$) is determined from the tangent formula, § 28,

$$\tan\theta = \sqrt{3}\frac{W_1 - W_2}{W_1 + W_2}$$

By referring to Fig. 3, power factor can be found directly from the ratio $W_2 \div W_1$.

§ 42. When there is an appreciable difference between the phase voltages (which we may term E_1 and E_2) across which the potential circuit is connected when W_1 and W_2 are read, a more accurate value of power factor will be obtained by correcting W_1 or W_2 by direct proportion to obtain values corresponding to equal voltages. The ratio $W_2 \div W_1$ then becomes $E_1W_2 \div E_2W_1$. The power factor thus determined is quite accurate, being independent of the calibration of any instrument and of any slight inequality in the phases. Even for an unbalanced load, it gives accurately the value of $\cos\theta$ for I_Z, where θ is the phase difference between I_Z and the voltage OZ (Fig. 2) midway in phase between XZ and YZ. The method is more accurate with one than with two wattmeters, § 28.

§ 43. *Power Factor by Sine Method.*—The power factor of a balanced 3-phase circuit can be determined by the sine method (§ 13) with only a single reading of voltmeter, ammeter and wattmeter. The method does not require the neutral to be available, nor does it require any auxiliary resistances or other devices.

Representing the three line wires as X, Y and Z, the ammeter and the current coil of the wattmeter are connected in one line, as Z. The voltmeter and the potential coil of the wattmeter are connected across the other two lines, X and Y. The wattmeter reading gives the wattless or quadrature volt-amperes,

$$W = E_{XY} I_Z \sin\theta,$$

from which θ and $\cos\theta$ are determined.

PART II. MEASUREMENTS.

§ 44. Many of the methods just described for measuring polyphase power and power factor can best be taken up as occasion arises for their use. Without undertaking in the present experiment to subject all of these methods to test, it will be well to select a few of them for trial in the laboratory in order to illustrate and make clear the methods as a whole. For this the following tests are suggested.

§ 45. **Verification of Methods for Measuring Polyphase Power.** —With a single-phase non-inductive load, forming a 2-wire system, measure the total power with two wattmeters. Each line is to contain the current coil of one wattmeter, the potential coil of which is connected from the line to a common point P, as in § 21. The experiment consists in connecting P to different parts of the circuit, of various potentials, and noting that the algebraic sum of the two wattmeter readings is constant.

When the power indicated by one wattmeter becomes greater, as P is changed, the power indicated by the other wattmeter becomes less.*

§ 46. For example, let the supply lines be a_1a_2, as in Fig. 6, Exp. 6–A. Connect P, successively, to points of different potential, as a_1, a_2, the neutral O, A_1 and A_2, these points being all on phase A. When phase B of a two phase supply is available, proceed, also, to connect P successively to points B_1, b_1, b_2, B_2 on phase B.

§ 47. Repeat with an inductive load.

§ 48. Repeat in some modified manner, as by using a_1b_1 as supply lines and connecting P, successively, to various points as described above.

§ 49. When points, as in Fig. 6, Exp. 6–A, are not available, a resistance can be bridged across the circuit and the point P

* A positive reading decreases; a negative reading increases.

connected to different points on this resistance. The load resistance itself can be thus utilized.

The experiment might be extended to using 3 wattmeters on a 3-wire system, 4 wattmeters on a 4-wire system, etc., but this seems hardly necessary. The method of n wattmeters, $n-1$ wattmeters and two wattmeters may, in this way, be experimentally verified.

§ 50. **Two-phase Power Factor.**—From one phase, *A*, of a 2-phase supply draw a single-phase load. Take measurements with a voltmeter, ammeter and wattmeter and determine the power factor by the "cosine method," § 14.

Transfer the voltmeter and potential coil of the wattmeter to the other phase, *B*, and determine the power factor by the "sine method," § 13, and by the "tangent method," § 12.

§ 51. **Three-phase Power and Power Factor.**—With a 3-phase balanced load supplied from a 3-phase circuit, take two readings of a wattmeter connected as in Fig. 7. Determine the total power; calculate the power factor by the tangent formula, § 28, and by the ratio of wattmeter reading, Fig. 3.

§ 52. Transfer the potential coil of the wattmeter to the third phase, so as to read the "quadrature" volt-amperes; take the necessary readings of the wattmeter, voltmeter and ammeter, and determine power factor by the "sine method," § 43.

APPENDIX I.

MISCELLANEOUS NOTES.

§ 53. **General Proof.**—In any system, with any number of conductors a, b, c, etc., let the instantaneous values of the currents in these conductors be i_a, i_b, i_c, etc. Designate by e_a, e_b, e_c, etc., the instantaneous values of the potentials of the several conductors. The currents and electromotive forces may vary in any manner whatsoever. There is no limitation as to the arrangement or method of connection of the generator and receiver circuits.

The total power at any instant is

$$(1) \qquad w = e_a i_a + e_b i_b + e_c i_c \ldots = \Sigma ei.$$

Let e_p be the instantaneous potential of any point P of the system. Since it is known that $\Sigma i = 0$, it follows that

$$(2) \qquad e_p i_a + e_p i_b + e_p i_c \ldots = e_p \Sigma i = 0.$$

Since (2) is equal to zero, it may be subtracted from (1) without affecting its value; hence

$$(3) \quad w = (e_a - e_p) i_a + (e_b - e_p) i_b + (e_c - e_p) i_c \ldots = \Sigma (e - e_p) i.$$

The total power at any instant is seen to be the sum of the products of the instantaneous currents in each conductor and the instantaneous differences of potential between the respective conductors and the point P.

The mean power W is found by integrating the instantaneous power over a time equal to one period, T, and dividing by T.

$$W = \frac{1}{T}\int_0^T w dt = \Sigma \frac{1}{T}\int_0^T (e - e_p) i dt,$$

$$W = \frac{1}{T}\int_0^T (e_a - e_p) i_a dt + \frac{1}{T}\int_0^T (e_b - e_p) i_b dt + \frac{1}{T}\int_0^T (e_c - e_p) i_p dt \ldots$$

But each one of these terms represents the power, as read by a wattmeter with current coil in series with one conductor and with potential coil connected from that conductor to the common point P, and the total power is the sum of the several wattmeters so connected. For an n-wire system, n wattmeters are required, the total power being

$$W = W_a + W_b + W_c + \ldots W_n.$$

When the point P coincides with one conductor, the wattmeter for that conductor reads zero and can be omitted; $n - 1$ wattmeters are then required.

The method for n wattmeters, for $n - 1$ wattmeters, and for two wattmeters, is, accordingly, proved without reference to wave form or the character of the load. This general proof was first given by A. Blondel, p. 112, Proceedings International Electrical Congress, Chicago, 1893.

CHAPTER VII.

PHASE CHANGERS, POTENTIAL REGULATORS, ETC.

EXPERIMENT 7–A. Polyphase Transformation.

§ 1. **Possible Kinds of Transformation.**—The transmission of power in a single-phase system is pulsating, no matter what the character of the load. This can be readily seen by sketching assumed curves for the instantaneous values of an alternating electromotive force and current, and plotting the products of the ordinates from instant to instant as a power curve.* In a balanced polyphase system, however, the pulsations of power in the different phases are seen to so combine that the total transmission of power is uniform,† without pulsation. (See § 2, Exp. 6–A.)

§ 2. *Polyphase to Single-phase Transformation not Possible.*—In a transformer, neglecting the slight modification due to losses, the power given into the primary at any instant is equal to the power taken out of the secondary at that instant. It is not possible, therefore, simply by means of transformers to change a

* (§ 1a). The area included between the power curve and time axis represents energy, this energy being positive (supplied to the line) or negative (returned from the line) according to whether the current and electromotive force have, at the time, like or unlike signs. It is instructive to sketch curves for currents differing in phase from the electromotive force by 0, 45 and 90 degrees.

† (§ 1b). This can be shown for a 2-phase system by drawing, for each phase, sine curves for electromotive force and current and plotting the product as a power curve. Two power curves are thus obtained, one for each phase, and it will be seen that the crests of one correspond to the hollows in the other, the algebraic sum of the two power curves being constant. The sum of the three power curves for a 3-phase circuit can be shown to be constant in the same way.

pulsating into a non-pulsating transmission of power, or *vice versa*.

It is accordingly not possible, by means of transformers, to draw single-phase current from a polyphase system and draw from the several phases equally, so that the flow of energy is non-pulsating. To accomplish such a transformation, use is made of a motor-generator* consisting of a polyphase motor driving a single-phase generator. The moving parts act as a flywheel, storing and restoring kinetic energy, thus accounting for the momentary difference between the pulsating output and non-pulsating input of electric energy. This method is advocated for running single-phase railway feeders from a polyphase transmission line.

§ 3. *Single-phase to Polyphase Transformation not Possible.*—It is likewise not possible, by means of transformers, to change a single-phase into a balanced polyphase system. This too can be done by means of a motor-generator, or by running a polyphase induction motor on a single-phase circuit,—a 2-phase or a 3-phase motor giving 2-phase or 3-phase currents. Various stationary phase-splitting devices will give difference in phase sufficient for starting induction motors on single-phase circuits, but such devices cannot give balanced polyphase currents.

§ 4. *Polyphase Transformation Possible.*—It is possible, however, by various arrangements† of transformers to change from one balanced polyphase system to *any other* balanced polyphase system, the flow of energy in each system being uniform. This is termed *polyphase transformation* and its study is the object of this experiment. The various methods of polyphase transformation are similar in principle, use being made of the fact‡

* Such a motor-generator has been installed in the chemical laboratory of Cornell University to supply 2,000 or 3,000 amperes of single-phase current for the electric furnace.

† (§ 4a). Two transformers, only, are necessary; but more than two are used in some arrangements, as Fig. 4.

‡ Fully discussed in Exp. 6–A.

that, if two coils with electromotive forces differing in phase are connected in series, the electromotive force across the two coils is the vector sum (or difference) of the two separate electromotive forces. A resultant electromotive force of any desired phase can thus be obtained from a polyphase supply by means of two transformers.

The transformation from 2-phase to 3-phase, or *vice versa,* is most important on account of the copper economy* in 3-phase transmission and the sometime advantage of 2-phase generation or utilization.

§ 5. **Two-phase to Three-phase Transformation (and vice versa) by T-connection.**—This method, first published by Mr. C. F. Scott, is shown diagrammatically in Fig. 1, in which *A* and *B* are the two phases of a 2-phase system; *X, Y* and *Z* represent the three line wires of a 3-phase system.

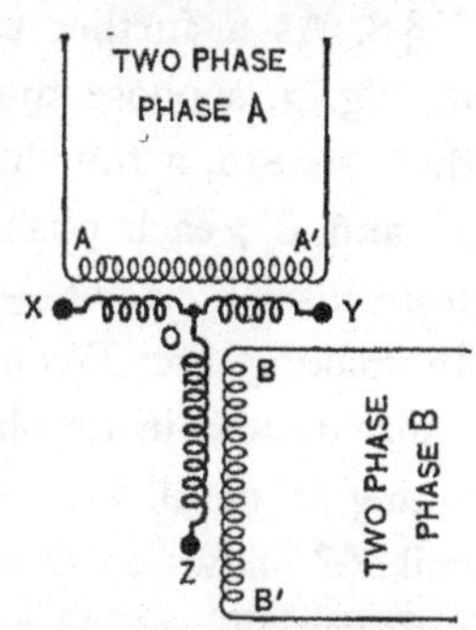

FIG. 1. Transformation from 2-phase (*AB*) to 3-phase (*XYZ*), or *vice versa.*

Let the transformation be from 2-phase to 3-phase. Two transformers are used. One has a primary *AA'* on phase *A* of the 2-phase system and has a secondary (*XY*) wound, let us say, for 100 volts with a middle tap at *O*, dividing the coil into two parts of 50 volts each. The second transformer has a primary *BB'* on phase *B* of the 2-phase system and has a secondary (*OZ*) wound for 86.6 volts ($86.6 = 100 \times \frac{1}{2}\sqrt{3}$), which has one end connected, as shown, to the middle tap of the first transformer. It will be found that the three voltages, *XY, YZ* and *ZX,* are equal and differ in phase from each other by 120°, thus making a 3-phase system. These voltages are represented in Fig. 2. They should be interpreted as in Exp. 6–A; see also Appendix I. to this experiment.

* See Appendix III., Exp. 6–A.

§ 6. This method of transformation is reversible; *i. e.*, if a 3-phase system be connected (see Fig. 1) to *XYZ* as primary, 2-phase circuits may be taken from *AA'* and *BB'* as secondary.

§ 7. *Double Transformation.*—In Fig. 3 is shown a double transformation, from the 2-phase generating circuits *A*, *B* to the 3-phase transmission circuits *X*, *Y*, *Z*, and from these to the 2-phase receiving circuits *A*, *B*. The receiving circuits, *A* and *B*, may be used together, as for operating polyphase motors, or separately as for lighting.

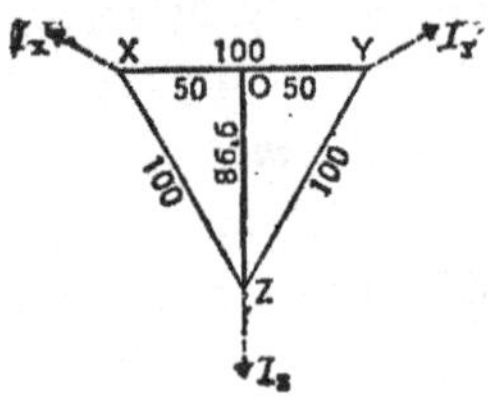

Fig. 2. Voltage and current relations.

§ 8. As a further explanation of the *T*-connection, referring to Fig. 3, suppose the connections *OZ'*, *OZ'* were left out and that, instead, a fourth wire *z'* (not shown) were used to connect *Z'* and *Z'*; each phase would then have its independent 2-wire transmission circuit,—wires *xy* for phase *A* and *zz'* for phase *B*. In making the *T*-connection of Fig. 3, the fourth wire *z'* is omitted* and in its place use is made of the two wires *x* and *y*, acting in parallel as a single conductor. The current from the coil *ZZ'* flows to *O* and divides, passing through *OX* and *OY* differentially, so as to have no magnetizing effect on the core of *XY*. With respect to the current from *Z'*, the two parts of the coil *XY* are wound non-inductively. They should be interspaced so as to have the least possible magnetic leakage and consequent leakage reactance, which would give poor regulation on phase *B*. *This precaution is necessary in winding any T-connected transformer.*

The regulation of phase *A* and of phase *B* are as independent of each other with three wires (Fig. 3) as they would be with four wires making separate circuits; phase *B* may have a heavy

* (§ 8a). There is obvious copper economy in this case in changing from a 4-wire 2-phase to a 3-wire 3-phase transmission; see Appendix III., Exp. 6-A.

motor load with 50 per cent. drop, while *A* has a lighting load with, say, 2 per cent. drop, unaffected by the starting and stopping of the motors on *B*. They are absolutely independent of each other.

§ 9. *Composite Transmission.*—If the phases *A* and *B*, Fig. 3, were generated and utilized separately, it would not be necessary for *B* to differ from *A* by ninety degrees; *B* could have any phase, even the same phase as *A*. Again *A* and *B* might be of different frequencies; in fact they can be treated as two independent transmission systems* whether of the same or of different frequencies. In the same way a direct and alternating current can be combined with economy of copper and independence of regulation.

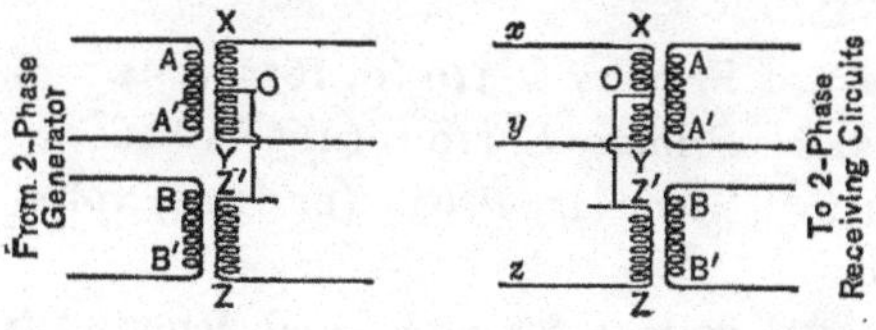

FIG 3. Two-phase generator and receiving circuits with 3-phase transmission.

§ 10. **Test.**—First note the single-phase transformations which can be made with the transformers to be used, and determine whether or not the transformers are suitable for the purpose. Connect the transformers so as to transform from a 2-phase system to a 3-phase system and make measurements of the primary and secondary line voltages, and the voltage of the *T*-connected coil, checking all by computation.

Make corresponding transformation from a 3-phase to a 2-phase system.

If the transformers are provided with two sets of coils, for parallel and series connection, make the polyphase transformations with all possible voltage ratios. Compute the volt-ampere

* (§ 9a). Various methods of composite transmission will be found in the following: *Elect. World,* February 28, 1903, pp. 347 and 351, Vol. XLI., No. 9; *Am. Electrician,* April, 1903, pp. 189 and 177, Vol. XV., No. 4; *Elect. Review* (New York), March, 1903, p. 362, Vol. 42, No. 11; *Elect. Age,* March, 1903, p. 179, Vol. XXX., No. 3; *Mill Owners,* April, 1903, p. 14.

capacity for the 3-phase side of each transformer when the total power output, on non-inductive load, is 100 watts; see § 24, Exp. 6–A.

§ 11. *Instructions for Special Transformers.*—These instructions relate to two transformers, *DEF* and $\alpha\beta\gamma\delta$. Each transformer has two primaries, which may be connected in series or in parallel. The windings are as follows:

Primary α, 110 (or 165) volts.
Primary β, 110 (or 165) volts.
Secondary γ, 36⅔ (or 55) volts.
Secondary δ, 36⅔ (or 55) volts.
Primary *D*, 110 (or 165) volts.
Primary *E*, 110 (or 165) volts.
Secondary *F*, 63.5 (or 95.25) volts.

The first number gives normal voltage for highest efficiency at 60 cycles; the number in parenthesis is 50 per cent. above normal voltage. These transformers were specially made for use at either voltage.

With *D* and *E* in parallel on one phase, and α and β in parallel on the other phase of a 2-phase system, connect *F* to the middle point of γ and δ connected in series, thus making a *T*-connection. From 2-phase circuits of 110 volts (and also 165 volts) obtain 3-phase secondary voltages by computation and measurement. Make the primary connections from 110-volt 4-wire 2-phase system, and also from 110-volt 3-wire 2-phase system.

Repeat with primaries in series instead of in parallel; compute and measure secondary 3-phase voltages.

Perform corresponding transformation from 110-volt 3-phase to 2-phase. What two 2-phase voltages can be thus obtained?

§ 12. **Monocyclic Transformation.**—In the monocyclic system (no longer being installed) a single-phase voltage is combined with a quadrature voltage of one fourth its value; thus, in Fig. 6, Exp. 6–A, a monocyclic voltage is obtained from $A_1A_2b_2$. It is an unsymmetrical 2-phase system. If two 1 : 1 transformers are used, the primary of one being connected to A_1b_2, of the other to A_2b_2, the secondaries (with two ends together for a 3-wire system) will give a monocyclic voltage the same as the primary. This secondary voltage is an open delta with one side reversed. Test this with a voltmeter and draw a diagram of voltages.

Reverse the primary or the secondary coil of one transformer; the secondary voltages now form an open delta, making very nearly an equilateral triangle, and hence forming a nearly symmetrical 3-phase system. Compute these voltages and verify by measurement. This method was used for obtaining polyphase current for operating 3-phase motors upon what was initially a single-phase circuit with a so-called "teazer" circuit (b_2) added. It was introduced by Steinmetz; the name indicated a pulsating (monocyclic) flow of energy instead of a non-pulsating (polycyclic) flow. The transformation is instructive, even though its introduction has been discontinued.

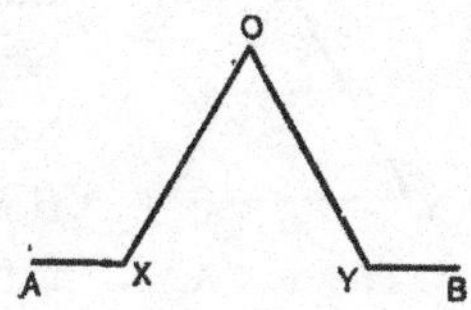

FIG. 4. Transformation from 3-phase (*XOY*) to 2-phase (*AOB*).

§ 13. **Miscellaneous Transformations.**—Several other transformations are here indicated. Try these by experiment, so far as time and facilities permit. Compute for each case the excess of volt-amperes over watts; see § 24, Exp. 6–A.

§ 14. Fig. 4 shows a method for transforming from 3- to 2-phase, with three transformers. The primaries are connected to a 3-phase supply. The secondaries of the first two transformers are *OX* and *OY*. The third transformer has two secondaries, whose voltages are *AX* and *YB*. What must be the values of these, in order that *AOB* will give true 2-phase voltages?

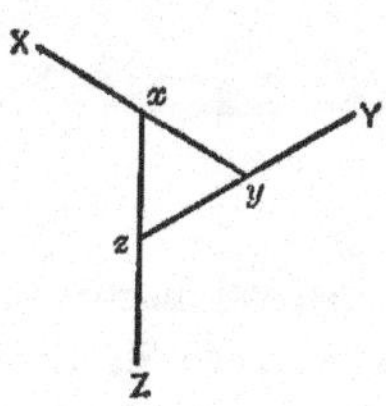

FIG. 5. Transformation from 3-phase (*XYZ*) to 3-phase (*xyz*).

§ 15. Fig. 5 shows a method, occasionally of laboratory use, of using three auto-transformers for 3-phase transformation from *XYZ* to *xyz*. What ratio of transformation will be obtained by using three auto-transformers with a middle tap?

§ 16. Fig. 6 shows two *V*-connected auto-transformers for 3-phase transformation from *XYZ* to *Xyz*. This is a method commonly used for obtaining a low starting voltage for 3-phase motors and converters. The taps y and z can be located where desired. It is to be noted that the voltage is changed, but not the phase. A third auto-transformer, YZ, might be used, with a tap at O. Although better for continuous operation, this would have the disadvantage of requiring an additional auto-transformer; furthermore, this arrangement could not give less than half voltage, and would make a reversal of phase in changing from low to high (starting to running) voltage, which is not desirable in starting a synchronous machine from the alternating current end.

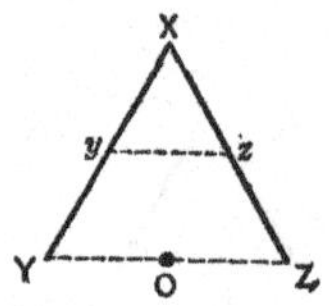

FIG. 6. Auto-transformers on 3-phase circuit.

§ 17. **Two-phase to Six-phase Transformation.**—The most practical method for this consists in transforming from a 2-phase primary circuit to two sets of T-connected secondaries, one set being inverted; (thus T and ⊥). Two or four transformers can be used. It is not necessary to make this transformation in the laboratory. (Detailed connections are given in McAllister's *Alternating Current Motors;* see also § 27, Exp. 6-A.)

APPENDIX I.

MISCELLANEOUS NOTES.

§ 18. **Further Interpretation of T-connection.**—A general discussion of the vector combination of electromotive forces is given in Exp. 6-A (particularly Appendix I.), and the general principles there given can be applied to the T-connection. The following is a more detailed discussion of this particular case.

The electromotive force of any alternating current coil may be represented by a vector in a certain direction. If this coil is the secondary of a transformer connected to a secondary line, as in the present case, the electromotive force impressed upon this line will be represented by the same vector. If in connecting the coil to the line the terminals are reversed, the vector representing the electromotive force impressed upon the line is likewise reversed. Thus, in Fig. 1, the electromotive force of the coil XY may be a vector XY; when it is connected in the opposite sense, the electromotive force of the coil YX is the vector YX. Fig. 7 shows these vectors for the secondary coils of Fig. 1. From these elementary principles can be shown the delta and the star equivalents of a T-connection.

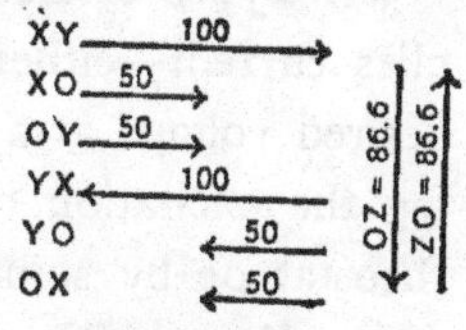

FIG. 7. Two senses in which vectors can be considered.

§ 19. *Delta Equivalent of T-connection.*—The electromotive forces between the terminals XYZ of Fig. 1 or Fig. 2, should be considered in a certain order, XYZ or ZYX. Let us consider them in the XYZ order, as shown in Fig. 8. Going from X to Y, we have the vector XY as shown. From Y to Z, we have the vectors YO and OZ (compare Fig. 7) which combine to give YZ. From Z to X, we have the vectors ZO and OX which combine to give ZX. The three resultant vectors are thus shown to be equal and to differ in phase by 120°. The line voltage, thus obtained by the T-connection, is accordingly the same as would be obtained by three 3-phase generator coils, XY, YZ, ZX, connected in delta.

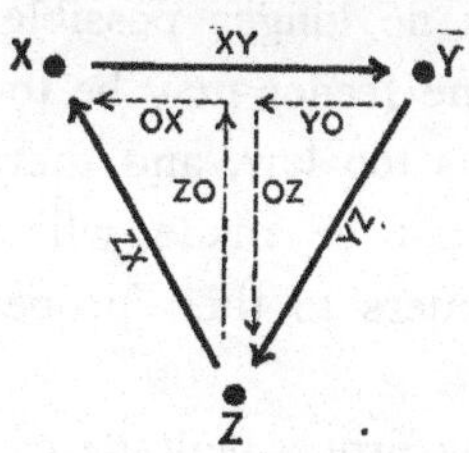

FIG. 8. Delta equivalent of T-connection.

§ 20. *Star Equivalent of T-connection.*—Suppose the neutral point N (either actual or imaginary) in the coil OZ divides its voltage into ⅓ and ⅔; thus, in Fig. 9, we have $ON = 28.9$ and $NZ = 57.7$, with arrows down; $NO = 28.9$, with arrow up as drawn. From the neutral N we have the vector NY, the resultant of NO and OY; and NX the resultant of NO and OX. It follows that NX, NY and NZ are each equal to 57.7 volts ($57.7 = 100 \div \sqrt{3}$) and differ in phase from each other by 120°. The line voltage, thus obtained by the T-connection, is the same as would be obtained by three 3-phase generator coils NX, NY, NZ, when star-connected.

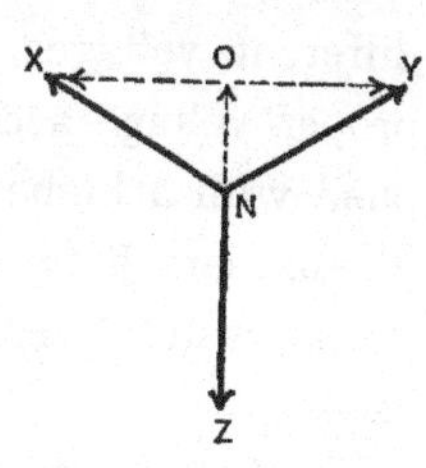

FIG. 9. Star equivalent of T-connection.

Experiment 7–B. Induction Regulators.*

§ 1. **Types of Potential Regulators.**—When a generator supplies current—either alternating or direct—to a single line, the desired voltage at a distant receiver can be maintained by varying the excitation of the generator, either by a hand-operated rheostat or by some automatic device as the Tirrell regulator (§ 3a, Exp. 1–B). This is also accomplished, to a certain extent, by a compound winding (§ 4, Exp. 1–B) or composite winding (§ 11a, Exp. 3–A) on the generator.

When, however, a generator (or several generators in parallel) supplies several lines or feeders, with independently varying loads, this simple method of regulation is no longer possible; for at any particular time the voltage on one feeder may be too high, while the voltage on another feeder is too low, and there is no change of station voltage which can be made which will bring the delivered voltages on all feeders to their proper values.

In direct current distribution systems, the proper voltage can be approximately maintained by inter-connecting the various feeders and proportioning the amounts of copper according to average load conditions. In large stations, a step further is taken by maintaining in the station several sets of bus bars at different voltages, so that feeders may be supplied with the proper voltage according to conditions, long feeders being supplied with a higher voltage than short ones. Use is also made of auxiliary batteries, motor-driven boosters, etc. In early stations, wasteful series resistances in each feeder were sometimes used.

§ 2. In an alternating current system, the most satisfactory results† are obtained by the use of a *potential regulator* in each

* Note that an induction motor with wound secondary can be used as an induction regulator; see § 4.

† (§ 2a). Series resistances and reactances have been used for this purpose. To use the former is not good practice on account of the energy

feeder. The potential regulator is a variable-ratio transformer or auto-transformer used to raise the voltage as a booster, or to lower the voltage as a negative booster. The regulator may be operated either manually, or automatically by means of a small motor which is controlled* by potential wires from any desired point in the system.

Alternating current potential regulators are of two types, the *step-by-step regulator* and the *induction regulator.*

§ 3. *The Step-by-step Regulator.*—The step-by-step regulator is an auto-transformer (or transformer) with switching arrangements for changing the number of turns. In principle it is the same as any auto-transformer (Exp. 5–A). In practice, the switching arrangement may consist of a number of individual knife switches, but more usually is either of a drum or a dial pattern, operated either manually or automatically. In the drum or dial pattern, resistance or reactance in the contact leads is sometimes used so that the contact arm can temporarily bridge two contact points without disastrous short circuit.

The step-by-step regulator is not easily made automatic. The contacts deteriorate, even when arcing tips are used, and hence this type of regulator is better for occasional than for constant adjustment. For continuous automatic adjustment, the induction regulator is generally† used.

§ 4. **The Induction Regulator.**—An induction regulator is a stationary transformer with a movable primary or secondary which may be set in different positions for obtaining potentials

wasted. Reactances are satisfactory for some cases; to be effective, however, they must be large and expensive.

* (§ 2b). The motor may be either direct or alternating and is usually controlled through a relay, one form of Tirrell regulator being made for this purpose.

† (§ 3a). In cases where *very rapid* continuous automatic adjustment is required, the induction type can not be used on account of the heavy mass to be moved. An automatic regulator of the step-by-step type is better for this rapid adjustment, because the moving part is only a light contact arm.

of different values or of different phase. A form* of apparatus in common use is essentially an induction motor with wound secondary brought out to terminals, and *any induction motor* so constructed can be used as an induction regulator.

Such an apparatus may be used:

1. As a single-phase potential regulator; used on lighting feeders.

2. As a phase shifter; used in laboratory testing.

3. As a polyphase potential regulator; used on polyphase lines, particularly in supplying current to synchronous converters.

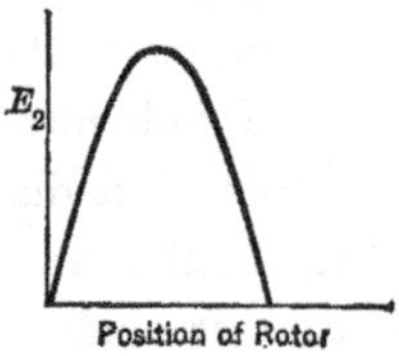

FIG. 1. Use as single-phase transformer; secondary voltage for different positions of rotor.

§5. (1) **Single-phase Potential Regulator.**—Supply the primary (or one phase of the primary if a polyphase induction motor is used) with a constant single-phase voltage not exceeding the normal voltage of the apparatus. The secondary voltage may be varied by turning the rotor by hand to any desired position, and current may be drawn up to the full-load rating of the secondary. The apparatus is used in two ways: (*a*) as a transformer with primary and secondary not connected together; (*b*) as an auto-transformer with the two coils connected, as in Fig. 2.

§6. (*a*) *Use as Transformer.*—Place a voltmeter across the open secondary and revolve the rotor step by step, so that the secondary potential changes between zero and a maximum. (On open circuit the data for methods (*a*) and (*b*) can be taken

* (§4a). An earlier form of regulator had stationary primary and secondary coils located at right angles, and a movable iron core which formed part of the magnetic circuit and permitted more or less of the primary flux to pass through the secondary. This device is sometimes referred to as a "magnetic shunt."

For a description of different forms of regulators, see: "Alternating Current Feeder Regulators," by W. S. Moody (a paper before the Toronto Section, A. I. E. E., February, 1908); "Alternating Current Potential Regulators," by G. R. Metcalfe, *Electric Journal,* August, 1908.

simultaneously.) Plot a curve, as Fig. 1, showing the secondary potential for various angular positions of the rotor. Note the ratio of maximum secondary to primary potential.

§ 7. (*b*) *Use as Auto-transformer.*—The second and commercially preferred method for use as a single-phase regulator is to supply the primary with single-phase constant voltage as before, and to connect the secondary in series with the load, as in Fig. 2,

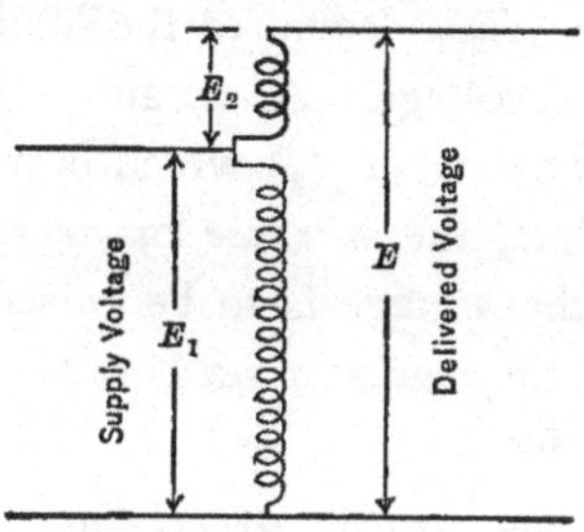

FIG. 2. Connections.

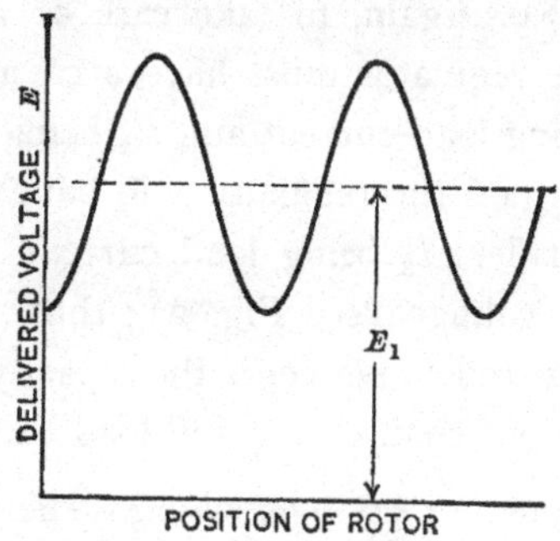

FIG. 3. Delivered voltage.

Single-phase potential regulator used as auto-transformer.

so that the delivered potential E taken from the machine, now acting as an auto-transformer, is equal to the primary potential E_1, either increased or decreased by the potential E_2 of the secondary. This is either additive or subtractive, the apparatus being a booster or a negative booster, according to the position of the rotor. In this manner, the potential may be varied between the limits of $E_1 + E_2$ and $E_1 - E_2$. Measure E_1, E_2 and E. Plot a curve, as in Fig. 3, showing the delivered potential for various positions of the rotor. Compare E with the algebraic sum of E_1 and E_2. What relation is there between the curves in Figs. 1 and 3?

§ 8. *A Comparison.*—In method (*a*), the output of the regulator is equal to the volt-ampere capacity of the secondary; in method (*b*), the output of the same regulator is much greater. Take, for example, a regulator with primary for 100 volts × 100 amperes, and secondary for 10 volts × 1000 amperes. In method (*a*), the secondary output is limited to 10 volts × 1000 amperes, or 10 kilowatts. In method (*b*), the potential may be varied from 90 to 110 volts, which with 1000 amperes gives an output of about 100 kilowatts. In practice, method (*b*) is therefore used; in the laboratory, method (*a*) is often convenient when the range of delivered voltage desired does not exceed E_2.

§ 9. Again, to take care of a certain load as a transformer, the regulator must have a capacity (E_2I_2) equal to the load, I_2 being load current and E_2 being load voltage. As an auto-transformer, the regulator will have a capacity (E_2I_2) which is much smaller, I_2 being load current and E_2 the *increase* or *decrease* of voltage (see Fig. 2); thus, if the voltage is to be raised or lowered 10 per cent., the capacity of the regulator needs to be only 10 per cent. of the full load of the feeder.

§ 10. *Further Experiments.*—The regulator may be tested under load, either inductive or non-inductive, as in Exp. 5–A; or its performance can be predetermined, as in Exps. 5–B and 5–C.

The air gap necessitates a larger magnetizing current than in a transformer with a closed magnetic circuit; and, on account of larger leakage reactance, gives poorer regulation and a smaller diameter to the circle diagram.

§ 11. *Tertiary Coil.*—As the secondary coil moves away from the influence of the primary and comes more nearly to the neutral position, it includes less and less of the primary flux; the secondary leakage flux now causes the secondary to act more and more as a choke-coil in series with the load, thus giving a low power factor. This has been overcome by a short-circuited *tertiary* coil, wound midway between the primary windings, so located that it is not cut by the primary flux. As the secondary moves away from the influence of the primary, the short-circuited winding comes into play, acting similarly to a short-circuited secondary on a transformer, so that the choking effect of the secondary or series coils becomes less and less, and is practically zero in the neutral position. (See citations, § 4a; also *Standard Handbook*, 6–158.) In a polyphase induction regulator no tertiary coil is needed (*Standard Handbook*, 6–161).

§ 12. (2) **Induction Regulator as Phase-shifter.**—Supply the primary with polyphase current at normal constant voltage. The secondary voltage will be found to be constant in value for all positions of the rotor (instead of varying as in the preceding tests), but to be of varying phase, having a definite phase position for each position of the rotor. This should be explained and demonstrated experimentally.

To do this, connect one primary circuit and one secondary circuit in series as before, measure E_1 and E_2 separately, and E the sum of the two for each position of the rotor, and construct triangles on a common base E_1, as in Fig. 4, which illustrates the varying phase* of E_2. Observe the relation between mechanical and electrical degrees, noting, for example, the mechanical angle through which the rotor is turned to shift the phase of E_2 by 45 electrical degrees.

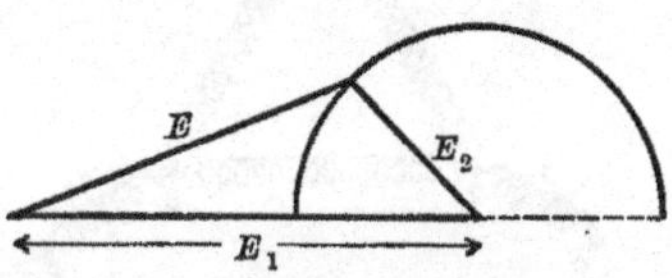

FIG. 4. Voltage relations as phase shifter.

Although of little commercial use, this method is extremely useful in the laboratory. If the primary supply is symmetrical and of constant voltage, the secondary voltage on open circuit will be constant and its phase angle will vary exactly with the position-angle of the rotor, which can be read with a suitable scale. A secondary load will, however, distort these conditions, so that the scale reading will not give the phase exactly.

The varying resultant potential E, in Fig. 4, shows that with polyphase supply the apparatus can also be used as a potential regulator, to be discussed in the next paragraph.

§ 13. (3) **Polyphase Potential Regulator.**—The primary is supplied, as in (2), with polyphase current at normal constant voltage. The secondary coils for each phase must be separate from each other, one secondary coil being connected in series with each delivery circuit. For a 3-phase regulator (or 3-phase motor used as a regulator) the connections are shown in Fig. 5. The supply circuit is connected to the terminals 1, 2, 3 of the primary—which may be star-connected, or delta-connected as

* This can also be shown by a phase-meter.

shown. The delivered currents are taken from *a, b, c.* The voltage relations are shown in Fig. 6, where *a', b', c'* gives the maximum delivered voltage. As the rotor is turned, this becomes *a'', b'', c''*, etc., until the minimum *a''', b''', c'''* is obtained.

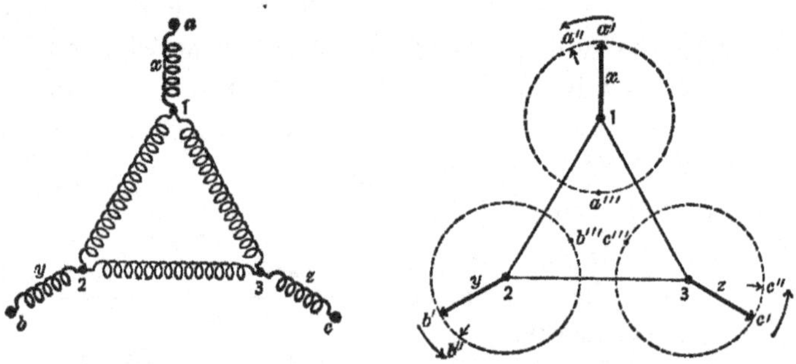

FIG. 5. Connections. FIG. 6. Voltage relations.
Polyphase potential regulator: supply voltage, 1, 2, 3; secondary coils, *x, y, z,* in series with load; delivered voltage, *abc.*

With a voltmeter, show that the secondary voltages *x, y* and *z* do not change in value with a change in position of the rotor; also show that the three delivered voltages, *ab, bc, ca,* are substantially equal for any one position of the rotor. (That *x, y, z* do not change and *ab, bc, ca* change simultaneously, as the rotor changes, is well shown by incandescent lamps.)

Measure one delivered voltage, as *ab,* for different rotor positions, noting particularly the positions for maximum and minimum values, and plot a curve, as in Fig. 3.

Construct a diagram to scale, as in Fig. 6, making the triangle 1, 2, 3 equal to the primary voltage; the circles have radii equal to the secondary voltages, *x, y* and *z.* From this diagram pick off values of delivered voltage, *a'b', a''b''*, etc., for different rotor positions and plot these values as a second curve, to be compared with the first curve already plotted from measurement. The limiting values of delivered voltage are shown to be $E_1 \pm 2E_2 \cos 30^\circ$.

CHAPTER VIII.

INDUCTION MOTORS.

EXPERIMENT 8–A. Preliminary Study of an Induction Motor and the Determination of its Performance by Loading.

PART I. INTRODUCTORY.

§ 1. **Use.**—The induction motor is the form of alternating current motor in most general use. It is practically a constant speed motor, the speed at full load being a few per cent. less than the speed at no load, and in this respect it resembles the shunt motor. On account of its simplicity and nearly constant speed, it is well adapted for operating machinery, but is not so well adapted for use in traction,* or similar service, where there is frequent acceleration. (For variable speed motors, see § 59.) Induction motors may be single-phase or polyphase, large motors being generally 3-phase or 2-phase.

§ 2. **Structure.**—An induction motor consists of two members, —a stationary member, or *stator,* and a revolving member, or *rotor,*—corresponding in a way to the field and armature of a direct current motor. (The terms field and armature should not be used, however, in referring to an induction motor, since they are inexact and confusing.)

* (§ 1a). For traction purposes the induction motor is best suited to service permitting long runs at uniform speed. The first such use in America, at the Cascade Tunnel of the Great Northern Railway Company, is described in a paper by C. T. Hutchinson read before A. I. E. E., Nov. 12, 1909 (see *Transactions,* Vol. XXVIII., p. 1281), in which the following advantages of the 3-phase induction motor are stated: simplicity, greater output for given space, uniform torque, constant speed, impossibility of excessive speed (either on down grade or when the wheels slip), regeneration on down grades, impossibility of excessive current, possibility of using 25 cycles. (See also discussion following the paper.) In Europe the 3-phase induction motor for traction is not uncommon.

Each of these members is built of laminated iron and has on it a winding disposed in slots. These windings are termed the *primary* and the *secondary*, for the induction motor is a form of transformer. The primary receives current directly from the line, while the secondary is *short-circuited* upon itself and has no electrical connection with the line or with the primary, the current in it being set up by *induction*, as in a transformer.

Usually, the primary winding is on the stator and the secondary is on the rotor. This gives the simplest form of motor; the primary can be connected directly to the supply line and there is no need of brushes or slip rings. Sometimes, however, the secondary is on the stator and the primary is wound on the rotor, the current from the line being in this case introduced through slip rings. The principle of operation is the same in the two cases.

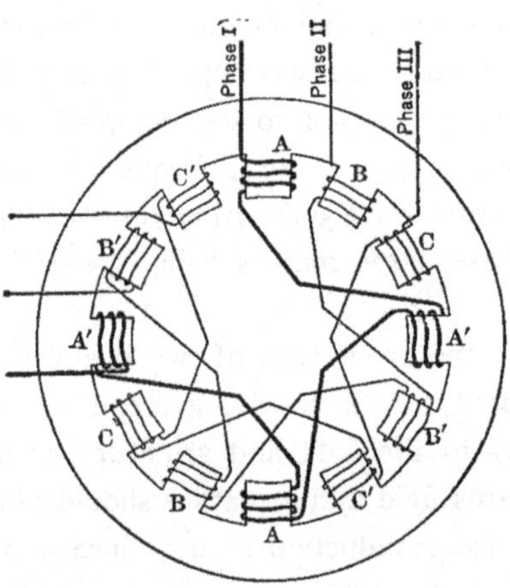

FIG. 1. Stator of a 3-phase motor with 4 poles per phase,—often termed a 4-pole motor. (This diagram is illustrative and does not represent construction.)

§ 3. **Rotating Field.**—The operation of an induction motor will be most readily understood by a consideration of its *rotating magnetic field*. Let us consider a polyphase* motor in which the primary is stationary.†

Fig. 1 illustrates the primary or stator of a 3-phase motor in which there are four poles per phase. When the current in Phase I. is a positive maximum, *AA* form north poles and *A′A′*

* (§ 3a). For a single-phase motor, see § 56.

† (§ 3b). In this case, with the primary stationary, the idea of the rotating field is most simply seen. The *relation* between primary and secondary, however, and the torque produced, will be the same, irrespective of which winding is stationary and which is moving.

form south poles. A little later (after 1/6 cycle) the current in Phase II. becomes a maximum, so that BB form north poles and $B'B'$ south poles. (A and C are now also north but weaker, the maximum field being under B.) Later (after 2/6 cycle) the current in Phase III. becomes a maximum and north poles are formed under CC. Each pole is accordingly seen to progress and to be successively under A, B, C, A', B', C', etc. The primary or stator winding thus produces a revolving or rotating field which tends to drag the rotor around with it, § 50. With the usual distributed winding, the field is uniform and revolves with uniform speed.

§ 4. In a 2-pole model (having two poles per phase) the field makes one revolution in one cycle; in a 4-pole model, as Fig. 1, one revolution in two cycles; in a 6-pole model, one revolution in three cycles, etc.

It is seen that, if n is the frequency in cycles per second and p is the number of pairs of poles (per phase), the rotating field makes $n \div p$ revolutions per second. This is known as the *synchronous speed* of the motor; compare § 1, Exp. 3–A.

§ 5. In revolutions per minute,

$$\text{Synchronous speed} = 60\,n \div p;$$

$$\text{Pairs of poles per phase} = 60\,n \div \text{synchronous speed.}$$

§ 6. **Speed and Slip.**—The rotor of an induction motor revolves at a little less than synchronous speed; for, at synchronous speed it would revolve at the same speed as the magnetic field, in which case there would be no cutting of lines of force, no secondary current and hence no torque.

The actual speed of an induction motor is, therefore, less than synchronous speed by a few per cent. called the *per cent. slip*. The slip increases with the load, thus increasing the cutting of lines of force, the current in both secondary and primary and the torque.

§ 7. **Primary Winding.**—An induction motor is not commonly constructed with separate poles, as in Fig. 1. The primary windings are usually distributed in slots and are so arranged in groups that a series of poles are produced which correspond to those shown in Fig. 1. There are a number of slots for each pole. A lap or a wave winding (see § 3b, Exp. 1–A) can be used, according to various winding schemes, as described in text-books.

✓ § 8. **Squirrel Cage Secondary.**—The secondary winding usually consists of a squirrel cage, made up of parallel copper bars set in slots with ends connected by two short-circuiting rings. Such a construction is strong and simple; it makes possible a low secondary resistance which gives high efficiency and good speed regulation but low starting torque. The squirrel cage secondary is self-contained and has no outside connection.

✓ § 9. **Phase-wound Secondary.**—To obtain a higher secondary resistance and hence a greater starting torque,* the secondary is sometimes *wound,* or *phase-wound* as it is called. Extra secondary resistance, either internal or external to the motor, may be included in the circuit on starting to increase the starting torque, this resistance being cut out as the motor speeds up so as to give higher efficiency and better speed regulation while run-

* (§ 9a). *Maximum Torque.*—Torque is proportional to secondary electrical input, $E_2I_2 \cos \theta_2$, see § 54. Starting torque is proportional to secondary input at standstill and is a maximum when $R_2 = X_2$, for I_2 then lags 45° behind E_2 and has a maximum power component in phase with E_2. (In any constant potential circuit with constant reactance, current has a maximum power component when the lag angle is 45°, as can be seen in Fig. 2, Exp. 4–B.) Increasing R_2 beyond a certain amount will decrease the starting torque.

When running with a slip s, the secondary reactance becomes sX_2 and the torque is a maximum when $R_2 = sX_2$. The secondary resistance can be given such a value as will give the motor its maximum torque at any desired slip, as for example at standstill when $s = 1.00$.

The maximum value which the torque can have, irrespective of speed, is independent of R_2, being dependent solely upon the input $E \div 2X$, except for the small effect of primary losses; see §§ 12a, 22, Exp. 8-B. Changing R_2 can not alter this maximum, but can cause it to occur at any desired speed.

ning. (High resistance in a squirrel cage armature is sometimes obtained by making the short-circuiting rings with high resistance.)

§ 10. The secondary resistance, besides increasing the torque, acts as a starting resistance (§ 14) and serves to reduce the starting current taken from the line, in the same manner as the starting resistance of a direct current motor.

The secondary resistance may also be used to vary the speed of an induction motor when running, but this is inefficient, as is the case when the speed of a shunt motor is varied by means of a resistance in series with the armature. See § 60.

§ 11. **Starting Polyphase Motors.**—Polyphase induction motors are self-starting and require no special starting devices in order to obtain a starting torque. There is, however, objection to starting a motor by connecting it directly to the line on account of the excessive current which would thus be taken. Small motors—say under 5 horse-power—may be started without load in this manner, inasmuch as such a motor forms but a small part of the total load of a system and cannot seriously affect its regulation. Furthermore, a small motor comes up to speed so quickly that the momentary excessive current does not overheat the motor. All large motors, however, and small motors when loaded, require some starting device to limit the starting current.

§ 12. *Starting Compensator or Auto-transformer.*—The most common method of starting is to connect the motor to a low voltage for starting and to throw it on to full line-voltage after it has reached, say, half speed. The low starting-voltage is obtained by means of auto-transformers connected across the line and provided with suitable intermediate taps, as in Fig. 2, Exp. 5-A. On a 2-phase circuit, two auto-transformers are used,—one on each phase. On a 3-phase circuit, two auto-transformers are also used, these being *V*-connected, as in Fig. 6, Exp. 7-A. The same starting box is, accordingly, suitable for either a 2-phase or a 3-phase motor. To save losses, the auto-transformers are cut out when the motor is running. The motor starters are often arranged so as to throw the motor step-by-step on successively higher voltages, thus avoiding too sudden acceleration and rush of current at any one step.

§ 13. Any auto-transformer serves to step up the current at the same time that it steps down the voltage and so requires less current from the line; thus, if at ⅓ voltage* a motor takes a starting current of 90 amperes, the current drawn from the line is only 30 amperes.† The auto-transformer, therefore, not only reduces the starting current taken by the motor itself on account of the reduced voltage but makes a further proportional reduction in the line current. It serves admirably as a motor starter for all cases except those in which a large starting torque is necessary.

§ 14. *Secondary Starting Resistance.*—When a large starting torque is necessary, it is best obtained by using a phase-wound secondary with additional resistance for starting; see § 9a. As the motor speeds up, this resistance is cut out, either all at once or gradually so as to control the acceleration and current of the motor. In motors with revolving secondaries, this extra resistance may be contained within the rotor and controlled by a lever bearing against a sliding collar, or it may be external to the motor with leading-in wires connected through slip rings.

§ 15. Some motors are provided with a centrifugal device for cutting out this resistance automatically when a certain speed is reached.

§ 16. **Starting Single Phase Motors.**—When supplied with single-phase current an induction motor has no starting torque, although it will run satisfactorily when once started (§ 56). In some cases small motors may be started by hand, but in gen-

* (§ 13a). At full voltage the current would be 270 amperes.

† (§ 13b). If a series resistance were used to reduce the voltage to 1/3, the current drawn from the line would be 90 amperes.

eral some special provision for starting a single-phase motor must be made. Whatever means are used for starting, the starting torque is small and it is preferable to start the motor without load. The load may be assumed after starting, by means of a loose pulley or clutch.

§ 17. *Shading Coils.*—One method for making small single-phase motors self-starting depends upon the use of shading coils. Each magnetic pole of the field of such a motor is divided into two parts. Around the leading portion of each pole is wound a short-circuit coil of low resistance, called a shading coil. When the flux is changing in any pole, its increase or decrease in the leading part is retarded by the short-circuited coil. In consequence of this action, the leading part of each pole attains its maximum magnetization *after* the other part, so that a revolving field is produced and the rotor is drawn around as in a polyphase motor.

§ 18. *Repulsion Motor.*—A single-phase motor is often made to start as a repulsion motor and is converted into an induction motor when nearly full speed is attained. Such a motor has a wound rotor provided with a commutator and brushes, as the armature of a direct-current generator or motor. The brushes are connected together by a heavy conductor. In a 2-pole model, the brushes are opposite each other.

In a 2-pole model, suppose the plane of the brushes (or rather the plane of the coils to which they are connected through the commutator) to be in the center line of the poles. There would be a large flow of current through the rotor windings and the brushes, but from symmetry there would be no torque; half the conductors would tend to turn in one direction and half in the other. If the plane of the brushes were turned at right angles, there would be no current flowing and therefore no torque. If, however, the brushes are set obliquely, a current will flow and there will be a resultant torque, for the current in the conductors under opposite poles will flow in opposite directions so that the conductors under both poles will tend to give rotation in the same direction.

After a certain speed is reached, a centrifugal device is commonly used to lift the brushes, thus saving friction and wear, and to short-circuit the rotor windings so that the motor runs as an induction motor. It then has the characteristics of an induction motor and not the characteristics of a repulsion motor, which are similar to those of a series motor.

§ 19. *Phase Splitters.*—A single-phase motor may be made to start as a 2-phase or 3-phase motor by means of polyphase currents temporarily derived from a single-phase circuit. Polyphase currents sufficient for this purpose can be obtained from a single-phase line by various arrangements of resistances and reactances; such devices, termed *phase-splitters,* are in common use and will be discussed later (§§ 26–32). Polyphase currents so derived can never be *balanced* (§§ 1–3, Exp. 7–A) and are only used for starting.

PART II. PRELIMINARY STUDY.

§ 20. **Structure and Rating.**—Study the general structure and windings* of the motor; note the rated full-load speed and output and the frequency, voltage and kind of circuit (single-phase, 2-phase 3-wire, 2-phase 4-wire, 3-phase, etc.) for which the motor is designed.

§ 21. Compute the watts input and the current per line at full load, assuming a certain efficiency (say 80 per cent.) and a certain power factor (say 80 per cent.). For a polyphase motor, com-

* (§ 20a). A detailed study of the windings may be made when the necessary data are obtainable. The study may include a diagram of winding, and the following data:—number of poles or groups of coils per phase; number of coils per group and turns per coil; size wire; current density for full-load current; total resistance; length of wire (per coil and total) computed from dimensions and from resistance measurement; type of secondary; number and size of conductors; number of slots, etc.

pare §§ 20, 21, Exp. 6-A. From the speed and frequency, compute the number of poles per phase, § 5.

Note the manner in which the motor is to be connected to the supply circuit and any special provision there may be for starting.

§ 22. **Polyphase Motor.**—Connect in one line circuit an ammeter with a range, say, 50 or 100 per cent. in excess of the full-load current. Start the motor in the regular manner, without load, and note the starting current and the current when the motor is running at full speed. Note the change in current as the motor gains speed and as the motor is changed over from the starting to the normal running conditions. Repeat the test with the motor belted to some load (belted, for example, to a generator) and note the current taken to start the motor and note that more time is required to attain full speed.

A 3-phase motor can be run on any 3-phase circuit, of proper frequency and voltage, irrespective of whether the circuit is star, delta, *T* or *V* connected.

A 2-phase motor with two independent circuits can likewise be operated on any 2-phase circuit. If, however, a 2-phase motor has its circuits brought out to only three terminals, as (*b*) in Fig. 1, Exp. 6-A, it can only be operated on a 3-wire circuit and cannot be operated on a quarter-phase system (as *c* or *d*). Conversely, if the motor is connected as *c* or *d*, it cannot be operated on a 3-wire circuit, as *b*.

§ 23. *Polyphase Motor Started with Secondary Resistance.*—If a secondary starting resistance is provided, start the motor with this resistance in circuit and cut it out, either gradually or in one step, as the motor attains full speed. Note that the primary current is increased by cutting out the starting resistance. Unless specially designed for continuous operation, this resistance will overheat if kept long in circuit.

If a half-voltage supply is available, start the motor, with no load, without the starting resistance, noting that the starting current is much greater than with the resistance in circuit. (This may be done at full voltage, if the motor is not too large.) Without the starting resistance, the starting torque is less; this can be shown* by means of scales and a lever arm fastened securely to the rotor, which is now stationary.

§ 24. *Polyphase Motor Started with a Compensator.*—If a starting compensator is provided, start the motor at the low starting voltage and throw over to full running voltage after the motor speeds up. The initial current would be too great to allow starting at full voltage, except in case of small motors without load. When started with a load, the current is greater and the acceleration less than when started without load.

§ 25. *Polyphase Motor Running on Single Phase.*—After the motor is in operation as a polyphase motor, disconnect all primary lines except the lines of one phase† and note that the motor will continue to run as a single-phase motor. While thus running, measure the various terminal voltages and note that the machine is also acting as a polyphase generator. In a 2-phase machine the voltage of the idle circuit will be found to be in quadrature with the voltage of the motor circuit which is connected to the line, showing that the machine is capable of delivering true 2-phase currents.

In a 3-phase machine, the three measured voltages will form a triangle. In this manner polyphase currents can be obtained from a single-phase system (§ 3, Exp. 7–A). (A polyphase motor may be *started,* as well as run, on a single-phase circuit by the methods shown in Figs. 2 and 3, discussed later.)

§ 26. **Single Phase Motor with Phase Splitter.**—To start with

* When an accurate measurement of starting torque is desired, it should be made with a Prony brake so adjusted that the motor just turns at a low speed, thus avoiding the error due to friction at standstill.

† (§ 25a). If an ammeter, voltmeter and wattmeter are connected in one circuit, it is instructive to take readings with the motor running single phase and polyphase and to compare the values of power factor, power component of current and quadrature component of current.

a phase-splitter, an induction motor must be wound as a 2-phase or 3-phase motor.

§ 27. *Starting as a Two Phase Motor.*—The stator is wound with two circuits, a main circuit *A* and an auxiliary starting circuit *B*, in series with which is a resistance or phase-splitter, *R*. The circuit *A* has high reactance and low resistance, while *B* has low reactance and high resistance. The starting circuit *B* is in parallel with *A*, as in Fig. 2, and is opened when the motor reaches about half speed.

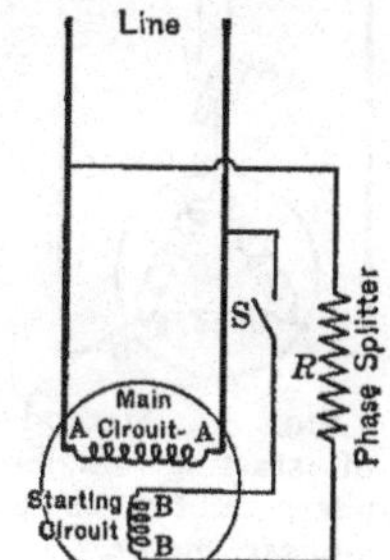

FIG. 2. Method of starting single-phase motor as a 2-phase motor.

§ 28. With the motor at rest, measure (at half voltage) the currents in *A*, in *B*, and in the line circuit. Plot the readings as a triangle to show the phase difference between branch currents. Note the effect of reversing the line terminals. In what way may the direction of rotation be reversed?

Connect the motor in circuit with the starting circuit open. Note that the motor remains at rest, but will start slowly in either direction when started by hand.

§ 29. *Starting as a Three Phase Motor.*—The motor is wound as a 3-phase motor. The starting device,* or phase splitter, con-

* (§ 29a). *Use of Condenser as Starting Reactance.*—The reactance *X*, Fig. 3, may be an inductance or a condenser. To reduce the size of the condenser, it may be connected through a transformer, the low voltage side of which is connected in the circuit; the high voltage side is connected to the condenser. The connections for an auto-transformer are shown in Fig. 4. The voltage is increased and the current is decreased as the ratio of transformation; the necessary reactance ($X = E \div I$) is thus decreased as the square of the ratio of transformation. A ratio of transformation of 1:3, as in Fig. 3, reduces the size of condenser to 1/9. See §16a, Exp. 5-B. Aside from the present application, this is a useful laboratory device for obtaining large capacity from small condensers. *The limit is the voltage that the condenser will stand.*

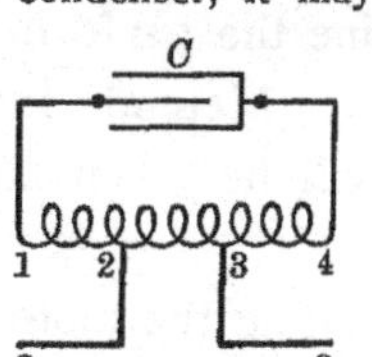

FIG. 3. Use of condenser with auto-transformer.

sists of a resistance R and a reactance X connected in series across the line, as in Fig. 4. These are thrown out of circuit by opening the switch S, when the motor reaches about half speed.

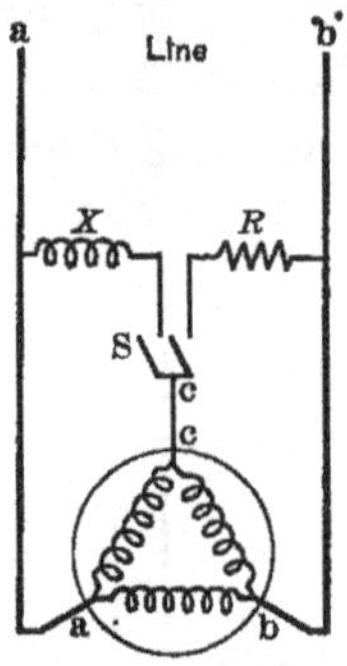

FIG. 4. Method of starting single-phase motor as a 3-phase motor.

§ 30. With R and X in circuit and the motor disconnected, measure the drop around R, the drop around X and the total line voltage. Plot these readings as a triangle to show how nearly a symmetrical 3-phase voltage is obtained.

§ 31. When starting the motor, take readings of the current in the third or starting circuit, c; also of the currents in the main and motor circuits.

With the motor running single phase, take readings of the voltage between each two of the three terminals of the motor and plot as a triangle.

§ 32. Note that the motor may be started with the X circuit open, by means of a single starting resistance between either line and c. Note also that any two terminals of the motor may be taken as the main terminal.

Note the manner in which the direction of rotation can be reversed.

PART III. LOAD TESTS.

§ 33. The purpose of these tests is to determine the performance of the motor in actual operation under load. Methods for predetermining the performance without load will be taken up in Exp. 8–B.

§ 34. Any available means may be used for loading the motor and for determining its mechanical output. A Prony brake may be placed on the motor pulley and the power computed from the torque measured by platform scales or spring balance. An objection to the Prony brake is the difficulty in holding the load constant while readings are taken; furthermore, except in the case of small motors, there is difficulty in disposing of the heat. In place of a Prony brake any form of transmission or absorption dynamometer can be used (see § 1a, Exp. 2–B).

A convenient method for conducting the test is to use a direct current generator for a load, and this method will be described in detail.

§ 35. **Load Run with Direct Current Generator as a Load.**—To load the motor, belt it to a shunt generator, preferably with constant separate excitation. The load of the motor is varied by varying the load of the generator and this is done by means of external load resistances. The motor input is measured electrically. The motor output is equal to the electrical output (EI) of the generator *plus* the generator and belt losses, which are separately determined as described later.

§ 36. **Measurements on Motor.**—For each load, read line voltage, line current,* watts input, frequency (unless this is constant and known) and speed; the slip should be measured directly when means are available. The line voltage should, if possible, be kept constant.

§ 37. For a single-phase motor, the electrical instruments to be read consist of one voltmeter, one ammeter and one wattmeter.

§ 38. For a polyphase motor, an ammeter and voltmeter may be used on each phase; or, one ammeter and one voltmeter may be shifted from phase to phase by means of suitable switches. Power and power factor† can be determined by any of the wattmeter methods described in Exp. 6–B. The 2-wattmeter method shown in Fig. 1, Exp. 6–B, is the best one to use in testing a 3-phase motor, or a 3-wire 2-phase motor.

§ 39. The results, on a polyphase motor, may be worked up in terms of (1) line current and line voltage, as measured; (2)

* If the secondary is phase-wound, read also secondary current.

† For a 2-phase motor, determine power factor by both the cosine and the tangent methods, §§ 11, 12, Exp. 6–B.

equivalent single-phase current and line voltage; or, (3) current per phase and voltage per phase. See §§ 28–30, Exp. 6–A.

§ 40. **Measurements on Generator.**—Measure terminal voltage, armature current and speed. Also measure field current, which is to be kept constant* during the run by adjusting the field rheostat. A strong field is desirable so as to minimize the effect of armature reaction on heavy load, which will change the core loss and so introduce error. The generator may be self excited when separate excitation from a constant source is not available; see § 40a. When the generator is self excited, the armature current is equal to the external current *plus* field current.

§ 41. **Readings.**—Readings should be taken as follows:

(1) Take readings with belt off. The watts input gives the no-load losses of the motor; these losses include the rotation losses W_0 (iron loss, friction and windage) *plus* the no-load copper losses. (Should it be desired to determine W_0 alone, the no-load copper losses are computed (§ 44) and deducted.)

(2) Take readings with belt on, the generator being without excitation. The increase† in the watts input of the motor gives the belt loss *plus* friction and windage of the generator.

(3) Take readings with the generator excited to the constant excitation (or constant terminal voltage) used throughout the test. (The difference between (3) and (2) would give the iron losses of the generator, a value, however, which is not used by itself in this test.)

§ 42. The difference between (3) and (1) gives the total rotation losses W_0 of the generator (iron loss, friction and windage),

* (§ 40a). The test may also be conducted by varying the field current for each load so that the terminal voltage of the generator is constant, § 42a. In some ways this is simpler, particularly with a self-excited generator, but is not quite as accurate.

† (§ 41a). This assumes that the motor losses remain constant, which will be practically true for such a small increase in the motor load. Should the increase in the copper losses of the motor seem to be appreciable, an allowance for it may be made.

including belt loss, for the particular speed and excitation. These losses may be taken as practically proportional to speed, for small variations in speed when the excitation is constant.* (Compare Fig. 2, Exp. 2-B.)

(4) Increase the load and take readings for a number of different loads up to say 25 per cent. overload, or until the motor comes to a standstill.†

§ 43. **Resistance Measurements on Motor.**—The resistance of the motor primary is found by the direct current fall of potential method, § 17, Exp. 1-A. This should be taken after the run is made and the motor is heated up. For commercial tests, standard specified conditions in regard to temperature should be followed; see A. I. E. E. Standardization Rules.

For a single-phase motor the resistance measurement is made as in any single-phase circuit.

For a 2-phase motor each phase is measured in this same manner.

For a 3-phase motor the resistance is measured between any two terminals. (For accuracy it is measured between each pair of terminals and averaged.) One-half the resistance thus measured gives the equivalent single-phase resistance, see § 29, Exp. 6-A, irrespective of whether the connection is star or delta. For a star connection, this is the resistance of one phase between one line terminal and the neutral. For a delta connection, it has no physical existence.

*(§ 42a). When the test is made by varying the field excitation so that the terminal voltage is constant, these losses may be taken as practically constant for different speeds through the small range of speed used in the test; the decrease in losses due to decrease in speed is roughly compensated for by the increase in losses due to the increase in excitation. This makes the test simpler but less accurate.

† (§ 42b). As motors are usually designed, the excessive current prohibits loading to standstill at full voltage, although it may be done at a lower voltage—say half voltage.

§ 44. The primary copper loss is computed* from the measurements of current and resistance.

§ 45. **Resistance Measurements on Generator.**—Immediately after the run, measure the resistance of the generator armature, including brushes and leads up to the voltmeter terminals.

§ 46. **Calculation of Motor Output.**—The output of the motor is equal to the generator output EI, *plus* the generator and belt losses. The losses are found as follows:

The armature copper losses RI^2 of the generator are calculated for the particular armature current, the resistance being determined as in § 45.

The field copper loss does not enter into the computations when the generator is separately excited. (When the generator is self-excited, the loss in the field circuit and in the field rheostat is determined by the product of field current and terminal voltage.)

The belt loss and the rotation losses of the generator are determined as in § 41, or by a separate test made as in Exp. 2–B (see particularly §§ 21, 24, Exp. 2–B). The method of § 41 is, how-

* (§ 44a). In a 3-phase motor, for example, in which the measured line current is 10 amperes and the measured resistance R between terminals is 1 ohm, the copper loss is computed from the equivalent single-phase current and resistance as follows:—

Equivalent single-phase resistance $R' = .5$ ohm.
Equivalent single-phase current $I' = 10\sqrt{3} = 17.3$ amperes.
Total copper loss $= R'I'^2 = .5 \times 300 = 150$ watts.

The computations may also be made by assuming the motor to be either star or delta connected, as follows.

Assuming a star connection, the resistance per phase is .5 ohm and the current per phase is 10 amperes. The RI^2 loss per phase is accordingly, 50 watts; the total loss is 150 watts as before.

Assuming a delta connection, the current per phase is $10 \div \sqrt{3} = 5.77$ amperes; the resistance per phase is 1.5 ohms. The RI^2 loss per phase is 50 watts and the total loss is 150 watts as above. This numerical example shows that the same result is obtained by the three methods of computation; it is unnecessary to know whether the primary is delta or star connected.

ever, more satisfactory on account of its convenience and the fact that belt loss is properly included in the results.

§ 47. **Curves.**—With power output of the motor as abscissæ, plot curves showing power input, voltage (if variable), primary current,* speed in per cent. of synchronous speed, efficiency, primary power factor and torque† in pounds at one foot radius or in synchronous watts. Plot also the primary copper losses, § 44. (Compare Fig. 3, Exp. 8–B.)

§ 48. Apparent power input of the motor is equal to the real power input divided by power factor. Apparent efficiency is power output divided by apparent power input; or it is equal to the real efficiency multiplied by power factor. These quantities may be computed when desired.

§ 49. **Circle Diagram.**—It is instructive likewise to compute the in-phase or power component of the primary current and the quadrature or wattless component of the current. The power component divided by the total current is equal to the power factor.

These results may best be shown as in Fig. 5, in which values of quadrature current are plotted as abscissæ and values of power component are plotted as ordinates. The curve obtained in this way forms the arc of a circle. The circle diagram for an induction motor, usually predetermined as in Exp. 8–B, is thus found by actual load test.

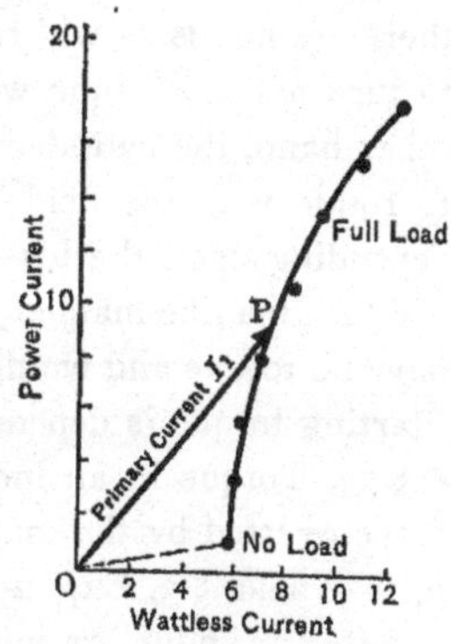

FIG. 5. Change in primary current from no load to full load; the curve is the arc of a circle.

The increase of wattless current with the load is due to leakage reactance, for it can be shown that the diameter of the circle locus

* Plot also the secondary current, if measured, and secondary copper loss.

† Torque is calculated from power output and speed, as in §3b, Exp. 2–A. In a brake test, torque is read directly.

is $E \div X$, as in a transformer, where X is the total leakage reactance of both primary and secondary; as X becomes smaller, the diameter becomes larger, the curve shown in Fig. 5 becoming more nearly a straight line.

APPENDIX I.

MISCELLANEOUS NOTES.

§ 50. **Operation of a Polyphase Motor.**—If a cylinder of glass, wood or other material in which there are no electric or magnetic losses is rotated in a stationary magnetic field, no torque is required—friction or mechanical load being neglected. If, however, the cylinder is constructed of a material as copper, in which there are losses due to the currents induced by the cutting of magnetic flux, it will require a torque to turn the cylinder depending upon the losses.

§ 51. Similarly, in a rotating magnetic field, the cylinder in which there are no losses will remain at rest; there will be no tendency for it to turn and no torque will be required to hold it stationary. On the other hand, the cylinder of copper in which there are losses will tend to rotate with the field; to hold it stationary would require a torque depending upon the losses.

§ 52. In a like manner, it may be seen that an induction motor would have no torque and would not start if there were no losses in the rotor. Starting torque is dependent upon rotor losses.

§ 53. Torque in an induction motor, as in any motor, is due to the force exerted by the magnetic field upon conductors carrying current in that field, § 3, Exp. 2-A.

After an induction motor starts, it continues to acquire speed until, when there is no load, nearly synchronous speed is reached. If the rotor did revolve at synchronous speed—the same speed as that of the rotating magnetic field—there would be no cutting of flux and hence no induced current in the secondary and no torque. The motor would then slow down and the cutting of flux would increase until the secondary current is sufficient to produce enough torque to maintain rotation.

§ 54. *Torque, Slip and Secondary Losses.*—When an induction motor is loaded, it slows down according to the load, in order to have the necessary increased torque; it has a greater slip, greater secondary current and greater secondary loss. The secondary input is $E_2I_2 \cos\theta_2$. A certain part of this, $R_2I_2^2$, supplies the secondary loss; the remainder is available for useful torque or mechanical output.

It can be shown * that slip depends directly upon the secondary copper loss; *per cent. slip is equal to secondary copper loss* $R_2I_2^2$, *divided by secondary input* $E_2I_2 \cos\theta_2$. Expressed in another way: the actual speed of an induction motor is the same percentage less than the synchronous speed as the useful mechanical output is less than the secondary input. The difference between the synchronous and the actual speeds is the slip; the difference between the secondary input and the mechanical output is the secondary loss.

Hence, since in general torque is equal to mechanical output divided by actual speed (§ 3b, Exp. 2–A), it follows that *the torque of an induction motor is equal to secondary input divided by synchronous speed.* Since synchronous speed is constant, torque is directly proportional to secondary input.

§ 55. A comparison between a shunt motor and an induction motor is of interest. In the rotor or armature of each there is an impressed electromotive force E_2, whether at standstill or rotating. At standstill this is all available for causing current to flow through the impedance of the windings and supplying RI^2 losses. When running, a counter-electromotive force is generated in proportion to the useful work or mechanical output; the electromotive force sE_2 which is available for overcoming the impedance and supplying losses is the *difference* between the impressed and counter-electromotive forces. For any given load, the rotor input is proportional to E_2 and the rotor losses are proportional to sE_2, or

$$\text{losses} \div \text{input} = sE_2 \div E_2 = s = \text{slip};$$

that is, slip is equal to rotor losses divided by input, as already shown. This is equally true for a shunt motor or for an induction motor.

---

*(§ 54a). This is shown as follows. For a slip s, the secondary current is

$$I_2 = \frac{sE_2}{\sqrt{R_2^2 + s^2X_2^2}} = \frac{sE_2 \cos\theta_2}{R_2}.$$

Hence, $R_2I_2^2 = sE_2I_2 \cos\theta_2$; or,

$$s = R_2I_2^2 \div E_2I_2 \cos\theta_2.$$

§ 56. **Operation of a Single-Phase Motor.**—At standstill a single-phase induction motor has no torque. Currents are induced in the secondary conductors, but from symmetry one half the conductors tend to give rotation in one direction and the other half, in the opposite direction. In a single-phase motor there is no rotating field at standstill. When running, however, there is a rotating field due to the combination of the main or field flux (which, referring to a 2-pole model, we will call vertical) and the cross or rotor flux (which we will call horizontal). On account of this rotating field the rotor is dragged around in the same manner as in a polyphase motor. The running of a single-phase motor becomes clear, therefore, when the production of the rotating field is understood.

§ 57. The main or vertical flux in a single-phase motor is set up directly by the field. When the motor is running, the conductors on the rotor under the field poles cut this field flux so that an electromotive force is generated in the rotor and this is a maximum when the field flux is a maximum. The rotor current which is caused to flow by this electromotive force sets up a flux in a horizontal direction.

Since this rotor flux must cause a counter-electromotive force proportional to its time rate of change to balance the impressed or generated electromotive force, it must come to its maximum a quarter of a period after the electromotive force generated in the rotor and therefore a quarter of a period after the field or vertical flux. The vertical flux and horizontal flux are out of phase with each other, as though set up by a 2-phase current; one is a maximum when the other is zero. They therefore combine to set up a rotating field.

§ 58. At synchronous speed the vertical and horizontal fluxes are equal,* thus producing a uniform or circular magnetic field. As the slip increases, the generated electromotive force and the horizontal flux become less and the field is elliptical; at zero speed the ellipse becomes a straight line.

§ 59. **Variable Speed and Multispeed Motors.**—It has been pointed out that the induction motor is nearly a constant speed motor and that the falling off of speed with load or slip is due to the secondary resist-

* (§ 58a). The generated electromotive force is proportional to speed times main flux; the counter electromotive force that balances this is proportional to frequency times rotor flux. When speed equals frequency, the two fluxes are equal, various constants being equal.

ance loss. Reducing the value of R_2 makes the speed more nearly constant. There is no generally satisfactory way for varying the speed; the following methods,* however, may be used.

§ 60. *Rheostatic Control.*—A variable speed may be obtained by varying the secondary resistance. As already stated, this method is inefficient, the operation being the same as that of a shunt motor with a resistance in series with the armature. When there is much reduction in speed, the speed will vary greatly with load.

§ 61. *Changing Number of Poles.*—A multispeed motor with a limited number of definite speeds can be obtained by changing the number of poles and this can be accomplished by re-grouping the coils of the primary winding or by using a separate winding for each speed.

§ 62. *Tandem or Cascade Operation.*—This involves the use of two motors, the rotors of which are mounted on the same shaft or otherwise mechanically connected, and arranged so that the secondary of the first motor supplies current to the primary of the second motor. The two motors may have the same number of poles, but usually the number of poles is different. The effect produced by the second motor is that of adding its poles to or subtracting them from the first motor so that four synchronous speeds are produced, two for the motors independently and two for the combinations.

§ 63. *By Varying Frequency.*—The synchronous speed of an induction motor may be varied by changing the supply frequency by means of apparatus external to the motor or constructed as a part of the motor itself. Such variation may also be obtained by supplying currents of constant frequency to the primary and of variable frequency to the secondary.

---

* For full discussion, see *Multispeed Induction Motors,* by Reist and Maxwell, A. I. E. E., Vol. XXVIII., p. 601; also articles by H. C. Specht, A. I. E. E., Vol. XXVII., p. 1177, and *Electric Journal,* Vol. 6, pp. 421, 492, 577, 611, 731.

Experiment 8–B. Predetermination of the Performance of an Induction Motor by Means of the Circle Diagram.

§ 1. **Data.**—To determine the complete performance of an induction motor without loading, two readings with voltmeter, ammeter and wattmeter are required: first, when the motor is running at normal voltage without load; second, when the motor is locked, and is supplied with reduced voltage so that the current is not more than, say, 1¼ the normal full-load current. These two tests correspond, respectively, to the open-circuit or core loss test and the short-circuit or copper loss test of a transformer.

In addition to these two sets of readings, the primary resistance is to be measured, as nearly as possible under normal working temperature conditions.

§ 2. **Example.**—A test is made of a 5 horse-power, 8-pole, 60-cycle, 3-phase motor having a wound secondary (rotor). The synchronous speed is 900 R.P.M. The primary is Y-connected* for a rated line voltage of 190 volts, or 110 volts from line to neutral. The test is made, however, at a measured line voltage of 185 volts, corresponding to 107 volts from line to neutral.

The average hot primary resistance, as measured from line to line, is 0.51 ohm; hence, the primary resistance per phase is $R_1 = 0.255$ ohm.

§3. **No-load Readings.**—The readings† are

* (§ 2a). In testing a 3-phase motor, it is not necessary to know how the primary or secondary is connected; as a matter of fact, it is usually impossible to tell. The results are not affected by assuming it either *Y*- or delta-connected. The results are here worked up as *Y*-connected. All results may be expressed in terms of equivalent single-phase quantities to advantage; see § 28, Exp. 6–A.

† (§ 3a). When there is a variation in the readings of current and voltage for the different phases, average readings are taken. The watts are the total watts of all phases, irrespective of whether the phases are alike or not, and are determined preferably by two wattmeter readings. (In using the 2-wattmeter method for testing a 3-phase motor, it is to be remembered that one reading may be negative; see §§ 25, 31, Exp. 6–B.)

For an investigation of motors on badly unbalanced voltage, see paper

Line Voltage.	Line Current.	Watts.	Apparent Watts.
E	I_0	W_0	$EI_0\sqrt{3}$
185	5.9	350	1890

Power factor $= W_0 \div EI_0\sqrt{3} = 0.185 = \cos 79°20'$.

No-load copper loss per phase $= R_1 I_0^2 = 0.255 \times \overline{5.9}^2 = 8.87$ watts.

No-load copper loss total $= 3R_1 I_0^2 = 3 \times 0.255 \times \overline{5.9}^2 = 26.6$ watts.

Iron loss, friction and windage $= 350 - 26.6 = 323.4$ watts.

§ 4. **Locked Readings.**—With the rotor locked in position, by any convenient means, the motor is a stationary transformer on short circuit. The locked or short-circuit readings, taken at reduced voltage, are

E.	*I.*	*W.*	Apparent Watts.	Power Factor.
54	16.5	581	1544	$0.375 = \cos 68°$

These readings were taken at about full load current and ⅓ normal voltage. At normal voltage the current would have been excessive, so the desired readings for normal voltage are found by proportion to be as follows:

Calculated Readings for Full Voltage.

E.	I_S.	W_S.	Apparent Watts.	Power Factor.
185	56.6	6810	18150	$0.375 = \cos 68°$

These calculations are made by taking current as proportional to voltage, and power as proportional to voltage squared; power factor remains the same. These relations are not strictly true (see § 32a) and it is better, therefore, to take readings for several currents and to determine the results by averages or by means of curves which are extrapolated for full voltage. These results depend upon the temperature at which the measurements are made; excessive current should, therefore, be avoided.

---

by Charters and Hillebrand, *Reduction in Capacity of Polyphase Motors due to Unbalancing in Voltage,* A. I. E. E., Vol. XXVIII., p. 559.

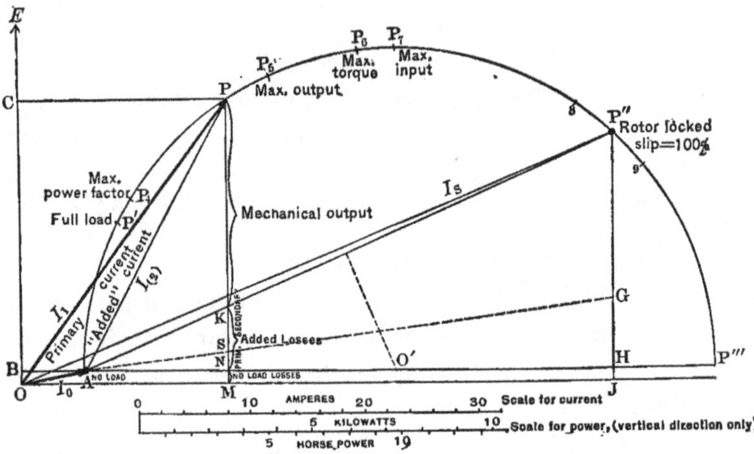

Fig. 1. Circle diagram locus of primary current for predetermining the performance of an induction motor.

Current.

Primary current at no load.................. $= I_0 = OA$
Added current due to load.................. $= I_{(2)} = AP$
Primary current at any load................ $= I_1 = OP$
Short-circuit or locked current............ $= I_S = OP''$
Power component of no-load current........ $= OB = MN$
Power component of primary current........ $= OC = MP$

Power. (For scale see § 10.)

Power input................................ $= MP$
Power loss at no load...................... $= MN$
Added loss at any load..................... $= NK$
Total loss at any load..................... $= MK$
Power output............................... $= KP$
Primary copper loss (approx.)............... $= NS$
Secondary copper loss (approx.)............ $= SK$
Secondary input .... $= SP$

Torque.

Torque is proportional to secondary input, SP.

In syn. horse-power $= SP$ in horse-power ÷ synchronous R.P.M.
In syn. watts $= SP$ in watts ÷ synchronous R.P.M.
In pound feet $= 7.04 \times SP$ in watts ÷ synchronous R.P.M.

§ 5. **Construction of Circle Diagram.**—As lines of reference, in Fig. 1 or 3, lay off line voltage E in a vertical direction and, at right angles to it, an indefinite line OJ, neither line being to scale.

Lay off to scale the no-load current, $I_0 = OA = 5.9$ amperes; this is to be laid off in its proper phase position, so that $\cos BOA$ = no-load power factor = 0.185.

§ 6. In a similar manner, lay off to scale the locked or short-circuit current at full voltage, $I_S = OP'' = 56.6$ amperes, in its proper phase position so that $\cos BOP''$ = power factor = 0.375.

§ 7. Two points, A and P'', of the circle diagram are thus located. The center of the circle lies on the diameter AP''' drawn parallel to OJ, and is located at O' by dropping a perpendicular from the middle of AP''. With the center located, the circle diagram is readily drawn; it is seen to be the same as the diagram Fig. 5, Exp. 8–A, found by loading the motor and to correspond to the circle diagram for a transformer, Fig. 11, Exp. 5–C, in which the lettering is similar. Compare, also, Fig. 2, Exp. 4–B.

§ 8. As the load increases, the primary current is increased from its no-load value, I_0, by the added current $I_{(2)}$ due to the load; $I_{(2)}$ is equal to the secondary current expressed in terms of the primary (§ 20a, Exp. 5–C). The total primary current is, accordingly, $I_1 = OP$. The point P moves around the semi-circle, as the load increases, until it reaches P'', which corresponds to such an over-load that the rotor is held at standstill. The full-load position of P is shown at P', maximum power factor at P_4, maximum output at P_5, maximum torque at P_6, maximum input at P_7, as will be seen later from the method of calculating these quantities.

§ 9. **Results.**—The diagram enables us to readily determine the various operating characteristics of the motor, as shown in Fig. 2.

Current values are scaled directly from the diagram. Power may be calculated from the current values or measured by a special scale.

§ 10. *Scale for Power.*—In any alternating current circuit, power is always proportional to the power component of current; thus, in a single-phase circuit, $W = E \times$ power component of I; in a 3-phase circuit, $W = E\sqrt{3} \times$ power component of I; etc.

In the induction motor diagram, all components of current in a vertical direction are power components and are, therefore, directly proportional to the corresponding power in watts.

For example, in Fig. 1, the power component of the no-load

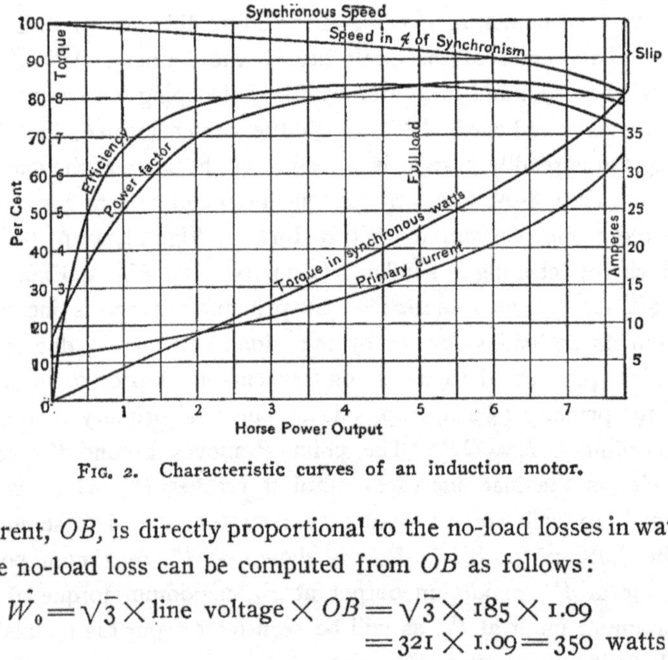

FIG. 2. Characteristic curves of an induction motor.

current, OB, is directly proportional to the no-load losses in watts. The no-load loss can be computed from OB as follows:

$$W_0 = \sqrt{3} \times \text{line voltage} \times OB = \sqrt{3} \times 185 \times 1.09$$
$$= 321 \times 1.09 = 350 \text{ watts};$$

$$\text{or, } W_0 = 3 \times \text{star voltage} \times OB = 3 \times 107 \times 1.09$$
$$= 321 \times 1.09 = 350 \text{ watts}.$$

§ 11. It will be seen that every ampere in a vertical direction in Fig. 1, *i. e.*, every ampere of power component of current, represents a definite number of watts and this number can be readily determined. We can then calculate watts from the scaled value of amperes, or can determine a separate scale for watts. In the present example, every ampere (vertical) represents 321 watts. To get horse-power, we have 1 horse-power = 746 watts.

Certain quantities,—as power factor, efficiency and slip—depend upon ratios and are, accordingly, independent of scale.

§ 12. **Input.**—For any load, the line MP is the power component of current and is, therefore, proportional to the total input in watts. The maximum* input is seen to be at P_7 where a tangent to the circle is parallel to the diameter AP'''.

§ 13. **Output.**—The useful output is proportional to KP, which is equal to the input MP less the total losses, MK, discussed in the next paragraph. The maximum output is seen to be at P_5, where a tangent to the circle is parallel to AP''.

§ 14. **Losses.**—The total losses MK, are equal to the no-load losses MN, increased by the "added" losses NK, which are copper losses due to the added current $I_{(2)}$ and vary as $I^2_{(2)}$. In the construction of Fig. 1, K is located on a straight line connecting A and P''. The point K moves toward P'' as the load increases, and, from the construction,† $NK \propto I^2_{(2)}$.

§ 15. The added losses NK consist of the secondary‡ copper loss SK, and the primary copper loss NS, which is practically||

* (§ 12a). Neglecting the no-load losses, the maximum input is the radius of the circle $O'P_7$, which may be taken as $E \div 2X$; see § 34. The input, and hence the output, of a motor on a given voltage is accordingly limited solely by the leakage reactance, which in design is made as small as possible.

† (§ 14a). In Fig. 1, the triangle ANP is similar to the triangle (not shown) APP'''. Accordingly, $AN:AP::AP:AP'''$ and $AN \propto (AP)^2$, since AP''' is constant. Knowing that $AP = I_{(2)}$ and that NK is proportional to AN, we have $NK \propto I^2_{(2)}$. Q. E. D.

‡ It will be remembered that R_2 and $I_{(2)}$ are the values of secondary resistance and current, respectively, in terms of the primary.

|| (§ 15a). The primary copper loss at no load is $R_1I_0^2$. For a primary current I_1, at any load, the primary copper loss is $R_1I_1^2 = R_1I_0^2 + R_1I_{(2)}^2 - 2R_1I_0I_2 \cos PAO$. Strictly speaking, therefore, the last two terms are the "added" primary copper loss due to $I_{(2)}$. We may consider, either (1)

equal to $R_1I^2_{(2)}$ and can be calculated. (For a 3-phase motor, this loss is $3R_1I^2_{(2)}$.)

§ 16. **Separation of Primary and Secondary Losses.**—If the primary copper loss is calculated and laid off as NS, we have SK equal to the secondary loss. This needs to be done for one point only, the line ASG being then drawn as a straight line. The point selected is usually the point G, corresponding to the locked position, and this can be located in several ways. The various methods for doing this are only approximate* and give slightly varying results,—sufficiently accurate, however, for practical results.

§ 17. *First Method.*—One procedure is to calculate the added primary loss for the locked position and lay this off as HG, thus locating the point G. The line AG is then drawn. For example, in the present test, $I_{(2)}$ on short circuit is $AP''=50.5$ amperes; $R_1=0.255$ ohms. The point G is located so that $HG=3\times0.255\times\overline{50.5}^2=1{,}950$ watts. This procedure was used in constructing Fig. 1.

§ 18. *Second Method.*—With the motor at standstill, JH has no particular significance; neither has I_0 (see § 32). Without involving either of these, G is readily, and perhaps more accurately, located from the short circuit current I_S, by laying off $JG=R_1I_S^2$, multiplying by 3 for a 3-phase motor. In the present example, $JG=3\times0.255\times\overline{56.6}^2=2{,}451$ watts.

§ 19. The same location for G as found in the preceding paragraph can be obtained by dividing JP'' in the ratio $R_1:R_2$, where R_2 is calculated as in § 33. Then $JG:GP''=R_1:R_2=0.255:0.453$. This does not involve any special scale for JG and GP'', and it is

that the last term is neglected as small in the working range of the motor (being zero when I_0 and $I_{(2)}$ are at right angles to one another), or (2) that the loss represented by the last term is included in MN and compensates for the decrease in friction and windage as the motor slows down with load.

* There is no method for determining secondary loss which is exact for all loads and all types of motors; compare § 32. The use of three straight lines AH, AG and AP'' radiating from A for defining the various losses is convenient but not exact.

not necessary to calculate their values in watts, which may prove a convenience.

§ 20. **Torque.**—Torque is equal to the secondary input divided by the synchronous speed; see § 54, Exp. 8–A. The secondary input, in Fig. 1, is seen to be SP, being the primary input MP, less the no-load losses MN and primary copper loss NS.

§ 21. In synchronous watts, torque is equal to SP, in watts, divided by the synchronous speed of the motor in revolutions per minute.

If this is multiplied by 7.04, we have the torque in pounds at one foot radius. (Compare § 3b, Exp. 2–A.)

To get torque in synchronous horse-power, SP is measured in horse-power and divided by the synchronous speed in revolutions per minute.

§ 22. The maximum or "pull-out" torque is seen to occur at P_6 where a tangent to the circle is parallel to AG. The maximum occurs at such a slip, s, that $R_2 = sX_2$; § 9a, Exp. 8–A.

§ 23. **Ratios.**—The following results, being ratios usually expressed as percentages, are independent of scale. They may be found by division, or graphically as in Appendix I.

§ 24. *Power Factor.*—The power factor at any load is equal to CO divided by OP; it is a maximum at P_4, where a line from O is tangent to the circle.

§ 25. *Efficiency.*—The efficiency is equal to the output divided by the input, namely, KP divided by MP.

§ 26. *Slip.*—The slip is equal to the secondary copper loss, divided by the secondary input (see §§ 54, 54a, Exp. 8–A); namely, SK divided by SP.

APPENDIX I.

GRAPHICAL CONSTRUCTION FOR OBTAINING POWER FACTOR, SLIP AND EFFICIENCY.

§ 27. Certain results—power factor, slip and efficiency—depend upon *ratios;* these are usually expressed in per cent. and are determined by dividing one quantity by another, as in §§ 23–26. Some prefer to obtain these ratios by using a slide rule; others, by using the graphical construction of Fig. 3 which is in common use, and is particularly convenient when one has many motors to test. The reader, however, should bear in mind that this construction is only a convenience for computing and not an essential for the proper understanding of the circle diagram.

§ 28. **Power Factor.**—In the direction of OE, lay off an arbitrary scale of 100 parts of any convenient length. From the 100th division, draw the quadrant of a circle with the center at O.

To obtain the power factor for any primary current OP, extend OP to R; from R draw a horizontal line to the point p, which gives the power factor. The construction is obvious; it may be used in any alternating current problem.

§ 29. **Slip.**—From the point A, draw AA' parallel to OE. The scale aa', of 100 equal parts, is drawn parallel to AG at any convenient distance. The point a on AA' is marked zero; the point a' on AP'' is marked 100.

To determine the slip corresponding to any primary current OP, locate the point s where the line AP cuts the slip scale. The per cent. slip* is sa. (If the measured value of the slip at no load is appreciable, the scale should be given this value, and not zero, at the point a. Ordinarily this refinement is unnecessary.)

---

*(§ 29a). *Proof of Slip.*—For any point P, the slip is $KS \div PS$. The triangles KAS and $Aa'a$ are similar; hence

$$KS : AS = Aa : a'a. \qquad (1)$$

The triangles APS and sAa are similar; hence

$$PS : AS = Aa : sa. \qquad (2)$$

Dividing (1) by (2), we have the slip

$$KS/PS = sa/a'a.$$

§ 30. **Efficiency.**—Extend the line AP'' back to L and draw LL' parallel to OE. The scale ll', of 100 equal parts, is drawn parallel to AP''' at any convenient distance. The point l on LL' is marked 100; the point l' on AP'' is marked zero.

To determine the efficiency corresponding to any primary current

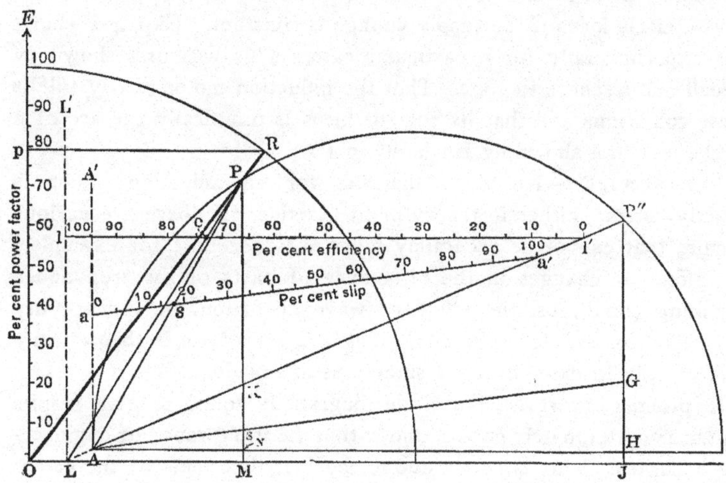

FIG. 3. Graphical method of obtaining power factor, slip and efficiency.

OP, locate the point e where the line LP cuts the efficiency scale. The per cent. efficiency* is $l'e$.

---

* (§ 30a). *Proof of Efficiency.*—For any point P, the efficiency is $PK \div PM$.

The triangles LKM and $l'Ll$ are similar; hence

$$KM : LM = Ll : l'l. \quad (1)$$

The triangles LPM and eLl are similar; hence

$$PM : LM = Ll : el. \quad (2)$$

Dividing (1) by (2), we have

$$KM/PM = el/l'l. \quad (3)$$

But

$$PM/PM = l'l/l'l. \quad (4)$$

Subtracting (3) from (4), we have

$$(PM - KM) \div PM = (l'l - el) \div l'l;$$

whence

$$PK/PM = l'e/l'l,$$

which is the efficiency.

APPENDIX II.

MISCELLANEOUS NOTES.

§ 31. **Basis of the Circle Diagram.**—In any circuit or apparatus with constant reactance and variable power consumption the current will have a circle locus if the supply voltage is constant. Exp. 4–B shows this experimentally for a particular case. This was first shown by Bedell and Crehore in 1892. That the induction motor nearly fulfills these conditions and that its current locus is practically the arc of a circle, was first shown by Heyland* in 1894.

§ 32. **Accuracy.**—No circle diagram for an induction motor is exactly correct, either in theory or in practice, for there are various factors that cannot be accurately taken into account; for example,† the effect of changes in the reluctance of parts of the iron under changing conditions, the effect of wave distortion, the uncertainty as to load losses, etc. The circle diagram, therefore, is theoretically correct only in case certain assumptions are made.

In practice, however, the circle diagram is found to give results that are approximately correct and within the usual range of engineering accuracy. This is partly due to the fact that some of the errors tend to cancel each other. The variations in the methods given by different writers for constructing and using the circle diagram have arisen from a difference in the selection of the errors to be eliminated or minimized, the remaining errors being neglected. If the no-load current I_0 were zero and there were no primary resistance loss, most of the errors would disappear and some of these different constructions would become identical.

Generally speaking the effect of errors in using the circle diagram becomes less as the size of motor increases, so that the method is reasonably accurate on motors larger than, say, 5 or 10 H.P. For motors under 5 H.P. the results, although not so accurate, are fairly

* *Elektrotechnische Zeitschrift,* Oct. 11, 1894; published later in book form and translated into English by Rowe and Hellmund.

† (§ 32a). Behrend shows, for example, that there is a departure from a true circle locus when the slots are bridged or closed; the short-circuit characteristic between volts and amperes, under these conditions, is a curve and becomes a straight line only if the leakage path contains no iron. See pp. 20, 21, *The Induction Motor,* by B. A. Behrend.

satisfactory, while for motors under 1 H.P. the results are of little value, unless refinements* are introduced in the construction of the diagram. It is, however, in testing large motors that a method of testing without load, as by the circle diagram, is particularly desirable; small motors can be readily tested by brake or other load methods.

§ 33. **Calculation of Secondary Resistance.**—The short-circuit watts W_S, are chiefly due to copper losses in the primary and secondary. In reality various load-losses are included,† which cannot be separately determined. These copper losses are $(R_1 + R_2)I_S^2$, per phase, where $R_1 I_S^2$ is the primary copper loss and $R_2 I_S^2$ is the secondary copper loss. Here R_2 is the secondary resistance in terms of the primary (§§ 16, 16a, Exp. 5–B; § 20a, Exp. 5–C) and is the quantity to be determined.

Per phase, we have

$$\text{Copper loss} = (R_1 + R_2)I_S^2;$$

hence,

$$R_1 + R_2 = \text{copper loss} \div I_S^2.$$

Since R_1 is known, R_2 is thus determined.

For a 3-phase motor,

$$W_S = 3(R_1 + R_2)I_S^2;$$

and,

$$R_1 + R_2 = 1/3(W_S \div I_S^2).$$

In the present test

$$R_1 + R_2 = 1/3(6{,}810 \div \overline{56.6}^2) = 0.708 \text{ ohms};$$

hence,

$$R_2 = 0.708 - 0.255 = 0.453 \text{ ohms}.$$

§ 34. **Leakage Reactance.**—The leakage reactance X, of an induction motor, both primary and secondary in terms of the primary, can

* (§ 32b). See a comprehensive article by H. C. Specht, *Elec. World*, p. 388, Feb. 25, 1905, in which it is said the modifications introduced give a diagram applicable to induction motors of all sizes, single-phase or polyphase. To correct for error due to primary resistance, Specht tips his diagram slightly, by dropping A and raising O' a small amount. Such a correction was pointed out by Heyland, p. 23 of the English translation.

† (§ 33a). This gives to R_2 a value somewhat greater than the value that would be determined by direct current resistance measurement.

be calculated* from the diameter of the circle locus AP''', which is equal to $E \div X$ amperes. Thus in Fig. 1, $AP''' = 55.4$ amperes; the leakage reactance per phase is $E \div AP''' = 107 \div 55.4 = 1.93$ ohms.

§ 35. **Leakage Coefficient.**—The leakage coefficient, or leakage factor, is defined by the ratio $BA \div AP'''$. To have this quantity small necessitates a small air gap. In Fig. 1, $BA \div AP''' = 5.8 \div 55.4 = 0.105$.

* (34a). The leakage reactance can also be calculated as in § 25, Exp. 5–B; thus,

$$Z = E \div I_S = 107 \div 56.6 = 1.89 \text{ ohms};$$
$$X = \sqrt{Z^2 - R^2} = \sqrt{(1.89)^2 - (.708)^2} = 1.75 \text{ ohms}.$$

The two methods of calculation agree only when I_0 is zero.

CHAPTER IX.

INDUCTION MACHINES: FREQUENCY CHANGERS AND INDUCTION GENERATORS.

EXPERIMENT 9–A. Operation and Test of a Frequency Changer (Secondary Generator).

§ 1. **Principles of Operation.**—The frequency of the current in the secondary of an induction motor depends upon the speed of rotation of the rotor; within limits any desired secondary frequency can be obtained by giving the rotor the proper speed. The usual form of frequency changer or frequency converter consists merely of an induction motor and a separate driving motor for driving the rotor at the proper speed.

The primary is furnished with polyphase current and produces a rotating magnetic field as in any induction motor. The secondary is phase-wound and delivers, usually, polyphase current to the receiving circuit. The driving motor may be of any type, but in commercial practice a direct-connected synchronous motor is commonly used, so that the delivered current has a definite fixed frequency. (Since the induction machine takes a lagging current of low power factor, the synchronous motor by taking leading current will raise the power factor of the set.)

§ 2. The secondary frequency is increased or decreased according as the rotor is turned in the opposite direction or in the same direction as the rotating field; the secondary voltage is increased or decreased in proportion to the secondary frequency. Aside from losses, the secondary current (with a 1 : 1 ratio) is equal to the primary current, either for an increased or decreased frequency.

Frequency changers are generally used to change from a lower to a higher frequency,—for example, from a low-frequency power circuit to a 60-cycle lighting circuit.

§ 3. The secondary frequency n_2 varies with the slip s; that is, $n_2 = sn_1$. At synchronous speed, $s = 0$ and $n_2 = 0$; at standstill, $s = 1$ and $n_2 = n_1$. When the rotor is driven against the rotating field, with a speed equal to synchronous speed, $s = 2$ and $n_2 = 2n_1$; etc. In this case the frequency is increased and the driving motor is supplying power proportional to the increase in frequency and voltage. Thus, when the frequency is increased 50 per cent., two thirds of the power is supplied by the primary of the induction machine and one third is supplied by the driving motor; losses are here neglected.

When the frequency is decreased, the rotor revolves in the same direction as the field at less* than synchronous speed; the secondary voltage and power are decreased so that the electrical power supplied to the primary is more than the power given out by the secondary. If the surplus power is more than enough to supply the losses, the induction machine runs as a motor and furnishes mechanical power. Frequency changers are not commonly used to decrease the frequency.

§ 4. Referring to the circle diagram, Fig. 1, Exp. 8–B, the short-circuit point has the position P'' when the rotor is at standstill ($s = 1$); the machine then acts as a stationary transformer, the range of working as the secondary external resistance changes being from A on open circuit to P' at full load and P'' on short circuit. The excessive current, however, prohibits going much beyond full load, as in any transformer.

When the rotor is turning with the field ($s < 1$), the short-circuit point P'' shifts to some point as 8 and when turning against the field ($s > 1$) to some point as 9, the full range of working being from A on open circuit to the short-circuit point

* (§ 3a). The rotor could be driven above instead of below synchronous speed, with a negative instead of a positive slip. Electrical power would then be given out by the primary as well as the secondary, the machine being simultaneously a primary and secondary generator (see Exp. 9–B), but it is doubtful whether there is any useful application for such operation.

P'' wherever located. The short-circuit point P'' may be located anywhere on the semi-circle, from A (when $s=0$) to P''' (when $s=\infty$), according to the slip. P''' would be reached only if the speed were infinite, or, if the secondary resistance, internal as well as external, were zero. See also Fig. 1, Exp. 9–B.

§ 5. **Apparatus.**—When a commercial frequency changer is available, it should be run under its rated conditions and appropriate measurements made of input and output.

In the laboratory, any phase-wound induction motor can be conveniently used as a frequency changer. It may be belt-connected to a direct-current shunt machine which may be driven at different speeds and will serve as a driving motor or as a generator.

§ 6. **Preliminary Test.**—With primary voltage constant, make a run at different speeds from synchronous speed in the same direction as the rotating field to the same speed in the opposite direction. (In changing the relative direction of rotation, it may be simpler to reverse the direction of the rotating field by changing primary connections than to reverse the direction of the driving motor.) For different speeds, note the frequency (which may be computed) and the voltage of the delivered currents. Plot voltage for different speeds, frequencies or slip.

§ 7. **Load Run at Constant Speed.**—(These tests may be curtailed or expanded as desired.) With the rotor driven against the field, so as to convert from a low to a high frequency, vary the load of the receiver circuit consisting of non-inductive resistances. Measure the primary input and secondary output of the induction machine, and the input of the driving motor. The mechanical power supplied by the driving motor is the motor input less losses determined as in Exp. 2–B. This mechanical power added to the measured primary power gives the total power supplied to the induction machine.

Compute the efficiency of the induction machine alone and of the complete set. Note the voltage regulation with load. Note how the relative amounts of power supplied by the driving motor and by the primary are related to the change in frequency.

§ 8. Repeat the test, converting from a high to a low frequency. The driving motor, when running as a generator, may pump power into the line.

Experiment 9-B. Operation and Test of an Induction Generator (Primary Generator).

PART I. INTRODUCTORY.

§ 1. **Principle of Operation.**—As the speed of an induction motor approaches synchronism, the slip decreases so that there is less cutting of magnetic flux by the secondary conductors. There is accordingly less electromotive force induced in the secondary, less secondary current and less torque. At synchronous speed the slip becomes zero, and there is no secondary current and no torque. When the rotor is driven above synchronous speed, the slip becomes *negative*; the secondary current and the torque are now reversed* and mechanical power is required to drive the rotor. The machine has become a generator and supplies electrical power to the line. (This condition occurs when an electric train, equipped with induction motors, runs down hill.) As the induction generator does not operate at synchronism, it is frequently described as *non-synchronous* or *asynchronous.*

§ 2. When the secondary current is reversed, the primary "added" current $I_{(2)}$ due to the secondary is also reversed (see § 1a), but it still follows the circle locus† as shown in Fig. 1. The point P follows the upper semi-circle (as P_M) when the machine is operating as an induction motor, the lower semi-circle (as P_G) when it is operating as an induction generator.

In Fig. 1, the impressed or line electromotive force is represented as E; the generated or counter electromotive force, as E'.

§ 3. As a motor, the primary current is $I_1 = OP_M$, consisting of the exciting current, $I_0 = OA$, and the added current $I_{(2)} = AP_M$. (The lines for a motor are not all shown in Fig. 1; see also Fig. 1, Exp. 8–B.) The primary current is always lagging with respect

* (§ 1a). Strictly speaking only the power component of current is reversed, as will be seen later.

† It will be understood that for an induction generator, as for an induction motor, the circle locus is approximate and not exact.

to E; for a definite load and slip it has a definite power component OC in phase with E and a definite wattless component CP_M lagging 90° behind E. The power factor, for a given load and slip, is likewise definite.

As a generator, the primary current is $I_1 = OP_G$, consisting of

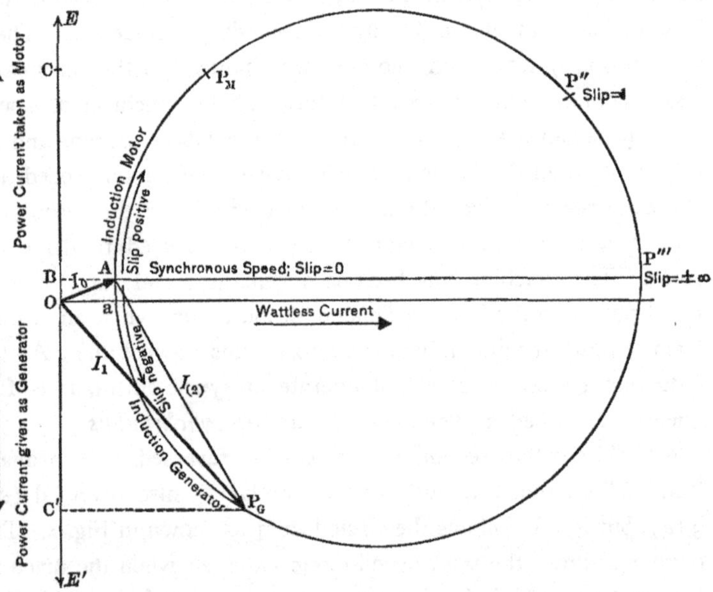

FIG. 1. Circle diagram showing the operation of an induction machine as a motor and as a generator. (In the range *Aa* the machine gives out no power.)

the same exciting current as before, $I_0 = OA$, and the added current $I_{(2)} = AP_G$. The primary current is *always leading* with respect to E'; for a definite load and slip it has a definite power component OC' in phase with E' and a definite wattless component $C'P_G$, which is 90° ahead of E'. The power factor for a given load and slip is definite.

§ 4. It is seen that in an induction generator, as well as in an induction motor, all the current can not be power current. There must be a flow of wattless magnetizing current* to set up the flux; otherwise the generator can produce no voltage.

This means that an induction generator will give voltage only when it is connected to a circuit that allows the proper wattless current to flow; it can not operate when connected to a resistance load (or other load taking power current only) unless there is connected in parallel some device, as a condenser or synchronous machine, that takes leading current.

§ 5. The size and cost of condensers† being prohibitive, the induction generator in practice is used (*a*) in parallel with a synchronous generator, or (*b*) in parallel with an over-excited synchronous motor or converter.

§ 6. When an induction generator is used in parallel with a condenser, synchronous motor or converter, the wattless current is a leading current supplied *by* the generator to the condenser or synchronous machine. Commonly but less logically, however, this wattless current is described as a lagging current supplied *to* the generator by the condenser or synchronous machine. The synchronous machine is said to "supply the excitation" for the generator.

When the synchronous machine is a generator, there are two generators in parallel and the current which circulates between them is due to the combination of their two electromotive forces, the current being lagging with respect to one and leading with respect to the other.

§ 7. **Uses.**—The induction generator has been but little used, due no doubt to its inability to supply lagging current and the

* (§ 4a). Aside from saturation, the voltage of an induction generator at no load is proportional to the wattless magnetizing current. As the load increases, the wattless current is increased from BA to $C'P_G$ on account of leakage reactance, as in a transformer or induction motor. The diameter of the circle, $E \div X$, becomes greater as X diminishes and would be infinite when $X=0$.

† For operation with condensers, see McAllister's "Alternating Current Motors."

necessity of using a synchronous machine in conjunction with it. It has the advantage* of rugged construction, with no commutator, brushes or slip rings. The squirrel-cage rotor requires no moving coils of wire and practically no insulation. The machine gives a smooth wave of electromotive force and tends to damp out, rather than to produce, harmonics and surges. On short circuit the machine gives no voltage, which is an advantage in operation.

§ 8. When an induction generator is operated in parallel with a synchronous generator, the frequency and voltage are determined by the latter. The load taken by the induction generator depends upon its slip,—that is, its speed with reference to the speed of the synchronous machine.

This characteristic may prove desirable or not according to circumstances. It would, for example, be obviously undesirable if the induction and synchronous generators were driven at constant speed; for, as the load increased, the induction generator would not take its share. On the other hand it would prove desirable if a station with induction generators driven by water power were connected in parallel with a station composed of synchronous machines driven by steam power. It could be so arranged that the induction generators would tend to speed up and take all the load up to the limit of the water power, the steam-driven synchronous machines carrying only the excess of load.

§ 9. When an induction generator is operated in parallel with a synchronous motor or converter, the frequency will depend upon the speed of the generator (see Fig. 5) but will vary also with the slip, that is, with the load. At constant speed the frequency would diminish with the load; or, for constant frequency, it would be necessary for the speed to increase with the load. The voltage

* For a discussion of the induction generator and its use, see a paper by W. L. Waters, A. I. E. E., Vol. XXVII., pp. 157–180 and the discussion pp. 217–254.

depends upon the speed of the generator, Fig. 5, and the field excitation of the synchronous machine, Fig. 4.

One synchronous machine, either in the station or in a substation, is sufficient for the operation of several induction generators. In the case of a long transmission line, it has been proposed to locate the synchronous machine at the receiving end so that the leading current supplied to it will improve the regulation of the line.

PART II. TESTS.

§ 10. (*a*) **Operation in Parallel with a Synchronous Generator or Supply Line.**—The induction generator is driven by mechanical power. As a driving motor, a shunt motor will be found con-

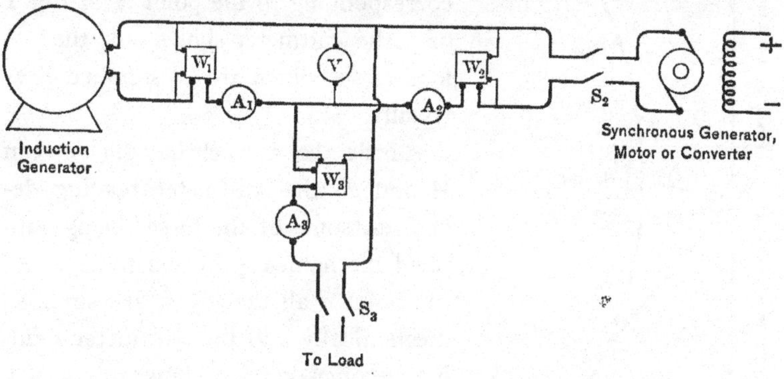

FIG. 2. Connections for operating an induction generator. The switch S_2 may connect to a supply line.

venient since its speed can be easily varied. The connections* are shown in Fig. 2.

§ 11. *Loading Back Test.*—No load is used, the switch S_3 being

* (§ 10a). The connections shown are for single phase; three ammeters and three wattmeters may be used, or one ammeter and one wattmeter can be switched from circuit to circuit. When polyphase apparatus is used, as a laboratory test it may be operated single phase for simplicity. When operated polyphase, the usual disposition of instruments should be made; care should be take that the polyphase connections are so made that the induction machine would run as a motor in the same direction it is driven as a generator.

open. The induction machine is driven at about normal speed and is then connected (by the switch S_2) in parallel with a synchronous generator or supply line. Note that the switch S_2 may be closed when the induction generator is running either above or below synchronous speed. After the switch is closed, the induction machine continues to run as a motor below synchronous speed or as a generator somewhat above synchronous speed.

§ 12. Vary the speed of the induction machine, by varying the speed of the driving motor, and note the wattmeter W_1.

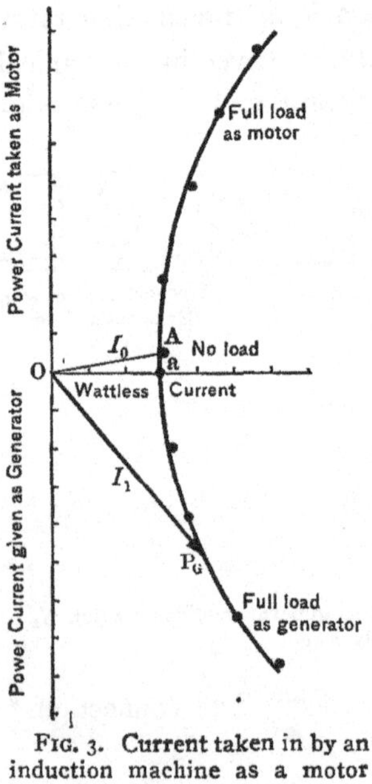

FIG. 3. Current taken in by an induction machine as a motor and given out as a generator.

Below synchronous speed, the induction machine takes power from the line as a motor. At synchronous speed, corresponding to the point A in Fig. 1 or 3, the wattmeter shows only the no-load losses which are all supplied electrically.

A little above synchronism (between A and a) the wattmeter reading decreases, some of the losses being supplied mechanically by the pulley. At the point a all the losses are supplied mechanically and the wattmeter reading becomes zero. Above this speed, the wattmeter reverses and shows electric power given to the line by the induction machine as a generator.

Note that the frequency and voltage in all cases are determined by the synchronous alternator or supply circuit.

§ 13. Take readings of volts, watts and amperes through the full range between no load and full load with the machine operating as a motor and as a generator. (It is instructive also to measure speed or slip and to plot slip—positive and negative—for different amounts of power.)

§ 14. For each reading, calculate

$$\text{Power current, } I_P = I\cos\theta = W \div E;$$
$$\text{Wattless current, } I_Q = I\sin\theta = \sqrt{I^2 - I_P^2}.$$

Plot results as in Fig. 3 by laying off wattless current as abscissae and power current as ordinates. Compare Fig. 5, Exp. 8–A.

§ 15. *Load Test.*—Connect to a non-inductive* load by closing the switch S_3, Fig. 2. With load constant, vary the speed of the induction generator and note instruments. The load receives power from both generators or from only one; $W_3 = W_1 + W_2$. When W_1 and W_2 are positive, both machines are supplying power as generators. When W_1 or W_2 is zero, the corresponding machine is neither supplying nor taking power. When W_1 or W_2 is negative, one machine is taking power as a motor, all the power being supplied by the other machine.

For several sets of readings, compare the values of $I\cos\theta$ and $I\sin\theta$ as calculated for each of the three circuits.

In commercial use, the machines would be so operated that both machines are supplying power, the division of the load depending upon their relative speeds.

§ 16. Tests can be made with variable load under any desired arrangement of conditions.

§ 17. (*b*) **Operation in Parallel with a Synchronous Motor or Converter.**—In this test the synchronous machine, hereafter referred to as the converter, may be either a motor or converter. The connections are as shown in Fig. 2. The converter can be readily brought up to speed with direct current, as a direct current motor.

After the induction generator and the converter† have been

*Loads that are not non-inductive can be made the subject of special investigation.

† (§ 17a). When a synchronous motor is used, a good procedure is to close S_2 and bring the machines to speed before exciting the motor field. The motor field current is then gradually increased until the induction generator gives the desired voltage.

brought up approximately to speed (the exact speed is not necessary) the two machines are connected together by closing the switch S_2, the power supply used in bringing the converter to speed being cut off.

§ 18. *No-load Excitation Curve.*—The load switch S_3 is open. Vary the field current of the converter and measure watts, amperes and volts. The induction generator is driven so as to give rated frequency at rated voltage; hold speed, or frequency, constant during the test.

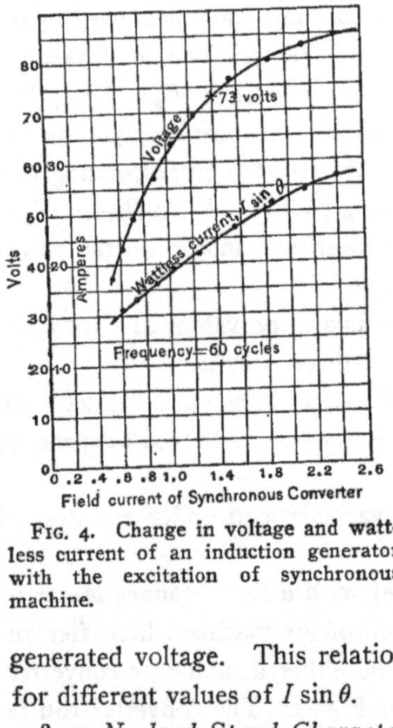

FIG. 4. Change in voltage and wattless current of an induction generator with the excitation of synchronous machine.

The converter may be separately or self excited. When the field current is reduced below a certain value, the converter goes out of step and stops; the induction generator then gives no voltage.

§ 19. For various field currents, plot voltage as in Fig. 4; also $I \sin \theta$, the wattless component of line current. It is the increase in $I \sin \theta$ that increases the excitation of the induction generator and so increases the generated voltage. This relation will be seen by plotting voltage for different values of $I \sin \theta$.

§ 20. *No-load Speed Characteristics.*—Separately excite the field of the converter and keep the field current constant. Vary the speed of the induction generator and measure line voltage and frequency. Begin with a speed of say 10 per cent. above normal and decrease speed until the converter stops. Converter speed may be measured instead of frequency.

§ 21. Results are plotted as in Fig. 5. The lower curve shows that, in order to give a frequency of 60 cycles, the machine to which the curve refers must be driven at 1823 R.P.M., corresponding to a negative slip of 1.3 per cent., the synchronous speed being 1800.

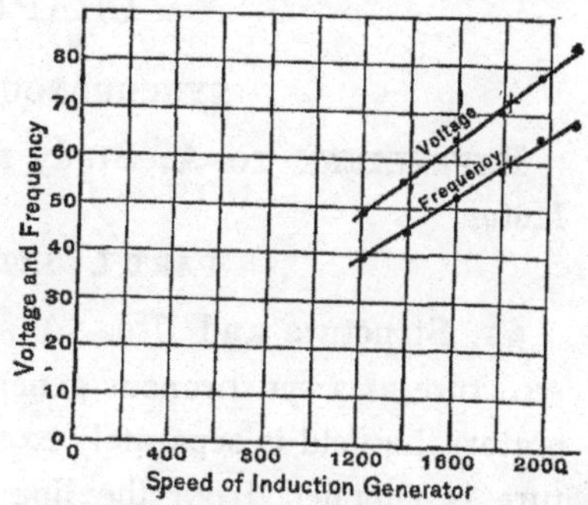

FIG. 5. Effect of change of speed of induction generator operated with a synchronous converter; field current of converter constant, 1.4 amperes.

§ 22. Repeat with the converter self excited.

§ 23. *Load Test.*—Connect a non-inductive load by closing S_3. Compare the values of $I \cos \theta$ and $I \sin \theta$ for the three circuits. All the power, and hence all the power current, is derived from the induction generator, some of this being used in driving the converter and the remainder being supplied to the load.

The wattless current, $I \sin \theta$, merely circulates between the two machines, since the non-inductive load takes none of it. (This test may be extended as desired.)

CHAPTER X.

SYNCHRONOUS MACHINES.

EXPERIMENT 10–A. Study and Operation of a Synchronous Motor.

PART I. INTRODUCTORY.

§ 1. **Structure and Use.**—A synchronous motor has the same structure as a synchronous generator (Exp. 3–A). As in a generator, the field is separately excited by direct current; the armature is supplied from the line with alternating current, either single-phase or polyphase. The speed is fixed by the frequency of the supply circuit, that is, the motor runs at synchronous speed or not at all. In starting a synchronous motor it is necessary to first bring it to speed, usually by external means,* and this limitation has no doubt prevented its more general use.

The fixed speed in many cases is a positive advantage, as for example in the operation of frequency converters (Exp. 9–A).

A synchronous motor has the important advantage that its power factor is adjustable and it can be made to take leading or lagging current by varying the field excitation; its leading current can be used to offset the lagging current taken by induction motors and other apparatus,† thus improving the power factor of the system. Synchronous motors are sometimes installed solely for this object.

A synchronous converter has the same general characteristics as a synchronous motor and most of the following statements will apply to a converter as well as to a motor.

*(§ 1a). An auxiliary starting motor is often provided. The means for starting may be within the motor; thus, some single-phase synchronous motors are provided with an extra winding and commutator so as to start as a series motor. For self-starting polyphase motors, see § 11.

† For operating an induction generator (Exp. 9–B) it is *necessary* to have a synchronous machine connected with the system.

§ 2. **Principle of Operation.**—In a direct-current motor the current in a particular armature conductor is reversed by the commutator and flows in one direction under a north pole and in the opposite direction under a south pole, so that the torque is always in the same direction. In a synchronous motor this reversal is caused not by a commutator but by the alternations of the supply and for the motor to run there must be one alternation for each pole passed, *i. e.*, the motor must run at synchronous speed; at any other speed the torque is alternately positive and negative and the mean torque is zero.

§ 3. In a shunt (direct-current) motor,* the counter-electromotive force E' varies with the speed and, as the load changes, the counter-electromotive force and speed so adjust themselves as to allow an armature current to flow which produces just the right torque for the particular load. The resultant of the impressed electromotive force E and the counter-electromotive force E' is their *algebraic* sum, and the armature current that flows is equal to this resultant divided by the resistance of the armature; or

$$I = (E - E') \div R.$$

§ 4. In a synchronous motor the speed is constant and the counter-electromotive force E' is constant in value for a particular field excitation. The phase of E', however, depends upon the running position (or mechanical phase position) of the armature and shifts with the load; as the load changes, the armature drops back or advances a few degrees and shifts the phase of E' so that an armature current flows with a power component that produces just the right torque (§ 8) for the particular load. In other words, the action of a synchronous motor depends upon the shifting of the phase of the counter-electromotive force rather than upon its change in value as in a shunt motor.

The resultant of the impressed electromotive force E and the counter-electromotive force E' is their *geometric* sum, E_z, and the

* Compare §§ 1–5, Exp. 2–A.

current which flows in the armature of a synchronous motor is equal to this resultant divided by the synchronous impedance Z of the armature; or,

$$I = E_Z \div Z.$$

This current lags behind E_Z by an angle θ_Z whose tangent is equal to the synchronous reactance of the armature divided by the armature resistance.

§ 5. **Synchronous Impedance.**—The synchronous impedance of the armature is made up of two components: true ohmic resistance and synchronous reactance, in quadrature. As in a generator, synchronous reactance includes *local reactance* due to inductance (as in any circuit) and the effect of *armature reaction* in strengthening or weakening the field, the two effects being so similar that for most practical purposes they may be considered as one. The *electromotive force method* of treating the synchronous motor, based upon the conception of synchronous reactance and synchronous impedance, is more fully discussed in Exp. 10–B, which can be read to advantage in connection with the present experiment.

§ 6. **Armature Reaction.**—The physical effect of armature reaction in a generator has already been discussed (§§ 1–15, 44–49, Exp. 3–B); a lagging current weakens the field and a leading current strengthens it. When a generator furnishes current to a synchronous motor, a current which is lagging with respect to the electromotive force of the generator is leading with respect to the counter-electromotive force of the motor, and *vice versa.* In the motor, therefore, a lagging current strengthens the field and a leading current weakens it,—the former being leading and the latter lagging with respect to the counter-electromotive force of the motor.

An under-excited motor takes a lagging current; this strengthens the field and increases the motor electromotive force until it just balances the line electromotive force. On the other hand, an over-excited motor takes a leading current; this weakens the field and decreases the motor electromotive force until it is equal to the electromotive force of the line. The operation of a synchronous motor is thus explained by the *magnetomotive force,* or *ampere-turn,* method.

§ 7. When the current in an armature conductor is in phase with the electromotive force, the current comes to its maximum value when midway under a pole piece, as does the electromotive force (see Fig. 9, Exp. 3–B) and the current does not strengthen or weaken the pole. When, however, the current is out of phase with the electromotive force, it comes to its maximum before or after reaching the middle of the pole and either strengthens or weakens it (see Fig. 10, Exp. 3–B). It is the quadrature or wattless component ($I \sin \theta$)—and not the in-phase or power component ($I \cos \theta$)—that has the magnetizing and demagnetizing effect.

The excitation of a synchronous machine depends not only upon the field current but also upon the wattless component of the armature current. Any change in the former makes a corresponding change in the latter, the combined effect of the two always being just sufficient to cause the counter-electromotive force of the motor to equal the electromotive force of the line.

§ 8. **Torque and Power.**—Torque is proportional to power divided by speed (§ 3b, Exp. 2–A). In a synchronous motor, since speed is constant, torque is proportional to power, and certain conclusions can be reached either from a consideration of torque or from a consideration of power. At constant voltage, the torque (neglecting losses) is proportional to the power component of current.

When the current and the electromotive force are in phase, their product—that is, the instantaneous power—is always positive; also, since in this case the current in an armature conductor remains in one direction while passing a pole, the instantaneous product of current and flux—that is, the instantaneous torque—is always positive. Both the current and electromotive force change sign when the conductor is between two poles.

Again, when the current and the electromotive force are not in phase, their instantaneous product—that is, the instantaneous power—is negative for a part of each cycle; also, since the current in this case reverses while a conductor is under a pole, the instantaneous product of current and flux—that is, the instantaneous torque—is negative for a part of each cycle. Power and torque are both seen to be pulsating.

The pulsating power and torque in a single-phase motor is taken care of by the fly-wheel effect of the moving parts; in a polyphase motor, the phases so overlap as to give uniform power and uniform torque (§2, Exp. 6–A, and § 1, Exp. 7–A).

§9. **Hunting.**—When the armature of a synchronous motor drops back or advances to assume a running position (§ 4) in which it will develop the power demanded by the load, the inertia of the armature causes it to go past the proper running position and then to approach it by a series of oscillations of a definite period which become damped only by losses which they occasion.

Any variation in the supply current or in the load will produce such oscillations and when the cause is periodic, with a period approximately equal to the natural period of the motor, the effect becomes cumulative and the oscillations become very great. Such a condition of oscillations is called *hunting*; when very bad it may make it impossible to keep the motor in step. The oscillation of the armature is accompanied by surging of the current and pulsation of the power. In the laboratory this makes testing unsatisfactory on account of the fluctuation of the instruments.

A remedy for hunting may sometimes be found in an adjustment of the prime-mover or its governor, thus changing the period of the cause.

§ 10. *Damping.*—Hunting may be effectively reduced by damping, commonly obtained by heavy copper grids imbedded in or surrounding the pole faces of the motor or by a complete squirrel cage structure. Any oscillations of the armature and armature flux induce currents in these copper structures and these currents, on account of RI^2 losses, tend to check or damp the oscillations.

§ 11. Starting a Poyphase Motor with Alternating Current.—If a polyphase synchronous motor with its field unexcited is connected to the line, the polyphase currents in the armature will set up a revolving field which, on account of the losses in the pole faces and damping coils, will bring the motor nearly to speed as an induction motor. (That the starting of an induction motor is produced by losses is shown in § 50, Exp. 8–A. The damping coils act as a short-circuited secondary and when they have the complete squirrel cage structure they are particularly effective in starting.) When the machine is nearly in synchronism as an induction motor, it will usually come into complete synchronism on account of the magnetic attraction between the separate poles of the field and the magnetic poles of the armature which finally stay together so that there is no slip. This attraction is due to the tendency of the lines of force in the air gap to shorten.

When the motor is in synchronism or nearly in synchronism, the field may be excited and the motor will lock in step; the armature current then decreases.

The disadvantages of alternating current starting are: (1) the large current drawn from the line; (2) the high induced voltage in the field winding. The starting current can be cut down by starting at a lower voltage than normal by means of auto-transformers (§ 12, Exp. 8–A). The danger of damage to insulation on account of the induced voltage is reduced by separating the field spools from each other on starting by a "break-up switch"; when synchronism is reached or nearly reached, the field is excited by closing this switch.

PART II. SYNCHRONIZING.*

§ 12. The motor (or other incoming synchronous machine) is brought up to speed by an auxiliary motor or by whatever means are provided. The field is excited with direct current which is adjusted by a rheostat until the electromotive force E', generated by the motor, is approximately the same as the line electromotive force E, a close adjustment being unnecessary. When the motor is running approximately at synchronous speed, the main switch connecting it with the supply line is closed† at a moment when the electromotive force E' of the motor is opposed to the electromotive force E of the line. The circuit should be protected by a circuit-breaker or other device, for a large current would flow if the switch should be closed when the two electromotive forces are not opposed, or if the motor should fall out of step and stop, in which case the line would be short-circuited through the armature.

§ 13. **Synchronizing with Lamps.**—There are various devices for synchronizing, the simplest consisting of lamps‡ which bridge the blades of the main switch with which the motor is to be connected to the circuit. Fig. 1 shows such an arrangement for a single-phase motor, or for one phase of a polyphase motor. When the motor is running at the proper speed and is in the proper phase for closing the switch, the lamps will be dark and may remain so for a considerable interval. When the motor is

* The same methods are used in synchronizing synchronous motors, synchronous converters and synchronous generators operated in parallel. Synchronizing by various methods should be practiced.

† The more inductance there is in the armature circuit, the less accurately is it necessary to synchronize a machine before closing the switch. Extra inductance, which may be cut out later, is sometimes introduced in the armature circuit so that the motor may be thrown into circuit without synchronizing, the inductance preventing excessive current while the motor is pulling into step.

‡ For each lamp mentioned in the following paragraphs, several lamps may be used in series when the voltage requires it.

running a little too fast or too slow, the line and motor electromotive forces are first opposed and then added, giving rise to "beats" shown by the periodic flickering of the lamps which are dark and bright alternately.

The lamps do not show whether the motor is running too fast or too slow; but, when a change of speed (increase or decrease) makes the flicker less rapid, the change is in the proper direction for bringing the motor into synchronism. It is not necessary, and usually would not be possible, to bring the motor to *exact* synchronous speed. When synchronous speed is practically reached,

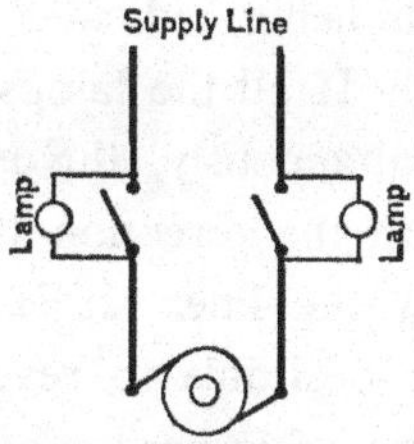

FIG. 1. Arrangement of synchronizing lamps (one phase).

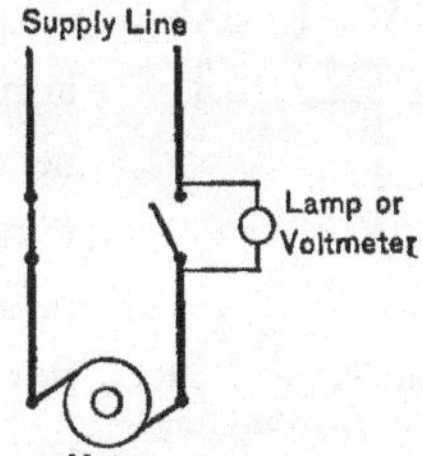

FIG. 2. Synchronizing with a voltmeter, or one lamp or other indicator.

shown by the slow changes of the lamps between dark and bright, the switch should be closed when the lamps are *dark*.

§ 14. *Synchronizing with Lamps Bright.*—To synchronize with the lamps *bright* instead of dark, interchange the two lamp terminals on the line or motor side (as the two upper lamps in Fig. 4).

§ 15. *Synchronizing with a Single Lamp or Voltmeter.*—By closing one switch-blade, one lamp (or series of lamps) of twice the voltage may be used instead of two, as in Fig. 2; a voltmeter, or other indicator, may be used in place of the lamp and will show more accurately the exact moment of synchronism.

§ 16. **Use of Transformers with Synchronizing Devices.**—Synchronous motors and converters are usually made for potentials so high that single lamps cannot be used directly across the switch blades. In this case step-down transformers are often used, these being connected as in Fig. 3 for synchronizing* with the lamp *dark* or with one transformer coil reversed for synchronizing with the lamp *bright*. A voltmeter or other indicator can be used in place of the lamp.

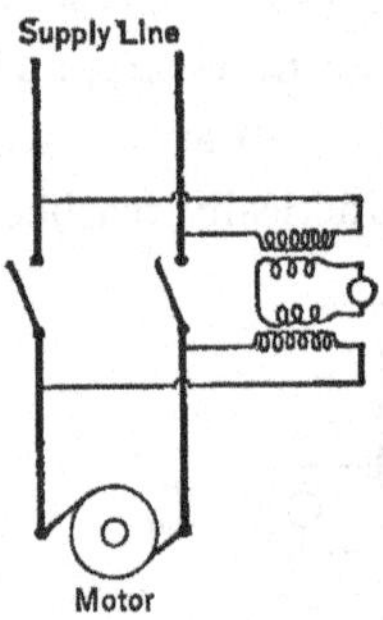

FIG. 3. Transformers for synchronizing.

§ 17. **Synchronizing a Polyphase Motor.**—If lamps are used, a lamp may be placed across each† switch blade as in the case of a single-phase motor, the proper time for closing the switch being indicated by the lamps being *dark*. If all the lamps do not become dark simultaneously, the motor is running counter to the direction it would have if driven by the line. It is usually not convenient or desirable to reverse the direction of rotation of generator or motor, but the leads may be interchanged in such a manner as to make the cyclic changes of the electromotive forces of line and motor similar. In the case of a 2-phase motor, this can be brought about by reversing the leads to one phase of the motor; in the case of a 3-phase motor, it is sufficient to interchange any two leads to the motor.

§ 18. In order to have the lamps *bright*—instead of dark—at the proper moment for closing the switch, in the case of a 4-wire 2-phase motor two terminals of the lamps on each phase can be interchanged, as will be seen by considering the 2-phase motor as consisting of two single-phase motors. A 3-phase motor can-

* In synchronizing with the lamp dark (or bright) the connections should be made so that the lamp will be dark (or bright) when the brushes on the motor are lifted and the switch is closed.

† Lamps across only two switch blades will be sufficient when it is known (as in permanent installations) that the connections and direction of rotation are correct.

not be synchronized with the lamps bright without the use of transformers.

§ 19. **Synchroscopes.**—There are various commercial synchroscopes* which not only indicate when the motor is running synchronously and is in the proper phase but also show whether the motor is running too fast or too slow when it is not in synchronism. Knowledge in regard to the latter saves time and annoyance in bringing the motor to the proper speed.

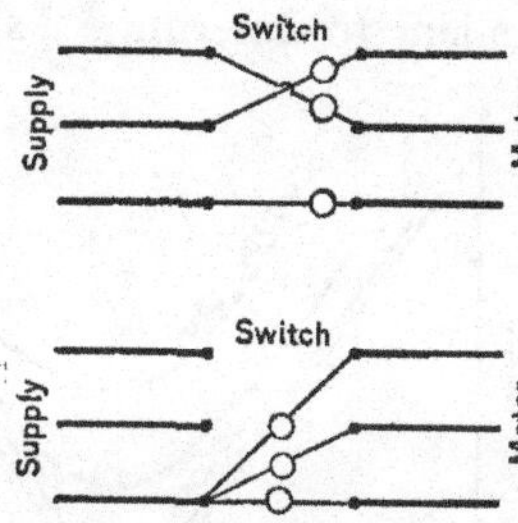

FIGS. 4 and 5.† Lamps for indicating "fast" or "slow."

§ 20. *Lamps for Indicating When Motor is Fast or Slow.*—When a polyphase motor is near synchronism, lamps can be used to show whether it is running too fast or too slow. (When the speed is much too high or too low, however, the flicker of the lamps will be too rapid to follow.)

Figs. 4 and 5 show two methods of connecting lamps for a 3-phase motor. When the motor is in the proper phase for closing the switch, a particular lamp (the lower one) will be dark and the other two will be of equal brightness. If the speed is not synchronous, one lamp after the other will be dark, the sequence depending on whether the motor is too fast or too slow. With the lamps arranged in a circle, or triangle, this gives the appearance of rotation in one direction or the other. Other arrangements of lamps can be made to give similar effects with polyphase motors of any number of phases.

* (§ 19a). Automatic synchronizers are also used. For description of synchroscopes and synchronizers, see hand and text books; also *Electric Journal,* Vol. V., p. 538, where other references are given.

† Fig. 5 shows the speed-indicating lamps only; although not shown, a lamp is required across each switch blade making five lamps in all.

PART III. EXCITATION CHARACTERISTICS.

§ 21. **No Load Run.**—Connect a circuit breaker in the armature circuit and an ammeter, voltmeter and wattmeter* to measure current, voltage, power and power factor; in the field circuit connect an ammeter and field rheostat. (It will often be convenient to determine $I \sin \theta$ directly by the method of §§ 13, 43, Exp. 6–B.) After synchronizing, run the motor at no load and vary the field current through as wide a range as is possible with the motor keeping in step. Take simultaneous readings of all instruments and, with field current as abscissæ, plot as ordinates: armature current, power factor ($\cos \theta$), phase angle (θ), power component of current ($I \cos \theta$) and wattless component ($I \sin \theta$).

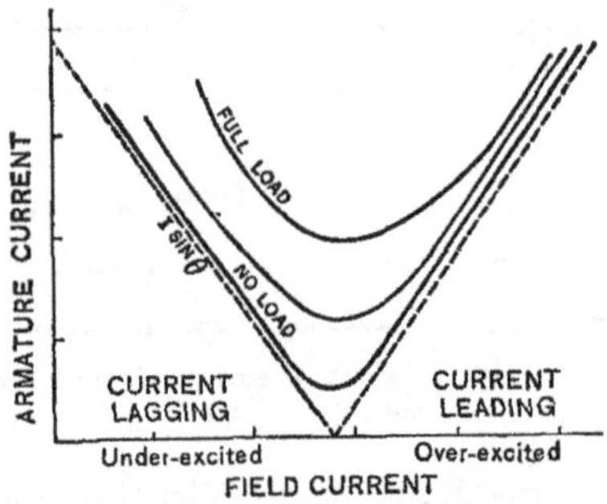

FIG. 6. Change of armature current with excitation for different loads.

§ 22. **Load Runs.**—Similar runs may be made at various loads.

§23. **V-curves.**—Fig. 6 shows the variation of the armature current with field excitation for no load, full load and one intermediate load. (These curves are often plotted with values of motor electromotive force E' as abscissæ instead of field current, the values of E' corresponding to particular values of field current being taken from the no-load saturation curve, § 6, Exp. 10–B.) Individual readings are likely to fall off the curves on account of hunting.

In an ideal case the curve for the wattless current $I \sin \theta$, would fall to a minimum value of zero, as shown by the dotted curve,

* (§ 21a). For a polyphase machine, determine power and power factor as in Exp. 6–B; or, make measurements on only one phase. In the laboratory, where the experiment is only illustrative, the motor may be operated as a single-phase machine.

and the curve for power factor would reach a corresponding maximum of 100 per cent.,* but in practical operation these ideal limits cannot be reached: $I \sin \theta$ never falls quite to zero and power factor never reaches 100 per cent.

The wattless current $I \sin \theta$ acts as an exciting current which strengthens the field when the motor is under-excited and the current is lagging and weakens the field when the motor is over-excited and the current is leading.

§ 24. **O-curves.**—For each load plot a polar curve for I, by laying off each reading of current to scale and with the proper phase angle. This gives a series of O-curves (each curve being like the bottom part of the letter O) corresponding to the series of V-curves. For constant power output and constant losses, with armature resistance zero, these curves would be parallel straight lines perpendicular to E; for an armature resistance R the curves are arcs of circles about a common center located in the direction of E at a distance $E \div 2R$ from the origin.

* (§ 23a). This would mean that the electromotive force and current are simple sine waves and that there is no hunting; otherwise currents of other than fundamental frequency would flow and the relations of plane vectors, upon which the derivation of the expression $I \sin \theta$ depends, would not hold. (See § 47, Exp. 6–A). Dissimilarity in the line and motor electromotive force waves will cause wattless currents of higher frequency to flow.

Experiment 10–B. Special Study of a Synchronous Motor.

§ 1. **Electromotive Force Method.**—The complete action of a synchronous motor can be simply explained* by considering that the line electromotive force E and counter-electromotive force E' combine to make a resultant electromotive force E_Z. The armature current I is equal to $E_Z \div Z$ and lags behind E_Z by an angle θ_Z whose tangent is $X \div R$, where Z is the synchronous inpedance, X the synchronous reactance and R the resistance of the armature circuit.

In Fig. 1, suppose the motor is thrown into circuit at a time when E' is exactly opposite to E in phase. E_Z is then a minimum, being the arithmetical difference between E and E'; the current I that flows is zero or small and generally will not produce enough power† for the motor to maintain rotation. The armature tends to stop or slow down, but as the armature drops back in its running position E' drops back in phase, thus increasing E_Z, I and the power. A balance is reached at some point as d, when the power developed is just sufficient to meet the demand.

§ 2. The electrical power input is the vector product of E and I, which is E times Oe, the projection of I upon E. The mechanical power developed is the vector product of E' and I, which is E' times Of, the projection of I upon the continuation of E'. Furthermore, the mechanical power developed is equal to the

* (§ 1a). Following a paper by Bedell and Ryan, *Journal of the Franklin Institute*, March, 1895, the first complete discussion of the synchronous motor with graphical diagrams and experimental verification, based upon preliminary papers in A. I. E. E. and *Sibley Journal*, May, 1894. A discussion of the circular current locus will be found in McAllister's *Alternating Current Motors.* For an elaborate discussion of the magnetomotive force or ampere turn method, see a series of articles by C. A. Adams, *Harvard Engineering Journal*, Jan., 1908, April, 1908, Jan., 1909. (A translation of Blondel's treatise on the synchronous motor is in preparation.)

† When E' is opposite to E, there will be no power at all as a motor unless E' is less than E, as the subsequent discussion will show; see § 4a.

electrical input less armature RI^2 loss; part of this mechanical power is used in friction and core loss, the remainder being available for useful work.

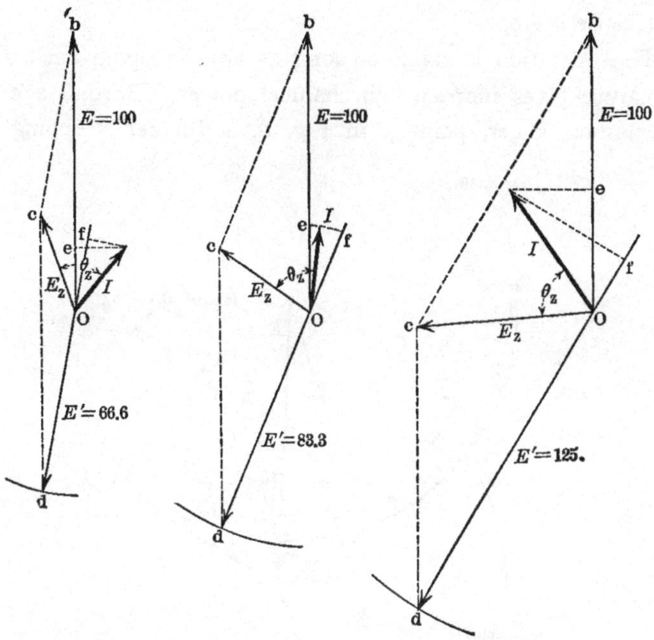

(*a*) under-excited. (*b*) under-excited. (*c*) over-excited.

FIG. 1. Electromotive force diagrams for a synchronous motor; three typical cases.

The projection *Oe* of *I* upon *E* is positive and is electrical power supplied by the line. The projection *Of* of *I* upon the continuation of *E'* is negative, being mechanical power developed by the motor. If *Of*, the projection of *I*, should fall upon *E'* itself and not upon its continuation, the machine would be operating as a generator and not as a motor.

The diagrams shown in Fig. 1 are for the three typical cases in which *I* lags behind, falls between and precedes *E* and *E'* prolonged.

§ 3. In normal operation, with any change of load or other

working conditions, E' drops back or advances until the mechanical power developed, namely $E' \times Of$, just equals the power demanded. The motor does this automatically, if it possibly can; otherwise it stops.

The operation is stable so long as any dropping back of the armature gives increased mechanical power. Beyond a certain maximum power, point 5 in Fig. 2, a further dropping back

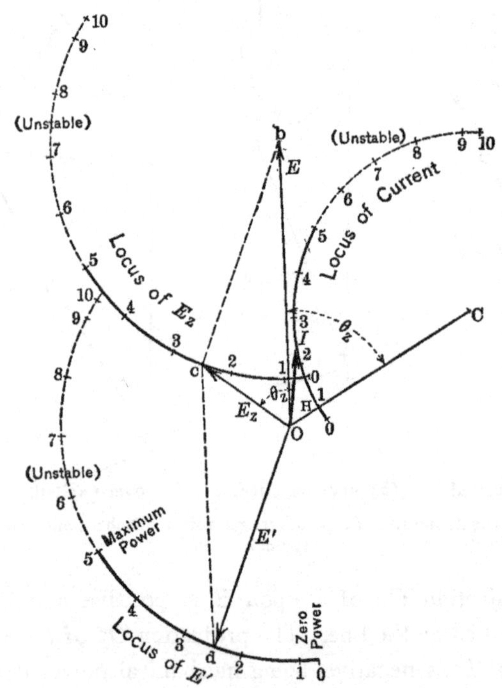

FIG. 2. Synchronous motor diagram; the solid part of each circle is the *operating range.*

of the armature means a decrease in power. Loaded beyond its maximum, the motor will stop. Even if not overloaded, unstable operation could not be maintained, for, assuming the power developed to be just equal to the demand, any momentary retardation of the armature due to hunting or variation in operating conditions would cause the motor to develop less power, insufficient for the demand, so that the motor would continue to drop back and would eventually stop.

§ 4. Fig. 2 shows the loci of E' and E_z, which are arcs of circles each with a radius E'. The numbers 0, 1, 2 . . . 10 indicate corresponding points on the different curves. For this particular case, 0 and 10 are points of zero mechanical power beyond which the machine acts as a generator and not as a motor; 5 is the point of maximum power. Stable operation is from 0 to 5; unstable operation for 5 to 10.

Practically E' must lie in the third quadrant,* for between 0 and 1 the power developed by any motor is not likely to be sufficient even to supply the iron and friction loss.

Maximum power occurs† when E' lags behind E by an angle $180° + \theta_z$. For stable operation, therefore, E' lags behind E by an angle that is less than $180° + \theta_z$ and is (practically) more than 180°. The larger the value of θ_z the wider is the range of stable operation, which means that the reactance of the armature circuit should be large compared with its resistance.

§ 5. **Current Locus.**—Since the locus of E_z is the arc of a circle, the locus of I—which is proportional to E_z—must likewise be the arc of a circle. The center C is on a line OC making an angle θ_z with E; the length OC is $E \div Z$, the radius CH is $E' \div Z$.

For different excitations the current loci consist of concentric circles with different radii, determined by the relation $CH:OC = E':E$. When $E' = E$, the current locus passes through O. For under-excitation (as in Fig. 2) the radius is less than OC; for over-excitation the radius is greater than OC, the point H falling on OC prolonged to the left.

* (§ 4a). For zero power, the lag of E' behind E is 180° when $E' = E$; it is more (or less) than 180° when E' is more (or less) than E.

† (§ 4b). This may be proved analytically, as in *Alternating Current Machines* by Sheldon, Mason and Hausman, or graphically, as in *Elements of Electrical Engineering* by Franklin and Esty.

§ 6. **Data.**—With the synchronous motor driven as a generator, determine the synchronous impedance (§§ 10–15, Exp. 3–B) and the no-load saturation curve (§ 5, Exp. 3–A).

§ 7. **Predetermination of Circular Loci.**—Predetermine the loci of Fig. 2 for the case when $E' = E$, for over-excitation $(E' > E)$ and for under-excitation $(E' < E)$.

§ 8. **Current Loci by Test.**—With constant excitation and varying load determine the current loci for several excitations and compare with the predetermined loci.

§ 9. **Further Investigations.**—The investigation may be extended to include a predetermination of *O*-curves and *V*-curves (as by McAllister), measurement of angular armature position (as by Bedell and Ryan), study of the frequency of hunting, limits of stability, etc. See § 1a; also § 24, Exp. 10–A.

EXPERIMENT 10–C. Study of a Synchronous Converter.

§ 1. **Methods for Obtaining Direct from Alternating Current.**—Direct current is usually distributed from substations and is obtained by means of synchronous converters which are operated by alternating current transmitted from a distance. For economy the transmission is usually 3-phase (§ 54, Exp. 6–A). The converters may be single-phase or polyphase, in practice being usually either 3-phase or 6-phase (§§ 27, 27a, Exp. 6–A). The growth of electric traction has been coincident with, if not indeed dependent upon, the general use of the synchronous converter. While more generally used on circuits of low frequency (25 cycles), its use at 60 cycles (§ 3, Exp. 3–A) is common.

§ 2. The synchronous converter is essentially a synchronous motor and direct-current generator combined in one machine;* it has one field, which is self-excited by direct current, and one armature winding which is provided with collector rings for receiving alternating current and with a commutator for delivering direct current. The armature connections for a 3-phase converter are shown in Fig. 1, which is the diagram for a 2-pole model. Each collector ring is tapped into the armature winding at one point in a 2-pole model, two equidistant points in a 4-pole model, three equidistant points in a 6-pole model, etc.

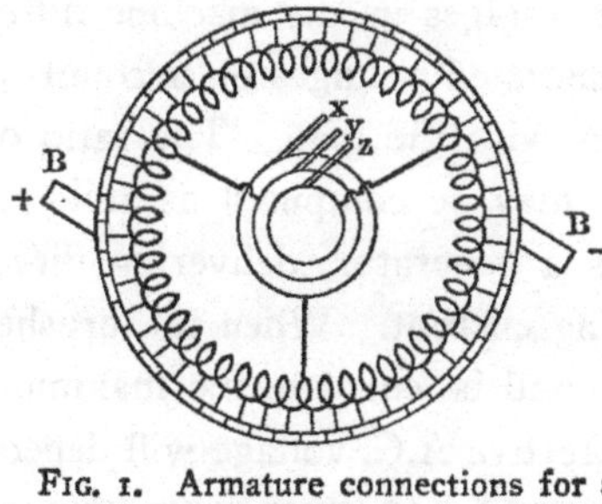

FIG. 1. Armature connections for a 3-phase converter, 2-pole model.

* (§ 2a). *Dynamotors and Motor-Generators.*—Provided with independent armature windings, but with a common field, the machine would be a *dynamotor;* with independent fields as well as armatures the machine would be a *motor-generator* with motor and generator separate,—a more flexible arrangement in regard to control and regulation but more costly in construction and less efficient in operation.

In a synchronous converter, the armature revolves while the field and brushes are stationary; for mechanical reasons the reverse arrangement—with revolving* field and brushes and with stationary armature—is undesirable. Connections for operating a synchronous converter are shown in Fig. 2.

§ 3. Other devices for deriving direct from alternating currents are: synchronous *commutators*† (which have proved short-lived both as individuals and as a class); and rectifiers that depend upon a *valve effect*, as the aluminum rectifier, mercury-arc rectifier, etc., the latter being the only one of these with high enough efficiency to warrant extensive use.

§ 4. A synchronous converter is normally used to receive alternating and to deliver direct current, but may be used as an *inverted converter*—to receive direct and to deliver alternating current—or, as a *double current generator* driven by power and delivering both direct and alternating currents.

§ 5. **Voltage Ratios.**—Terminal voltages in any machine differ somewhat from the induced or generated voltages on account of drop in the windings which varies with the load. The ratio of generated voltages in a converter may be computed as follows:

Consider a converter driven as a generator, delivering direct current and single-phase alternating current. When the brushes are properly set, the D.C. voltage will be equal to the maximum value of the A.C. voltage. The effective A.C. voltage will depend upon wave form, being for a sine wave $1/\sqrt{2}$ times the maximum, or direct current, value.

* (§ 2b). *Permutators.*—The rotating field may be produced *electrically*, in which case both field and armature windings are stationary, the brushes being driven at synchronous speed by a light driving mechanism. A modification of this arrangement is the *permutator*, the introduction of which has no doubt been prevented by the difficulties introduced by the revolving brushes. See: *Elektrotech. Zeit.* (Vienna), Aug. 28, 1898; *L'Industrie Electrique*, Feb. 10, 1902, Nov. 25, 1905; *Lond. Electrician*, Dec. 9, 1905, Dec. 10, 1906; *Elect. Age*, Nov., 1908; *Sibley Journal of Eng.*, June, 1909.

† For a test of such a commutator, see paper by J. B. Whitehead and L. O. Grondahl, *Elec. World*, pp. 896 and 914, April 15, 1909.

Hence, for a single-phase (or 2-phase) machine, the A.C. voltage is

$$E_{A.C.} = (1/\sqrt{2})E_{D.C.} = .707E_{D.C.}$$

The star voltage E_S, measured between one alternating current line and the neutral, is

$$E_S = \frac{1}{2}(1/\sqrt{2})E_{D.C.} = 0.354E_{D.C.},$$

which is true for a polyphase, as well as for a single-phase, machine. From the star voltage, the line voltage in any case is readily computed; thus,—

In a single-phase or 2-phase machine, the line voltage is twice the star voltage.

In a 3-phase machine (§ 19, Exp. 6–A), the line voltage E_3 is

$$E_3 = \sqrt{3}E_S = \frac{\sqrt{3}}{2\sqrt{2}}E_{D.C.} = .612E_{D.C.}.$$

In a 6-phase machine, the line voltage E_6 is

$$E_6 = E_S = 0.354E_{D.C.}.$$

§ 6. **Current Ratios.**—Assuming a certain efficiency and power factor, the alternating current in each supply line corresponding to any particular value of direct current output can be computed. Thus, if power factor and efficiency are 1.00, each ampere of direct current requires an alternating current of $\sqrt{2}$ amp. (=1.414) single-phase, $\frac{1}{2}\sqrt{2}$ amp. (=0.707) 2-phase, $\frac{2}{3}\sqrt{2}$ amp. (=0.943) 3-phase, $\frac{1}{3}\sqrt{2}$ amp. (=.472) 6-phase.

§ 7. **Rating.**—In a converter, each armature conductor carries an alternating current and a rectified direct current, giving an irregular wave form, with a chopped up appearance, differing in the various conductors according to the time that has elapsed since each conductor has passed under a brush. The rating* of a con-

* For a good discussion, see paper by O. J. Ferguson, *Elect. World,* p. 214, Jan. 21, 1909, where ratings are derived for various power factors, based upon hottest and coolest coils as well as upon average heating. See also paper by W. L. Durand, *Elect. World,* p. 235, Jan. 26, 1911.

verter depends upon armature heating and may be based upon several assumptions. Based upon average armature heating and the assumption of unity power factor, the relative capacities of a converter are

single-phase, 0.85; 3-phase, 1.33; 6-phase, 1.93;

the capacity as a direct current generator being unity.

The advantage of the 6-phase converter is obvious (§ 27a, Exp. 6-A), but there is little advantage in more than six phases, the capacity for infinite phases being only 2.3 as compared with 1.93 for six phases.

§ 8. **Voltage Control.***—The D.C. voltage of a converter is usually controlled by altering the A.C. voltage supplied to the collector rings, and this is commonly done: (1) by means of an *induction regulator* (or some other form of potential regulator, Exp. 7-B); (2) by means of *reactance* placed in series with the converter on the A.C. side, as discussed later; or, (3) less commonly by a *synchronous booster.*† Another method, used in the split-pole‡ converter, controls the D.C. voltage without altering the A.C. voltage.

* (§ 8a). For a discussion of converter construction and operation, with particular reference to voltage control, see the following papers and their discussions: A. I. E. E., Vol. XXVII., C. W. Stone, p. 181; J. E. Woodbridge, p. 191; C. A. Adams, p. 959; also *Elect. Journal,* Vol. V., F. D. Newbury, pp. 615, 616.

† (§ 8b). *Synchronous Booster.*—A small auxiliary alternator, mounted on the same shaft as the converter is connected in series with it as a booster on the A.C. side. The A.C. voltage supplied to the converter depends, therefore, upon the excitation of the booster, which may be controlled by a suitable regulator. Some of the field windings of the booster may be put in series with the D.C. load, thus giving an increasing excitation with load.

‡ (§ 8c). *Split-pole Converter.*—In this converter each pole is divided into sections which can be given different excitations so as to vary the flux distribution. This shifts the flux, which is practically the same as shifting the brushes, and so changes the D.C. voltage; or, it alters the wave form of electromotive force and so changes the ratio of the D.C. to the A.C. voltage; or, it both shifts the flux and alters the wave form.

§ 9. *Reactance Control.*—A lagging current through a reactance always causes a drop in voltage; but a leading current, when sufficiently in advance of the electromotive force, will cause a rise in voltage. With reactance in a circuit, therefore, the voltage delivered will depend upon how much the current is lagging or leading, and this—in a synchronous motor or converter—is controlled by the excitation.

§ 10. Fig. 2 shows a reactance X located in one line; in practice, an equal reactance is located in each line (§ 20) either as separate coils or as leakage reactance* in transformers. Let E_O be the line voltage and E_T be the terminal voltage at the collector rings of the converter.

When the current I is in phase with E_T, the reactance causes a voltage drop in quadrature with I so that E_T is somewhat less than E_O, as shown in Fig. 3, Exp. 3–B. This corresponds to a particular excitation of the converter.

When the current I is lagging, the drop through the reactance is more effective and E_T is much less than E_O, as in Fig. 4, Exp. 3–B.

When the current I is leading—corresponding to an increased converter excitation—E_T is increased, as in Fig. 5, Exp. 3–B.

It is seen, therefore, that the A.C. voltage supplied to the converter (and so the D.C. voltage delivered) can, when series reactance is used, be increased or decreased by increasing or decreasing the motor excitation.

§ 11. In the laboratory, and sometimes in practice, this change in excitation can be made by a hand adjustment of the field rheostat. In practice, however, it is usually made automatic by *compounding* the converter, by means of series turns carrying the D.C. load current. As the load increases or decreases, this increases or decreases the excitation and the voltage. By properly proportioning the reactance and the series winding (or a shunt

* (§ 10a). The unavoidable reactance of the generator, line, transformers and converter may be sufficient.

around it) the converter may be over-, under-, or flat-compounded in the same manner as a D.C. generator (§§ 1, 23, Exp. 1–B).

§ 12. **Derived Neutral.**—An interesting feature of a converter or double-current generator is that the potentials of the neutral of the A.C. system and of the D.C. system are alike.

The neutral may be considered as a (usually fictitious) middle point of the armature. Taking this neutral potential as zero, the positive D.C. brush is at a constant positive potential and the negative D.C. brush is at a constant negative potential, the mean of these being the potential of the neutral. Referring to a single-phase machine, the two A.C. brushes have alternating potentials that are equal but of opposite sign, the mean of which is the potential of the neutral.

The A.C. neutral is readily obtained from a tap in the middle of a transformer or choking coil across the A.C. lines of a single-phase system, or opposite lines of a 2-phase or 6-phase system.

The D.C. neutral, otherwise not easily obtained, can be readily "derived" from the A.C. neutral, *i. e.*, the neutral of the A.C. side is used as the neutral for the D.C. system. This is one advantage of a 2-phase or 6-phase converter with diametral connections (§ 27, Exp. 6–A).

§ 13. Direct-current generators are often constructed with A.C. collector rings across which are placed choking coils for the purpose of deriving a neutral for a 3-wire D.C. system. Any direct current returned to the generator by the neutral or third wire, due to an unbalanced load, passes into the middle tap of such a choking coil and thence differentially through the two halves of the coil with no magnetizing effect upon the core.

Special 3-wire generators are constructed with the choking coils contained within the armature, the outside connections being made through the commutator and one slip ring.

PART II. OPERATION AND TESTS.

§ 14. **Synchronizing.**—A synchronous converter can be brought up to speed and synchronized by any of the means used for starting and synchronizing a synchronous motor (Exp. 10–A). It can also be brought to speed as a direct-current shunt motor (Exp. 2–A) by means of direct current supplied to the commutator end of the converter, a starting resistance being used in series with the armature to limit the starting current; the field rheostat regulates the field current and so controls the speed in synchronizing. In the laboratory, direct current starting is usually the most convenient.

In practice, alternating current starting with low starting voltage is most common; a "break-up" switch is used for separating the field spools (see § 11, Exp. 10–A). A D.C. voltmeter on the D.C. side shows when synchronism is reached by ceasing to beat and by assuming a steady reading, either positive or negative. If the polarity is not the one desired, the machine must be synchronized again or allowed to slip a pole by opening and closing the main switch. Another way to slip a pole is to reverse the field connections and, after the D.C. voltmeter has come to rest near zero, to again reverse the field; the converter then locks in step with the proper polarity.

§ 15. **Voltage Ratio.**—On open circuit, measure the A.C. and D.C. voltage and compare their ratio with the calculated ratios, § 5.

§ 16. **Tests (Without Series Reactance).**—On the A.C. side, connect* a circuit-breaker, voltmeter, ammeter and wattmeter; on the D.C. side, connect an ammeter and voltmeter and a variable resistance for a load. No reactance X is in the line; otherwise the connections are as shown in Fig. 2. Line voltage should, if possible, be kept constant. The following scheme of tests may be followed or modified as seems desirable.

§ 17. *No-load Excitation Test.*—Take the same no-load exci-

* For a polyphase machine, see § 21a, Exp. 10–A.

tation characteristics as for a synchronous motor (§§ 21, 23, 24, Exp. 10–A).

§ 18. *Full-load Excitation Test.*—Repeat the test, keeping the direct current constant at rated full-load value, or other selected value.

§ 19. *Load Run; Excitation Constant.*—With constant excitation* (for which separate excitation may be convenient), vary the D.C. load and take simultaneous readings of all instruments. Make runs with over-excitation, normal-excitation and under-excitation.

§ 20. **Tests with Series Reactance.**—Place a reactance in series with each line (for a single-phase converter, reactance in one line is sufficient†) and repeat the preceding tests.

Read all instruments, as shown in Fig. 2; also read voltage

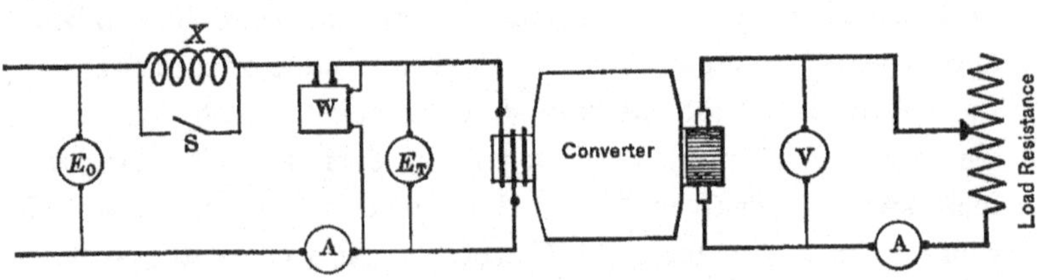

FIG. 2. Connections for operating a synchronous converter.

drop around the coil *X*, which should be so designed that the drop does not exceed, say, 25 per cent. of the terminal voltage.

Measure the resistance and reactance of the coil, plotting results as in Fig. 7, Exp. 5–B.

§ 21. **Results.**—Show all results by curves and compare curves with and without series reactance,—particularly the curves showing voltage variations. Efficiencies are computed from input and output. Construct diagrams, as Figs. 3, 4, 5, Exp. 3–B.

* As a modification, with the converter self-excited, leave the field rheostat in one position.

† When supplied from independent transformers, 2 reactances are sufficient for a 2-phase converter; 3 reactances, for a 6-phase converter.

§ 22. **Inverted Converter.**—When a converter is driven by direct current, its speed depends upon the field excitation, as in the case of a D.C. motor. If the field is weakened through any cause (by a decrease in field current or by the demagnetizing effect of armature reaction), the speed increases; likewise, if the field is strengthened (by an increase in field current or by the magnetizing effect of armature reaction), the speed decreases.

When alternating current is being delivered by the converter at unity power factor, the magnetizing or demagnetizing effect of the armature current is insignificant.

At other power factors, however, a lagging current weakens and a leading current strengthens the field, as in a generator, and makes a corresponding change in speed and in frequency. A machine designed for operation as an inverted converter should, therefore, be designed with a magnetically weak armature; or, some device should be provided for controlling the excitation and maintaining the speed constant. This is sometimes done by using an exciter mounted on the same shaft so that any increase of speed of the converter is checked by the increase of exciting current which it produces.

§ 23. *Test.*—Operate an inverted converter with an inductive load* and with a non-inductive load adjusted for the same value. (CAUTION: Be careful to avoid excessive speed.) With constant field current, compare the speeds in the two cases. (Complete curves from no load to full load may be taken when desired.)

§ 24. Note the change in field current necessary to produce the same speed in the two cases.

§ 25. **Compounding with Series Reactance.**—With a given series reactance, determine the number of series turns needed to give the same D.C. voltage at full load as at no load (§ 11); proceed as with a D.C. generator (§ 28, Exp. 1–B).

* An induction motor, locked, may be found convenient for this.

§ 26. With a given series winding, flat-compounding can be obtained by trial by adjusting the series reactance or adjusting a shunt around the series turns (§ 23, Exp. 1–B).

§ 27. *Derived Neutral.*—Obtain a derived neutral (§ 12) and test as a converter or as a generator with unbalanced load.

CHAPTER XI.

WAVE ANALYSIS.

EXPERIMENT 11–A. Analysis of a Complex Wave by the Method of 18 Ordinates.

§ 1. **Introductory.**—An alternating current or electromotive force is rarely an exact sine wave; in addition to the fundamental wave or first harmonic it usually comprises odd harmonics of 3, 5, 7, etc., times the fundamental frequency. (Even harmonics* are never present when the negative half-wave is a repetition of the positive half-wave.) If we are given the ordinates—or certain ordinates—of a complex wave, we can "analyze" it into its components, that is, we can find the fundamental and the harmonics of higher frequency of which it is composed, each component wave being defined by its amplitude and its phase position with respect to the fundamental.

§ 2. Any complex wave in which there are no even harmonics (the negative half-wave being a repetition of the positive) can be represented by a Fourier's series consisting of the following sine and cosine terms:

$$y = A_1 \sin x + A_3 \sin 3x + A_5 \sin 5x \cdots$$
$$+ B_1 \cos x + B_3 \cos 3x + B_5 \cos 5x \cdots. \qquad (1)$$

x is an angle varying with time; thus $x = \omega t$, $3x = 3\omega t$, etc., where ω is 2π times the fundamental frequency.

By combining† the sine and cosine terms, (1) may be written

* For the analysis of waves with even harmonics, see Appendix II.

† (§ 2a). To prove equation (2), expand as follows:

$$\sqrt{A^2+B^2} \sin (x + \phi) = \sqrt{A^2+B^2} (\cos \phi \sin x + \sin \phi \cos x)$$
$$= A \sin x + B \cos x,$$

where

$$A = \sqrt{A^2+B^2} \cos \phi; \quad B = \sqrt{A^2+B^2} \sin \phi; \quad B \div A = \tan \phi.$$

This will be seen more clearly by constructing a right triangle with C as the hypothenuse and A and B as the two sides.

$$y = C_1 \sin(x + \phi_1) + C_3 \sin(3x + \phi_3) + C_5 \sin(5x + \phi_5) \cdots, \quad (2)$$

where*

$$C_1 = +\sqrt{A_1^2 + B_1^2}; \quad C_3 = +\sqrt{A_3^2 + B_3^2}; \text{ etc.} \quad (2a)$$

$$\phi_1 = \tan^{-1}\frac{B_1}{A_1}; \quad \phi_3 = \tan^{-1}\frac{B_3}{A_3}; \text{ etc.} \quad (2b)$$

The first term in (2) represents the fundamental; the remaining terms represent the harmonics of 3, 5, 7, etc., times the fundamental frequency. Their amplitudes are given by C_1, C_3, C_5, etc., and their relative phase positions by ϕ_1, ϕ_3, ϕ_5, etc. The absolute values of ϕ_1, ϕ_3, ϕ_5, etc. (and the corresponding values of A_1, A_3, A_5, etc., and B_1, B_3, B_5, etc., but not of C_1, C_3, C_5, etc.), depend upon the origin or point of reference from which angles are measured. In the following analysis the origin from which the angles ϕ_1, ϕ_3, ϕ_5, etc., are measured is determined by the selection of the initial ordinate (see Fig. 2) where $y = y_0$ when $x = 0°$.

It is convenient, when plotting, to measure time or angle from the zero of the fundamental wave. We therefore rewrite (2) by substituting $x - \phi_1$ for x; thus,

$$y = C_1 \sin x + C_3 \sin 3\left(x + \frac{\phi_3}{3} - \phi_1\right) + C_5 \sin 5\left(x + \frac{\phi_5}{5} - \phi_1\right) \quad (3)$$

or,

$$y = C_1 \sin x + C_3 \sin 3(x + \alpha_3) + C_5 \sin 5(x + \alpha_5) \quad (4)$$

where

$$\alpha_3 = \frac{\phi_3}{3} - \phi_1; \quad \alpha_5 = \frac{\phi_5}{5} - \phi_1; \text{ etc.} \quad (4a)$$

* (§ 2b). In computing ϕ, note the signs of B and A; thus

with $+B$ and $+A$, we have $\phi = +\tan^{-1}\frac{B}{A}$;

with $-B$ and $-A$, we have $\phi = +\tan^{-1}\frac{B}{A} \pm 180°$;

with $-B$ and $+A$, we have $\phi = -\tan^{-1}\frac{B}{A}$;

with $+B$ and $-A$, we have $\phi = -\tan^{-1}\frac{B}{A} \pm 180°$;

where $+\tan^{-1}\frac{B}{A}$ is a positive angle (from 0° to +90°) and $-\tan^{-1}\frac{B}{A}$ is a negative angle (from 0° to −90°).

§ 3. Equation (4) is the most convenient for plotting and for general use. The phase angles α_3, α_5, etc., are measured from the zero of the fundamental in the same angular scale as the fundamental wave, as in Fig. 1; that is, 180° always represent half a wave of the fundamental and not half a wave of each particular harmonic. (On the other hand, ϕ_3, ϕ_5, etc., are measured each to the scale of the particular harmonic; the scale for ϕ_3 measures 180° for a half-wave of the third harmonic, the scale for ϕ_5 measures 180° for a half-wave of the fifth harmonic, etc.)

A positive phase angle indicates a leading wave, as the third

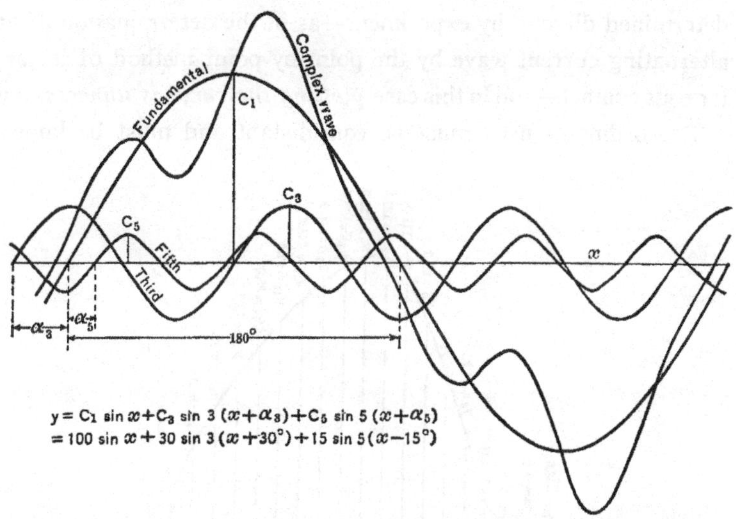

FIG. 1. Complex wave composed of a fundamental, a leading third and a lagging fifth harmonic. Phase angles α_3 and α_5 are measured from zero of the fundamental.

harmonic in Fig. 1; a negative phase angle indicates a lagging wave, as the fifth harmonic in Fig. 1, this being the usual notation in alternating currents.

Note that in measuring to or from a zero of a wave the zero selected is always one where the wave changes from negative to positive.

§ 4. We are given the ordinates of a complex wave. The process of analysis consists first in determining the values of A_1, A_3, A_5, etc., and B_1, B_3, B_5, etc. These values may then be substituted in (1); or, as is more useful, the values of C_1, C_3, C_5, etc., ϕ_1, ϕ_3, ϕ_5, etc., and α_3, α_5, etc., are computed and substituted in (2), (3) or (4).

§ 5. When the wave to be analyzed is given in the form of a curve—as an oscillograph record, for example—the values of the ordinates necessary for computation are measured from the curve. In some cases, however, the values of the ordinates are determined directly by experiment—as in the determination of an alternating current wave by the point-by-point method of instantaneous contact—and in this case *plotting the curve is unnecessary.*

The ordinates used must be equi-distant and must be known

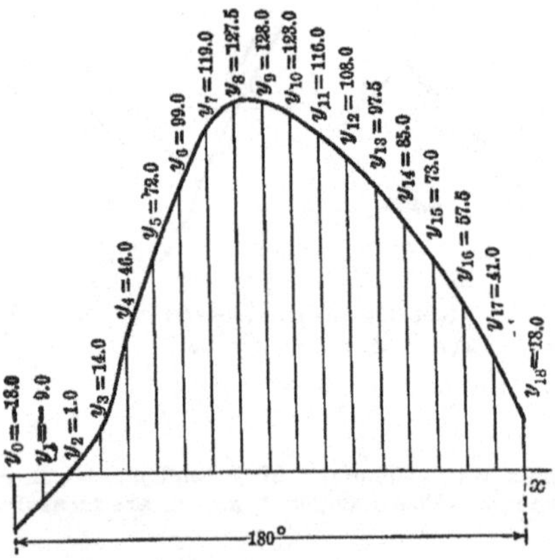

FIG. 2. Showing 18 ordinates taken in a half wave.

for one half of a wave. The following method is based upon 18 ordinates for one half wave and is sufficient for determining the amplitude and the phase of the odd harmonics up to and including the seventeenth; even harmonics are assumed to be absent.

For the origin of the method, see Appendix I.; for the determination of even as well as odd harmonics, see Appendix II.

§ 6. **Procedure.**—Ascertain the values of 18 equidistant ordinates distributed over any interval of 180°, as $y_0, y_1, y_2 \cdots y_{17}$ in Fig. 2, the ordinates being determined for every 10 degrees. The initial ordinate y_0 may have any position whatsoever, without reference to the zero or maximum of the curve; y_{18}, y_{19}, etc., are repetitions of y_0, y_1, etc., and are not used, unless y_0 and y_{18}, y_1 and y_{19}, etc., are averaged for greater accuracy. In all cases care should be taken to note the algebraic sign; all additions, subtractions and multiplications are *algebraic*.

§ 7. *Scheme.*—Arrange the 18 ordinates in a scheme as shown.

SCHEME (18 ordinates).

	y_0	y_1	y_2	y_3	y_4	y_5	y_6	y_7	y_8	y_9
		y_{17}	y_{16}	y_{15}	y_{14}	y_{13}	y_{12}	y_{11}	y_{10}	
Sums		s_1	s_2	s_3	s_4	s_5	s_6	s_7	s_8	s_9
Differences	d_0	d_1	d_2	d_3	d_4	d_5	d_6	d_7	d_8	

Write the algebraic sums and differences as indicated, where

$$s_1 = y_1 + y_{17}; \cdots s_9 = y_9;$$
$$d_0 = y_0; \cdots\cdots\cdots d_8 = y_8 - y_{10}.$$

Certain of the values thus obtained are further combined, algebraically, in the following manner:

$$s_1 + s_5 - s_7 = s_1'; \qquad d_0 - d_6 = d_0';$$
$$s_2 + s_4 - s_8 = s_2'; \quad d_1 - d_5 - d_7 = d_1';$$
$$s_3 - s_9 = s_3'; \quad d_2 - d_4 - d_8 = d_2';$$
$$s_1' - s_3' = s''; \qquad d_0' - d_2' = d''.$$

§ 8. *Tabulating.*—These values should be placed in the Table, *after being multiplied by the sine of the angle shown in the first column;* thus, $\mathbf{s}_1$ denotes the algebraic product of s_1 and the sine of 10°, or, $\mathbf{s}_1 = s_1 \times 0.1736$, etc.

Write the algebraic sums of the first columns (1, 3, 5, 7, 9) on line I., and of the second columns (17, 15, 13, 11) on line II.

The line (I. + II.) is found by adding line II. to line I.; the line (I.—II.) is found by subtracting. Dividing these results by 9 gives the values of $A_1, A_3, \cdots A_{17}$ and $B_1, B_3, \cdots B_{17}$, as shown in the last two lines, which may be substituted in equation (1).

§ 9. *Check.*—As a check on the computation, the following relations should hold, each constant being given its proper sign:

$$A_1 - A_3 + A_5 - A_7 + A_9 - A_{11} + A_{13} - A_{15} + A_{17} = y_9;$$
$$B_1 + B_3 + B_5 + B_7 + B_9 + B_{11} + B_{13} + B_{15} + B_{17} = y_0.$$

Table (18 ordinates).

	Sine Components.									Cosine Components.								
Harmonic.	1	17	3	15	5	13	7	11	9	1	17	3	15	5	13	7	11	9
sin 10°	$\mathbf{s}_1$				$-\mathbf{s}_7$		$-\mathbf{s}_5$			$\mathbf{d}_8$				$-\mathbf{d}_2$		$\mathbf{d}_4$		
sin 20°		$\mathbf{s}_2$				$-\mathbf{s}_4$		$-\mathbf{s}_8$			$\mathbf{d}_7$				$-\mathbf{d}_5$		$\mathbf{d}_1$	
sin 30°	$\mathbf{s}_3$		$\mathbf{s}_1'$		$\mathbf{s}_3$		$-\mathbf{s}_3$			$\mathbf{d}_6$		$\mathbf{d}_2'$		$\mathbf{d}_6$		$\mathbf{d}_6$		
sin 40°		$\mathbf{s}_4$				$\mathbf{s}_8$		$\mathbf{s}_2$			$\mathbf{d}_5$				$\mathbf{d}_1$		$-\mathbf{d}_7$	
sin 50°	$\mathbf{s}_5$				$\mathbf{s}_1$		$\mathbf{s}_7$			$\mathbf{d}_4$				$\mathbf{d}_8$		$-\mathbf{d}_2$		
sin 60°		$\mathbf{s}_6$		$\mathbf{s}_2'$		$-\mathbf{s}_6$		$\mathbf{s}_6$			$\mathbf{d}_3$		$\mathbf{d}_1'$		$-\mathbf{d}_3$		$-\mathbf{d}_3$	
sin 70°	$\mathbf{s}_7$				$-\mathbf{s}_5$		$\mathbf{s}_1$			$\mathbf{d}_2$				$-\mathbf{d}_4$		$-\mathbf{d}_8$		
sin 80°		$\mathbf{s}_8$				$\mathbf{s}_2$		$-\mathbf{s}_4$			$\mathbf{d}_1$				$\mathbf{d}_7$		$\mathbf{d}_5$	
sin 90°	$\mathbf{s}_9$		$\mathbf{s}_3'$		$\mathbf{s}_9$		$-\mathbf{s}_9$		$\mathbf{s}''$	$\mathbf{d}_0$		$\mathbf{d}_0'$		$\mathbf{d}_0$		$\mathbf{d}_0$		$\mathbf{d}''$
I. Sum 1st col.	......		......		....		....		......	....		......		....		....		......
II. Sum 2d col.	......		......		....		....			....		......		....		....		
I.+II.	......		......		....		....		......	....		......		....		....		......
I.—II.	......		......		....		....			....		......		....		....		
$\frac{1}{9}$(I.+II.)	A_1		A_3		A_5		A_7		A_9	B_1		B_3		B_5		B_7		B_9
$\frac{1}{9}$(I.−II.)	A_{17}		A_{15}		A_{13}		A_{11}			B_{17}		B_{15}		B_{13}		B_{11}		

Calculate C_1, C_3, C_5, etc., as in equation *2a*.
Calculate ϕ_1, ϕ_3, ϕ_5, etc., as in equation *2b*.
Calculate α_3, α_5, etc., as in equation *4a*.

§ 10. **Example.**—Required to analyze a wave of which the following 18 ordinates at 10° intervals are known (see Fig. 2):

$y_0 = -\ 18.0 \quad y_1 = -\ 9.0 \quad y_2 = +\ 1.0 \quad y_3 = +\ 14.0 \quad y_4 = +\ 46.0$
$y_5 = +\ 72.0 \quad y_6 = +\ 99.0 \quad y_7 = +119.0 \quad y_8 = +127.5 \quad y_9 = +128.0$
$y_{10} = +123.0 \quad y_{11} = +116.0 \quad y_{12} = +108.0 \quad y_{13} = +\ 97.5 \quad y_{14} = +\ 85.0$
$y_{15} = +\ 73.0 \quad y_{16} = +\ 57.5 \quad y_{17} = +\ 41.0.$

These are arranged according to the scheme of § 7.

SCHEME.

	− 18.0	− 9.0	+ 1.0	+ 14.0	+ 46.0
		+ 41.0	+ 57.5	+ 73.0	+ 85.0
Sums		+ 32.0	+ 58.5	+ 87.0	+ 131.0
Differences	− 18.0	− 50.0	− 56.5	− 59.0	− 39.0
	+ 72.0	+ 99.0	+ 119.0	+ 127.5	+ 128.0
	+ 97.5	+ 108.0	+ 116.0	+ 123.0	
Sums	+ 169.5	+ 207.0	+ 235.0	+ 250.5	+ 128.0
Differences	− 25.5	− 9.0	+ 3.0	+ 4.5	

$s_1' = +32.0 + 169.5 - 235.0 = -33.5 \qquad d_0' = -18.0 + 9.0 = -9.0$
$s_2' = +58.5 + 131.0 - 250.5 = -61.0 \qquad d_1' = -50.0 + 25.5 - 3.0 = -27.5$
$s_3' = +87.0 - 128 = -41.0 \qquad d_2' = -56.5 + 39.0 - 4.5 = -22.0$
$s'' = -33.5 + 41 = +7.5 \qquad d'' = -9.0 + 22.0 = +13.0$

TABLE.

SINE COMPONENTS.

	1	17	3	15	5	13	7	11	9
sin 10°	+ 5.56				− 40.81		− 29.43		
sin 20°		+ 20.01				− 44.80		− 85.67	
sin 30°	+ 43.50		− 16.75		+ 43.50		− 43.50		
sin 40°		+ 84.21				+ 161.02		+ 37.60	
sin 50°	+ 129.84				+ 24.51		+ 180.02		
sin 60°		+ 179.27		− 52.83		− 179.27		+ 179.27	
sin 70°	+ 220.83				− 159.28		+ 30.07		
sin 80°		+ 246.69				+ 57.61		− 129.01	
sin 90°	+ 128.00		− 41.00		+ 128.00		− 128.00		+ 7.50
I. Sum 1st col.	+ 527.73		− 57.75		− 4.08		+ 9.16		+ 7.50
II. Sum 2d col.	+ 530.18		− 52.83		− 5.44		+ 2.19		+ 7.50
I. + II.	+ 1057.91		− 110.58		− 9.52		+ 11.35		+ 7.50
I. − II.	− 2.45		− 4.92		+ 1.36		+ 6.97		
$\frac{1}{9}$ (I. + II.)	$A_1 = +117.55$		$A_3 = -12.29$		$A_5 = -1.06$		$A_7 = +1.26$		$A_9 = +0.83$
$\frac{1}{9}$ (I. − II.)	$A_{17} = -\ 0.27$		$A_{15} = -\ 0.55$		$A_{13} = +0.15$		$A_{11} = +0.77$		

Check: $A_1 - A_3 + A_5 - A_7 + \cdots + A_{17} = 128.01$; $y_9 = 128.00$.

COSINE COMPONENTS.

	1	17	3	15	5	13	7	11	9
sin 10°	+ 0.78				+ 9.81		− 6.77		
sin 20°		+ 1.03				+ 8.72		−17.10	
sin 30°	− 4.50		−11.00		− 4.50		− 4.50		
sin 40°		−16.39				−32.14		− 1.93	
sin 50°	−29.87				+ 3.45		+43.28		
sin 60°		−51.09		−23.81		+51.09		+51.09	
sin 70°	−53.09				+36.65		− 4.23		
sin 80°		−49.24				+ 2.95		−25.11	
sin 90°	−18.00		− 9.00		−18.00		−18.00		+13.00
I. Sum 1st col.	−104.68		−20.00		+27.41		+9.78		+13.00
II. Sum 2d col.	−115.69		−23.81		+30.62		+6.95		
I. + II.	−220.37		−43.81		+58.03		+16.73		+13.00.
I. − II.	+ 11.01		+ 3.81		− 3.21		+ 2.83		
$\frac{1}{9}$ (I. + II.)	$B_1 = -24.49$		$B_3 = -4.87$		$B_5 = +6.45$		$B_7 = +1.86$		$B_9 = +1.44$
$\frac{1}{9}$ (I. − II.)	$B_{17} = +\ 1.22$		$B_{15} = +0.42$		$B_{13} = -0.36$		$B_{11} = +0.31$		

Check: $B_1 + B_3 + B_5 + B_7 \cdots + B_{17} = -18.02$; $y_0 = -18.00$.

RESULTS.

$C_1 = 120.07$ $\quad$ $C_3 = 13.22$ $\quad$ $C_5 = 6.54$
$\phi_1 = -11.8°$ $\quad$ $\phi_3 = +21.6° \pm 180°$ $\quad$ $\phi_5 = -80.7° \pm 180°$
$\alpha_1 = 0°$ $\quad$ $\alpha_3 = +79.0°$ $\quad$ $\alpha_5 = +31.7°$

$C_7 = 2.25$ $\quad$ $C_9 = 1.66$ $\quad$ $C_{11} = 0.83$
$\phi_7 = +55.9°$ $\quad$ $\phi_9 = +60.0°$ $\quad$ $\phi_{11} = +21.9°$
$\alpha_7 = +19.8°$ $\quad$ $\alpha_9 = +18.5°$ $\quad$ $\alpha_{11} = +13.8°$

$C_{13} = 0.39$ $\quad$ $C_{15} = 0.69$ $\quad$ $C_{17} = 1.25$
$\phi_{13} = -67.4°$ $\quad$ $\phi_{15} = -37.4° \pm 180°$ $\quad$ $\phi_{17} = -77.5 \pm 180°$
$\alpha_{13} = +\ 6.6°$ $\quad$ $\alpha_{15} = +21.3°$ $\quad$ $\alpha_{17} = \pm 17.7°$

APPENDIX I.

ORIGIN AND PROOF OF METHOD

§ 11. **Origin.**—Many* methods have been used for analyzing a complex wave, either accurately or approximately, but until Runge† devised a simplification the accurate methods have been exceedingly laborious, sometimes involving hundreds of multiplications. Runge found that by combining terms the number of multiplications can be much reduced.

The method here given is based‡ upon that of Runge, being further simplified by assuming that the negative half-wave is a repetition of the positive, *i. e.*, that there is no constant term and that even harmonics are absent (see Appendix II.). No assumption, however, is made that the initial ordinate is zero ($y_0 = 0$); when analyzing a wave for which the ordinates are obtained by the method of instantaneous contact such a limitation adds to the labor and reduces the accuracy, for it involves plotting the curve according to the ordinates found by experiment and then measuring from the curve a second set of ordinates, beginning with zero, in order to make the analysis. Evidently it is more accurate, as well as easier, to base the analysis upon the original set of ordinates.

§ 12. **Number of Ordinates Used.**—If m ordinates in one half-wave are used, the method determines the harmonics to and including the

* (§ 11a). For a description of some of these, see Sir William Thomson, *Proc. Roy. Soc.*, XXVII., 1878, p. 371; J. A. Fleming, *Lond. Elect.*, Jan. 22–9, 1892; J. Perry, *Lond. Elect.*, Feb. 5, 1892, and June 28, 1895; S. Berson, *Ecl. Electrique*, 15, 1898; Houston and Kennelly, *Elec. World*, XXXI., 1898; Michelson and Stratton, *Am. Jour. Sc.*, 5, 1898; A. S. Langsdorf, *Physical Rev.*, XII., 1901, p. 184; Fischer-Hinnen, *Elektrotech. Zeitschrift*, May 9, 1901; S. M. Kintner, *Elec. World*, XLIII., 1904, p. 1023; J. Harrison, *Engineering*, Feb., 1906, p. 201; P. M. Lincoln, *Elec. Journal*, V., 1908, p. 386; C. S. Slichter, *Elec. World*, July 15, 1909; C. A. Pierce, *Elec. World*, Apr. 13 and Oct. 21, 1911.

† *Zeitschrift für Mathematic und Physic*, 1903, Vol. XLVIII., p. 443. A discussion of Runge's method is given by S. P. Thompson, *Proc. London Phy. Soc.*, Vol. XIX., p. 443, and by E. B. Tuttle, *Iowa Engineer*, Sept., 1906. Thompson introduces the condition that $y_0 = 0$.

‡ Runge's method is followed in combining and tabulating results but not in developing the equations nor in determining phase angles ϕ and α.

Added Note.—For a convenient method of constructing schedules for even as well as odd harmonics, see H. O. Taylor, *Phys. Rev.*, 1915.

$(m-1)$ harmonic, assuming that higher harmonics are negligible; thus, 18 ordinates determine harmonics to the seventeenth, 6 ordinates determine harmonics to the fifth, etc., assuming that no higher harmonics are present to an appreciable extent.

Reducing the number of ordinates used, however, not only reduces the number of harmonics that can be determined but reduces the accuracy with which these are determined. For example, if the seventh or ninth harmonic has considerable amplitude, the method with 6 ordinates in general would not accurately determine even the third and fifth harmonics. The more ordinates used, therefore, the more accurate is the method, but the labor is likewise increased.

Except in work of the greatest precision, the use of more than 18 ordinates in a half-wave is hardly worth while, for the accuracy of the data will rarely warrant it. On the other hand, it does not pay to make an analysis with too few ordinates and to risk errors in the result, unless only an approximate analysis is desired.

For simplicity, the following discussion is limited to the method using 18 ordinates, but it can be readily modified so as to apply when more ordinates, or less, than 18 are used.

§ 13. **Development of Method.**—For practical use the infinite series in equation (1) must be limited to a finite number of terms; thus, excluding all harmonics above the seventeenth, we have

$$y = A_1 \sin x + A_3 \sin 3x + A_5 \sin 5x \cdots A_{17} \sin 17x$$
$$+ B_1 \cos x + B_3 \cos 3x + B_5 \cos 5x \cdots B_{17} \cos 17x. \qquad (5)$$

Substituting for x 18 known consecutive values (0°, 10°, 20°, etc.) and for y the corresponding 18 known values (y_0, y_1, y_2, etc.), we have 18 simultaneous equations of the first degree which may be solved for the 18 unknown coefficients A_1 to A_{17} and B_1 to B_{17}.

The coefficients of the nth harmonic may be written as summations, in which the values of k vary from 0 to 17; thus

$$A_n = \frac{2}{18} \sum_{k=0}^{k=17} y_k \sin nk\, 10^\circ, \qquad (6)$$

$$B_n = \frac{2}{18} \sum_{k=0}^{k=17} y_k \cos nk\, 10^\circ.$$

§ 14. *Proof.*—The foregoing expressions may be derived* as follows. To determine a coefficient A_n, multiply the first of the 18 equations by $\sin 0°$, the second by $\sin n\,10°$, the third by $\sin 2n\,10°$, etc., and add. The sum† of all terms that contain A_n on the right hand of the equations (after multiplying) is $9A_n$; the sum of the other terms is zero. The sum of the left hand terms may be written as a summation. Thus,

$$\sum_{k=0}^{k=17} y_k \sin nk\,10° = 9A_n. \quad (7)$$

Transposing, we have the value of A_n, as in (6). The value of B_n is similarly found by multiplying by cosines instead of sines.

§ 15. **Determining Values for Individual Coefficients.**—Given the general expression (6) for A_n and B_n, the next step is to find particular expressions for A_3, A_5, A_7, etc., which may be conveniently used in a numerical solution. It will suffice to determine A_3 as an illustration. In (6) or (7), let $n=3$ and assign values for k from 0 to 17. We then have

$$\left.\begin{aligned} 9A_3 = &+y_0 \sin 0° + y_1 \sin 30° + y_2 \sin 60° + y_3 \sin 90° \\ &+y_6 \sin 180° + y_5 \sin 150° + y_4 \sin 120° \\ &\quad + y_7 \sin 210° + y_8 \sin 240° + y_9 \sin 270° \\ &+y_{12} \sin 360° + y_{11} \sin 330° + y_{10} \sin 300° \\ &\quad + y_{13} \sin 30° + y_{14} \sin 60° + y_{15} \sin 90° \\ &\quad + y_{17} \sin 150° + y_{16} \sin 120° \end{aligned}\right\} \quad (8)$$

Since $\sin 150° = \sin 30°$, $\sin 210° = -\sin 30°$, $\sin 330° = -\sin 30°$, $\sin 120° = \sin 60°$, $\sin 240° = -\sin 60°$, $\sin 300 = -\sin 60$, $\sin 270° = -\sin 90°$, we may write (8) as follows:

$$\begin{aligned} 9A_3 = &(y_1 + y_{17} + y_5 + y_{13} - y_7 - y_{11}) \sin 30° \\ &+ (y_2 + y_{16} + y_4 + y_{14} - y_8 - y_{10}) \sin 60° \\ &+ (y_3 + y_{15} - y_9) \sin 90° \end{aligned}$$

* (§ 14a). *General Expression for Coefficients.*—Determined more generally for an infinite number of terms, the coefficients of the nth order are

$$A_n = \frac{2}{\pi}\int_0^{\pi} y \sin nx \cdot dx; \quad B_n = \frac{2}{\pi}\int_0^{\pi} y \cos nx \cdot dx.$$

(See Byerly's *Fourier's Series and Spherical Harmonics,* Chap. II.; Todhunter's *Int. Calculus,* Chap. XIII.; Greenhill's *Int. Calculus,* § 183; etc.)

† If there are m terms and m equations instead of 18, this sum is $\frac{1}{2}mA_n$, the average value of the sine of an angle, squared, being ½.

$A_3 = 1/9\,(s_1' \sin 30^\circ + s_2' \sin 60^\circ + s_3' \sin 90^\circ),$

which is the form used in the Table, § 8.

§ 16. A_3 is developed in terms of $\sin k\, 30^\circ$ while A_{15} is developed in terms of $\sin k\, 150^\circ$. The latter can be written $\sin k\,(180^\circ - 30^\circ)$, which expanded gives $\sin k\, 180^\circ \cos k\, 30^\circ - \cos k\, 180^\circ \sin k\, 30^\circ$. It is evident that, when k is odd, $\sin k\, 30^\circ = \sin k\, 150^\circ$, while when k is even $\sin k\, 30^\circ = -\sin k\, 150^\circ$; hence, A_{15} can be determined from A_3 by changing the signs of all terms in the third column of equation (8). Hence

$$A_{15} = 1/9\,[s_1' \sin 30^\circ - s_2' \sin 60^\circ + s_3' \sin 90^\circ].$$

The other coefficients are determined in a similar manner.

§ 17. *Check.*—In equation (5), $y = y_9$ when $x = 90^\circ$; and $y = y_0$ when $x = 0^\circ$. Substituting these values, we have

$$A_1 - A_3 + A_5 - A_7 \cdots + A_{17} = y_9;$$
$$B_1 + B_3 + B_5 + B_7 \cdots + B_{17} = y_0.$$

After analyzing a wave, these equations may be used to check the values of A_1, A_3, A_5, etc., and B_1, B_3, B_5, etc.

APPENDIX II.

ANALYSIS OF WAVE WHICH MAY HAVE EVEN AND ODD HARMONICS AND A CONSTANT TERM.

§ 18. **Method of 12 Ordinates.**—In the most general case,* when the negative part of the wave is not equal to the positive and is not a repetition of it, both even and odd harmonics may be present and also a constant term, B_0. To analyze such a wave, see Fig. 3, equidistant ordinates must be taken over an *entire period,* or 360°, and not merely for a half period, as in the preceding pages when odd harmonics only were considered.

Let ordinates be taken† at intervals of 30°, *i. e.,* there are 12 known ordinates in a complete wave.

* This general method, which applies whether there is a constant term or not and whether odd or even harmonics are present or not, is taken directly from Runge's article, where the method is given in detail for 12 and for 36 ordinates.

† For greater accuracy (§ 12) more ordinates must be used.

Arrange these in the scheme, as shown; then compile the table, after multiplying by the sine of the angle indicated: thus, $\mathbf{a}_1 = a_1 \sin 30°$.

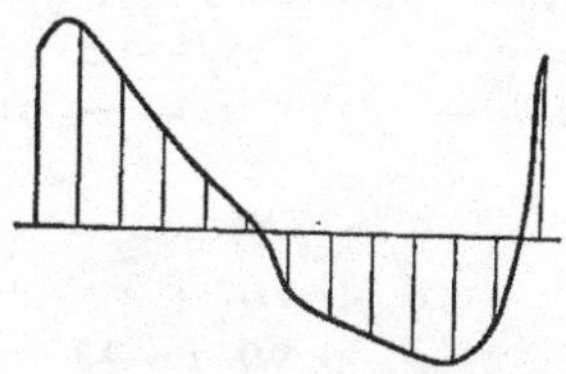

FIG. 3. Wave with even and odd harmonics and a constant term; 12 ordinates used.

SCHEME (12 ordinates).

Ordinates:	y_0	y_1	y_2	y_3	y_4	y_5	
		y_{11}	y_{10}	y_9	y_8	y_7	y_6
Diff.:		d_1	d_2	d_3	d_4	d_5	
Sum:	s_0	s_1	s_2	s_3	s_4	s_5	s_6

$$\begin{array}{llll} & d_1 & d_2 & d_3 \\ & d_5 & d_4 & \\ \hline \text{Sum:} & a_1 & a_2 & a_3 \\ \text{Diff.:} & a_1' & a_2' & \end{array} \qquad \begin{array}{lllll} & s_0 & s_1 & s_2 & s_3 \\ & s_6 & s_5 & s_4 & \\ \hline \text{Sum:} & b_0 & b_1 & b_2 & b_3 \\ \text{Diff.:} & b_0' & b_1' & b_2' & \end{array} \qquad \begin{array}{lll} & a_1 & b_0' \\ & a_3 & b_2' \\ \hline \text{Diff.:} & a'' & b'' \end{array} \qquad \begin{array}{lll} & b_0 & b_1 \\ & b_2 & b_3 \\ \hline \text{Sum:} & c_0 & c_1 \end{array}$$

TABLE (12 ordinates).

	SINE COMPONENTS.			COSINE COMPONENTS.			
	1 5	2 4	3	1 5	2 4	3	0 6
sin 30° = 0.500	$\mathbf{a}_1$			$\mathbf{b}_2'$	$-\mathbf{b}_2$ $\mathbf{b}_1$		
sin 60° = 0.866	$\mathbf{a}_2$	$\mathbf{a}_1'$ $\mathbf{a}_2'$		$\mathbf{b}_1'$			
sin 90° = 1.000	$\mathbf{a}_3$		$\mathbf{a}''$	$\mathbf{b}_0'$	$\mathbf{b}_0$ $-\mathbf{b}_3$	$\mathbf{b}''$	$\mathbf{c}_0$ $\mathbf{c}_1$
I. Sum 1st col.	....	....	....	....	....	....	....
II. Sum 2d col.	....	....		....	....		....
I. + II.	....	....	....	....	....	....	....
I. − II.	....	....		....	....		....
$\frac{1}{6}$(I. + II.)	A_1	A_2	A_3	B_1	B_2	B_3	$2B_0$
$\frac{1}{6}$(I. − II.)	A_5	A_4		B_5	B_4		$2B_6$

Check: $A_1 - A_3 + A_5 + B_0 - B_2 + B_4 - B_6 = y_3;$

$-A_1 + A_3 - A_5 + B_0 - B_2 + B_4 - B_6 = y_9;$

$B_0 + B_1 + B_2 + B_3 + B_4 + B_5 + B_6 = y_0;$

$B_0 - B_1 + B_2 - B_3 + B_4 - B_5 + B_6 = y_6.$

§ 19. **Example.**—Required to analyze the wave shown in Fig. 3, the following 12 ordinates at 30° intervals being known:

$y_0 = +14.0$	$y_1 = +15.8$	$y_2 = +12.0$	$y_3 = +7.7$
$y_4 = +\ 4.3$	$y_5 = +\ 1.4$	$y_6 = -\ 4.6$	$y_7 = -6.8$
$y_8 = -\ 8.2$	$y_9 = -\ 9.3$	$y_{10} = -\ 9.9$	$y_{11} = -6.8$

SCHEME.

Ordinates:	+ 14.0	+ 15.8	+ 12.0	+ 7.7	+ 4.3	+ 1.4	
		− 6.8	− 9.9	− 9.3	− 8.2	− 6.8	− 4.6
Diff.:		+ 22.6	+ 21.9	+ 17.0	+ 12.5	+ 8.2	
Sum:	+ 14.0	+ 9.0	+ 2.1	− 1.6	− 3.9	− 5.4	− 4.6

	+ 22.6	+ 21.9	+ 17.0		+ 14.0	+ 9.0	+ 2.1	− 1.6
	+ 8.2	+ 12.5			− 4.6	− 5.4	− 3.9	
Sum:	+ 30.8	+ 34.4	+ 17.0	Sum:	+ 9.4	+ 3.6	− 1.8	− 1.6
Diff.:	+ 14.4	+ 9.4		Diff.:	+ 18.6	+ 14.4	+ 6.0	

	+ 30.8	+ 18.6		+ 9.4	+ 3.6
	+ 17.0	+ 6.0		− 1.8	− 1.6
Diff.:	+ 13.8	+ 12.6	Sum:	+ 7.6	+ 2.0

TABLE.

	SINE COMPONENTS.			COSINE COMPONENTS.			
	1 5	2 4	3	1 5	2 4	3	0 6
sin 30° = 0.500	+15.4			+ 3.0	+0.9 +1.8		
sin 60° = 0.866	+29.8	+12.5+8.1		+12.5			
sin 90° = 1.000	+17.0		+13.8	+18.6	+9.4 +1.6	+12.6	+7.6+2.0
I. Sum 1st col	+32.4	+12.5	+13.8	+21.6	+10.3	+12.6	+7.6
II. Sum 2d col.	+29.8	+ 8.1		+12.5	+ 3.4		+2.0
I. + II.	+62.2	+20.6	+13.8	+34.1	+13.7	+12.6	+9.6
I. − II.	+ 2.6	+ 4.4		+ 9.1	+ 6.9		+5.6
$\frac{1}{6}$ (I. + II.)	$A_1 = +10.4$	$A_2 = +3.4$	$A_3 = +2.3$	$B_1 = +5.7$	$B_2 = +2.3$	$B_3 = +2.1$	$B_0 = +0.8$
$\frac{1}{6}$ (I. − II.)	$A_5 = +\ 0.4$	$A_4 = +0.7$		$B_5 = +1.5$	$B_4 = +1.1$		$B_6 = +0.5$

Check: $A_1 - A_3 + A_5 + B_0 - B_2 + B_4 - B_6 = \quad 7.6; \quad y_3 = \quad 7.7;$

$-A_1 + A_3 - A_5 + B_0 - B_2 + B_4 - B_6 = -\ 9.4; \quad y_9 = -\ 9.3;$

Check: $B_0 + B_1 + B_2 + B_3 + B_4 + B_5 + B_6 = \quad 14.0; \quad y_0 = \quad 14.0;$

$B_0 - B_1 + B_2 - B_3 + B_4 - B_5 + B_6 = -\ 4.6; \quad y_6 = -\ 4.6.$

CHAPTER XII.

PROBLEMS.

Many who are efficient in carrying out standardized experiments are not so efficient in carrying out experiments for which no instructions are given. It is very important to possess such ability and it can be acquired only by attacking problems which demand initiative and responsibility.

It is futile to prepare a schedule of problems of this sort with any expectation of its being adequate or complete; some of the problems here given, however, may prove useful or suggestive.

Various reference books and periodicals should be consulted in most cases before proceeding with experimental work. A familiarity with original sources and the ability to give proper weight to different authorities is highly desirable.

§ 1. Given an over-compounded D.C. generator. Determine a shunt to go in parallel with the series coils to produce a definite regulation (as flat compounded or, say, 5 per cent. over compounded) for a certain speed and voltage. Determine how the result would be affected, and the cause for it, if the generator is operated at the same speed, but at a different voltage (say 100 instead of 125 volts); or at the same voltage but different speed. If the source of power is an induction motor, how will its slip enter into the problem?

§ 2. Determine the relation between electromotive force and speed in a separately excited generator and in a self-excited shunt generator.

§ 3. Determine the relation between line voltage and speed in a shunt motor.

§ 4. Operate two D.C. generators in parallel, first as shunt and then as compound machines, and ascertain how any desired division of the load is obtained. In the case of compound generators, an equalizing bus-bar is necessary connecting the two brushes (one on each machine) to which the series coils are connected.

§ 5. Explore the field of a dynamo-electric machine by determining the distribution of the flux in the air gap.

§ 6. Determine the relation between the total flux set up by the field windings of a dynamo-electric machine and the useful flux that passes through the armature. The ratio of the former to the latter is the dispersion, or leakage, coefficient.

§ 7. Given a separately excited D.C. motor the armature of which is supplied with current from a series generator. Investigate and explain the conditions affecting the direction of rotation of the generator and the conditions under which the direction will periodically reverse. The motor should have brushes which will not damage the commutator when the direction of rotation is reversed.

§ 8. In a separately excited motor in which the armature resistance drop is so large that the counter-electromotive force is practically negligible (as in certain watthour meters), determine the relation between speed and field excitation when line voltage is constant.

§ 9. Design, have made, and test commutating interpoles for some machine which commutates badly.

§ 10. Find the relation of potential drop to current density between brushes of various materials and slip rings at usual speeds; also at very high speeds.

§ 11. Analyze all the losses in a given machine or apparatus.

§ 12. Make a study of the temperature rise in a machine or apparatus by thermometer and by resistance measurements.

§ 13. Given a differential D.C. motor which runs too fast at full load. Determine a shunt to go in parallel with the series coil that will give a certain speed at no load and full load. Determine whether the same shunt will do for a different speed, and report as to why it will or will not.

§ 14. Take some point concerning which you find your knowledge inadequate, on some subject you have already studied, and if possible plan an experiment to settle the matter to your own satisfaction.

§ 15. Take a technical article which proves of interest (as the paper or papers of some A. I. E. E. meeting) and investigate such points as you can in the laboratory. The Digest of the *Electrical World* and the Question Box of the *Electric Journal*, can be used to advantage as a source of timely practical problems.

§ 16. Given a patent specification and claims. Investigate the invention by experiment and study, and report on one or more of the following: (1) Its usefulness (from the standpoint of a possible user or purchaser); (2) its apparent novelty, including points which differentiate this invention from other methods or apparatus for securing similar ends; (3) its operativeness without further invention. (To be valid, a patent must be new, useful and operative.)

§ 17. Determine the insulation resistance of a machine or line by means of a voltmeter.

§ 18. Make a study of a Tirrell or other voltage regulator.

§ 19. Study the electrolytic or "pail forge" method of heating rods for welding. The following solution may be used: 10 gal. water; ¾ lb. borax; 3¼ lb, sal soda; ¼ lb. salt. A D.C. dynamo of 200 to 300 volts has one terminal connected to a submerged lead plate in the solution. The other terminal is connected to the rod or to a horizontal piece of copper upon which the rod rests when contact is desired. The rod becomes heated when submerged, if the current flows in the proper direction.

§ 20. Determine the torque of a machine by the electrical method of McAllister, using a shunt motor as load; see *Standard Handbook* and McAllister's *Alternating Current Motors.*

§ 21. Investigate various methods for obtaining a neutral on a 3-wire D.C. system.

§ 22. Operate two alternators in parallel and study the conditions that determine the division of the load between the two machines.

§ 23. Determine the characteristics of a high frequency alternator and note the effect of lagging and leading currents upon the terminal voltage. With condensers in parallel with the load, no field excitation may be necessary. Care is necessary in this test as there is danger of excessive voltage.

§ 24. Select and use one or more methods for determining inductance and capacity with a fair degree of accuracy.

§ 25. Connect in series two electromotive forces, one alternating and the other direct (or alternating of a different frequency). Measure the combined voltage and determine the relation between it and the separate voltages.

§ 26. Superpose in a conductor with resistance R an alternating current I_1 and a direct current I_2 (or an alternating current of different frequency). Determine the relative values of the copper loss (RI_1^2 and RI_2^2) for each current alone and (RI^2) for the total current, I. Determine the relation between the effective values of I, I_1 and I_2.

§ 27. Given a transformer with an open magnetic circuit, as the now obsolete "Hedgehog" transformer. Investigate the transformer with a view to bringing out the characteristic differences between it and a transformer with a closed magnetic circuit. Report on the relative advantages and disadvantages of the two types. At one time this subject was much debated.

§ 28. Required to find at what frequency, current and voltage a given transformer will give the highest efficiency. The voltage should not exceed a certain specified value; the temperature rise should not exceed the limit set by A. I. E. E. Standardization Rules; assume that any frequency, from say 20 to 125 cycles, is available. Outline completely the method of procedure before making tests.

§ 29. Make a comparison of loading-back methods for testing transformers.

§ 30. Make a general study of a series current transformer; make a particular study of the accuracy of its ratio at different frequencies and of phase errors when used with a wattmeter.

§ 31. Determine the effect of different wave forms upon transformer losses, regulation and magnetizing current.

§ 32. Study the effect of wave form upon circle diagrams and other vector diagrams.

§ 33. Given a 2-phase 4-wire supply from two independent transformers. Insert between the two neutrals an additional source of electromotive force of different frequency (or, direct current). Measure all voltages and construct a vector model. What must be the value of the inserted electromotive force to cause all voltages between line wires to be equal (see § 9a, Exp. 7–A).

§ 34. Determine the variation in the starting torque of an induction motor for different values of secondary resistance.

§ 35. An induction motor was purchased for a certain frequency. The central station equipment has been changed to a different frequency. Can this motor be used? If so what voltage will be best? Will any benefit come by changing its 3-phase primary from star to delta or *vice versa?*

§ 36. Test an induction generator with condenser excitation.

§ 37. Test the method of obtaining polyphase from single-phase current by means of an induction motor (§ 3, Exp. 7–A) and a similar method using a synchronous motor.

§ 38. Make a study of devices for indicating synchronism and of methods for synchronizing.

§ 39. Construct a shunt for an A.C. Wattmeter so as to extend its range; it should be designed so as to be correct at different frequencies.

§ 40. Determine the losses of a machine by the retardation method; see *Standard Handbook*.

§ 41. Determine the hysteresis loss in different dielectrics.

§ 42. Set up and test an electrolytic or a mercury arc rectifier.

§ 43. Make a study of potentiometer methods for measuring alternating currents and voltages.

§ 44. Make a comparison of the behavior of various types of instruments at commercial frequencies and at high frequencies.

§ 45. Determine the current efficiency and energy efficiency of a storage cell.

§ 46. Adapt the Ryan-Braun Tube to the measurement of power—particularly of small power.

§ 47. With an oscillograph, study the behavior of fuses and circuit breakers with D.C. and A.C. loads, inductive and non-inductive.

§ 48. Determine the instantaneous values of resistance, for example, of a lamp with alternating current, or with direct current at brief intervals of time after the circuit is closed.

§ 49. Determine the instantaneous values of flux in a transformer, either indirectly (from instantaneous values of current and voltage) or directly by some special device; see *Bulletin of the Bureau of Standards*, Vol. IV., p. 467.

§ 50. Make a study of different methods and apparatus for measuring one of the following quantities, with a view to determining their relative advantages and developing modifications of new methods: (*a*) slip; (*b*) frequency; (*c*) speed; (*d*) phase; (*e*) form factor; (*f*) wave form; (*g*) power factor; (*h*) reactive factor; (*i*) very small (or large) alternating current, voltage or power.

§ 51. Make a study of an interpole motor, repulsion motor, series A.C. motor, or other particular machine.

§ 52. Make a study of "concatenation," or other method of speed control for induction motors.

§ 53. Determine the wave form of alternating electromotive forces and currents by means of the Pierce Analyzer (§ 11a, Exp. 11-A) and by other methods.

INDEX.

Made in the USA
Monee, IL
13 September 2026

62976737R00225